Mathematical Aspects of Paradoxes in Cosmology

Michal Křížek • Lawrence Somer

Mathematical Aspects of Paradoxes in Cosmology

Can Mathematics Explain the Contemporary Cosmological Crisis?

Springer

Michal Křížek
Institute of Mathematics
Czech Academy of Sciences
Prague 1, Czech Republic

Lawrence Somer
Department of Mathematics
Catholic University of America
Washington, DC, USA

ISBN 978-3-031-31767-5 ISBN 978-3-031-31768-2 (eBook)
https://doi.org/10.1007/978-3-031-31768-2

This Springer imprint is published by the registered company Springer Nature Switzerland AG
The registered company address is: Gewerbestrasse 11, 6330 Cham, Switzerland

Never identify any mathematical model with reality.

—THE AUTHORS

Preface

Einstein's General Theory of Relativity represents a very complex theoretical discipline which intersects mathematics and physics. Its results are applied in relativistic cosmology, which is a branch of astronomy that deals with the greatest spatial and temporal distances. For some time there has been a growing crisis in cosmology. The contemporary cosmological model possesses at least 20 serious paradoxes, see e.g. Sect. 1.1. Nevertheless, it is still defended by most cosmologists. We show that the main reason for this is that the simple mathematical models used in cosmology are identified with reality and that various incorrect extrapolations are being made.

In 1999, the first author of this monograph published the paper [157] about delay differential equations: *Numerical experience with the finite speed of gravitational interaction* which indicated that the laws of conservation of energy and angular momentum can be slightly violated in our universe. This was the beginning of a detailed analysis of mathematical and numerical methods used in current astrophysics and relativistic cosmology.

This treatise is based on our more than 50 works in Special and General Relativity and related topics that we published mostly in prestigious international journals such as *Journal of Computational and Applied Mathematics, Gravitation and Cosmology, Communications in Computational Physics, Astrophysics and Space Science, Mathematics and Computers in Simulation, International Journal of Astronomy and Astrophysics, New Astronomy, Neural Network World, Astronomische Nachrichten, and the Czech journal Pokroky (=Advances) matematiky, fyziky a astronomie* (see e.g. `dml.cz`). Most of our results were reported at many international conferences (see e.g. [162, 175, 177–180, 195]) and also at our regular seminars *Current Problems in Numerical Analysis* and *Cosmological Seminar* which take place at the Institute of Mathematics of the Czech Academy of Sciences in Prague [387, 398].

Prague stood at the birth of two revolutions in building the theory of gravity:

1. At the beginning of the seventeenth century in Prague, Johannes Kepler formulated his first and second laws of planetary motion with the help of Tycho

Brahe's observational data. This discovery played a key role in Newton's theory of gravitation.
2. In the period 1911–1912, Albert Einstein received his first full professorship at Charles-Ferdinand University in Prague. His colleague Georg Pick worked on non-Euclidean geometries and taught Einstein the foundations of tensor calculus. In Prague, Einstein found the calm atmosphere necessary to formulate the basic ideas of his General Theory of Relativity. Here he wrote several important papers on this topic.

There are many books on General Relativity and cosmology. Let us mention, e.g., [107, 184, 249, 270, 272, 332, 337, 369, 371]. However, our book is written in a different style. Most of the important results are formulated in the form of mathematical theorems with precise assumptions and statements. Our book also contains some nonstandard topics such as: the violation of the principle of causality in Newtonian mechanics, critical mathematical and numerical analysis of Mercury's perihelion shift, and the large computational complexity of Einstein's equations which renders them factually inapplicable for the classical two-body problem. Further topics include the nonuniqueness of the notion universe, topology of the universe, various descriptions of a hypersphere, regular tessellations of hyperbolic spaces, flight around the universe, local Hubble expansion of the universe, neglected gravitational redshift in detection of gravitational waves, possible distribution of mass inside a black hole. We would also like to dispel some myths appearing in the theory of relativity and contemporary cosmology.

Our book is intended for a general audience—especially for those who can appreciate the beauty of both mathematics and cosmology, who are open to new ideas, and who want to break out of the rut. We only assume that the reader is familiar with the basic rules of arithmetic and has no problem with adjustments of algebraic formulas. Only very rarely it is necessary to understand some deep relationships from linear algebra or calculus. Most chapters can be read independently from one another. Some parts are quite simple, others more complicated. If some part is too difficult, there is no problem in skipping it.

For inspiration, there are also several links to websites, although we are well aware that they are not subjected to any review and change quite frequently. Newly defined terms are highlighted in italics in the text for the convenience of the reader. They can also be found in the Index.

In order to read the individual chapters, it is not necessary that the reader understands all assertions. Mathematicians formulate their ideas in the form of theorems that contain only what is relevant in the problem in question. To avoid any discussion whether a particular statement is true or not, we provide proofs of most statements so that everyone can verify their validity. For more complicated proofs, we only give suitable references to the corresponding literature.

In a number of discussions, many researchers helped us to improve the content of this book, in particular, Jan Brandts, Jan Chleboun, Caroline Griffis, Jiří Grygar, Vesselin Gueorguiev, Ivan Hlaváček, Kurt Koltko, Pavel Křížek, Monika Lanzendörferová, František Lomoz, André Maeder, Jan Maršák, Jaroslav Mlýnek,

Vladimír Novotný, Luboš Pick, Mirko Rokyta, Karel Segeth, Antonín Slavík, Václav Vavryčuk, Tomáš Vejchodský, Emil Vitásek, and Vladimír Wagner. We really appreciate their help and they deserve our great thanks. Furthermore, we are deeply grateful to Attila Mészáros for his excellent regular lectures on cosmology at Charles University in Prague, and to Hana Bílková and Eva Ritterová for their technical assistance in the final preparation of the manuscript. Preliminary LaTeX typesetting was prepared by Michal Křížek.

Finally, we are profoundly indebted to Elena Griniari, Francesca Ferrari, and Harini Rajesh from Springer-Verlag for their helpful cooperation in the preparation of this book. Our great thanks go to the Springer Publishing House for its care in the graphical design of the book and to the referees for invaluable suggestions. We are also grateful to our families for patience and understanding.

This book was supported by the Czech Academy of Sciences (RVO 67985840). Some of our results contained in Chaps. 3–8 were developed in part by our work supported by Grant No. 23-06159S of the Grant Agency of the Czech Republic.

Prague 1, Czech Republic — Michal Křížek
Washington, DC, USA — Lawrence Somer

Contents

Glossary of Symbols

$\mathbb{N} = \{1, 2, 3, \ldots\}$	Set of natural numbers
$\mathbb{E}^n$	n-dimensional Euclidean space
$\mathbb{S}^3$	Three-dimensional sphere (hypersphere)
$\mathbb{H}^3$	Three-dimensional hyperbolic pseudosphere
k	Curvature index from $\{-1, 0, 1\}$
$\mathbb{C}$	Set of complex numbers
(a, b)	Open interval
$[a, b]$	Closed interval
$\lvert \cdot \rvert$	Absolute value
$\log_b$	Logarithm to the base b
ln	Natural logarithm
e	Euler number 2.718 281 828...
$\forall$	For all
$\exists$	There exist(s)
diag	Diagonal matrix (tensor)
det	Determinant
vol	Volume
$\top$	Transposition
$\mathcal{O}(\cdot)$	Landau symbol: $f(\alpha) = \mathcal{O}(g(\alpha))$ if there exists $C > 0$ such that $\lvert f(\alpha)\rvert \le C\lvert g(\alpha)\rvert$ for $\alpha \to 0$ or $\alpha \to \infty$
$\emptyset$	Empty set
π	Ludolph number 3.141 592 653...
$\prod$	Product
$\sum$	Sum
$\cap$	Intersection
$\cup$	Union
$\setminus$	Set subtraction
$\subset$	Subset
$\in$	Is element of
$\notin$	Is not element of
$\{x \in A;\ \mathcal{P}(x)\}$	Set of all elements x from A which possess property $\mathcal{P}(x)$

$f: A \to B$	Mapping (function) f from the set A to the set B
$x \mapsto f(x)$	Function which assigns value $f(x)$ to x
$\Longrightarrow$	Implication
$:=$	Assignment
$a = a(t)$	Expansion function
$\dot{a}$	Time derivative of $a = a(t)$
$\square$	Halmos symbol
$\odot$	Sun
$R_\odot$	Radius of the Sun 695 700 km
$M_\odot$	Mass of the Sun $1.988547 \cdot 10^{30}$ kg
m_p	Mass of the proton $1.67 \cdot 10^{-27}$ kg
z	Redshift
c	Speed of light in vacuum 299,792,458 m/s
G	Gravitational constant $6.674 \cdot 10^{-11}$ $\mathrm{m^3 kg^{-1} s^{-2}}$
Λ	Cosmological constant $\approx 10^{-52}$ $\mathrm{m^{-2}}$
σ	Stefan-Boltzmann constant $5.669 \cdot 10^{-8}$ $\mathrm{Wm^{-2}K^{-4}}$
h	Planck constant $6.626\,070 \cdot 10^{-34}$ Js
au	Astronomical unit 149,597,870,700 m
pc	Parsec 3.262 ly = 206 265 au = $3.086 \cdot 10^{16}$ m
H_0	Hubble-Lemaître constant $\approx$ 70 km/(s Mpc)
$H(t)$	Hubble parameter
yr	Sidereal year 365.256 36 days = 31 558 149.54 s $\approx \pi \cdot 10^7$ s
ly	Light-year 63,242 au = $9.46 \cdot 10^{15}$ m
cy	Century
eV	Electronvolt = $1.602 \cdot 10^{-19}$ J

Chapter 1
Mathematical Modeling

1.1 Introduction

Cosmology is a branch of astronomy that deals with the greatest spatial and temporal distances and the questions of the origin of the universe as a whole. However, the current standard cosmological model leads to a wide range of serious paradoxes (or even catastrophes—see Index) and does not give acceptable results (see e.g. [24, 25, 79, 184, 205–207, 217, 226, 360]). Therefore, in this book we will analyze mathematical models used in cosmology and investigate their theoretical background.

The main problems of the standard ΛCDM (i.e. Lambda-Cold Dark Matter) cosmological model, which should not be ignored, can be summarized as follows:

1. The problem of the existence of dark matter (see Sect. 8.1).
2. The problem of the existence of dark energy (see Sect. 9.1).
3. The problem of isotropy of the universe (see Remark 6.1).
4. The problem of homogeneity of the universe (see Remark 6.2).
5. The flatness problem (see Remark 6.19).
6. The problem of the existence of the cosmological constant (see Remark 3.15).
7. The problem of the accelerated expansion of the universe and the energy non-conservation problem (see Remark 3.16 and Sect. 10.6).
8. The problem with observed values of the cosmological parameter σ_8 (see Remark 6.22).
9. The problem that the standard cosmological model admits a division by zero and that cosmological parameters have a strange behavior in time (see Sect. 6.7).
10. The problem with inconsistency of different values of the Hubble-Lemaître constant obtained by different methods (see Remark 6.23).
11. The problem of hierarchical structures (see Remark 1.12).

M. Křížek, L. Somer, *Mathematical Aspects of Paradoxes in Cosmology*,
https://doi.org/10.1007/978-3-031-31768-2_1

12. The problem of the existence of stars that are older than the universe according to the standard cosmological model (see Remark 7.28).
13. The problem of setting up accurate initial conditions (see Remark 10.2).
14. The problem of the existence of an infinite universe (see Remarks 6.27 and 6.28).
15. The inflation problem (see Remark 7.7).
16. The problem of the existence of giant black holes in the early universe (see Sect. 3.6).
17. The horizon problem (see [122]).
18. The problem of the large angular momnetum and velocities of all spiral galaxies (see Remark 9.7).
19. The problem of a 120-order-of-magnitude discrepancy between the density of fluctuations in a vacuum, which should have gravitational effects, and theoretically derived density of dark energy (see [11] for the vacuum catastrophe).
20. The Big Bang problem itself (see Remarks 6.8 and 6.21).

Some of these problems are overlapping, but their number is growing. Most mainstream cosmologists do not mind that these problems exist and still defend the standard cosmological model. Their frequently used argument: *"We have nothing better"* does not hold up.

Why is this so? The main reason is that physical reality is identified with very simple mathematical models (see e.g. (7.21) below) ignoring the modeling error.

The second reason is that people like sensationalism: Our universe consists of 25 % of some invisible mysterious dark matter and 70 % of even more mysterious dark energy, and nobody knows what it is, etc. However, these mysterious entities could be mostly the above-mentioned modeling errors.

The third reason is that cosmology is developing too rapidly and spontaneously. It usually does not rigorously use exact mathematical procedures and methods.

From 1–20 it is clear that the standard cosmological model, which is a direct mathematical consequence of Einstein's equations, does not approximate the physical universe well [200]. In our book we will investigate mathematically some of the above-mentioned problems and puzzles.

Mathematical models of physical phenomena, in general, are typically described by algebraic equations, ordinary or partial differential equations, integro-differential or integral equations, systems of these equations, variational inequalities, systems of differential-algebraic equations, etc. Their exact analytical solution is usually not known and there is no universal method capable of solving the above-mentioned models. Therefore, each class of problems has to be investigated individually. This consists of several steps including, as a rule, the proof of the existence (and, if applicable, also uniqueness) of the true and approximate solution, numerical solution of approximate problems and its stability, the treatment of convergence questions, continuous dependence of solutions on data, etc.

We should have in mind several important features in the mathematical modeling of physical problems. First, one should realize that no physical relation (except for definitions) holds entirely exactly, since the universe is composed from elementary particles and their behavior can be predicted only with some probability (due to the Heisenberg uncertainty principle and the Born probabilistic interpretation of wave functions). That is why each physical equation can only approximately describe reality. Any model problem thus includes several simplifications. The most common simplification is a "continuous approach", where discrete objects (such as molecules or atoms) of the universe are replaced by continuous ones. This idea allows us to employ the theory of differential or integral equations.

Second, we must never overestimate mathematics itself. Do not forget that mathematics is only a "game" (even though this game has precisely defined notions and rules). Mathematical models do exist only in our minds. They are invented mainly for better understanding of the universe.

Third, validation and verification of problems of mathematical physics and their computer implementation is a very important part of numerical analysis. We always encounter two basic types of errors: modeling error and numerical errors such as discretization errors, computational errors (iteration errors, algebraic errors, rounding errors) and also undiscovered programing bugs. Validation tries to estimate the modeling error and to answer the question:

Do we solve the correct equations?

On the other hand, verification tries to quantify the numerical errors and to answer the question:

Do we solve the equations correctly?

We will return to these questions in Sects. 1.5 and 1.6. Now we will focus on mathematical modeling in celestial mechanics.

1.2 Kepler's Laws

Already since the seventh century AD Chinese astronomers were assembling detailed tables of positions of the planets, a thousand years before Johannes Kepler (1571–1630). They even knew that the trajectories of the planets form peculiar loops on the celestial sphere. However, they did not discover Kepler's laws, since they did not have a good geometric model of the functioning of the Solar system and their measurements of positions of planets were one order of magnitude less accurate than the measurements of the Danish astronomer Tycho Brahe (1546–1601). It must also be taken into account that Kepler knew the Copernican heliocentric model of the Solar system very well and that Kepler's ingenuity certainly played a big role as well.

Johannes Kepler analyzed the very accurate data of Tycho Brahe on the movements of the planets during his stay in Prague from 1600 to 1612. He found that the planets follow elliptic orbits and that the areal velocity of the radius vector that originates at the Sun is constant for each planet. In this way the first and second laws of Kepler were discovered. They were first published in 1609 in one of Kepler's most important works, *Astronomia nova*. At present they are formulated as follows:

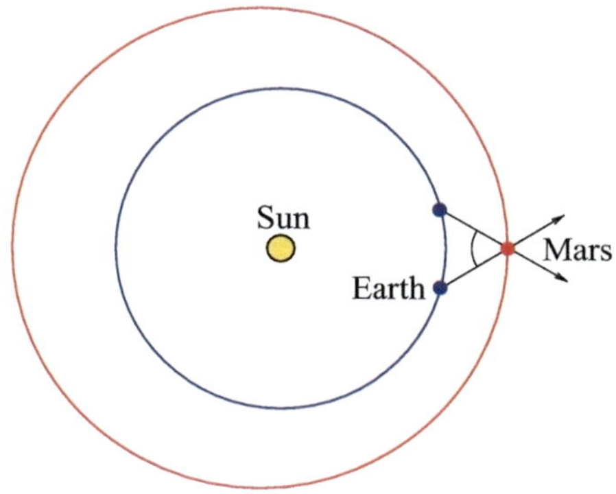

Fig. 1.1 Illustration of Kepler's method for the determination of the elliptic orbit of Mars, which after one Martian year projects to a different part of the sky. Mars is located at the intersection of the two directions

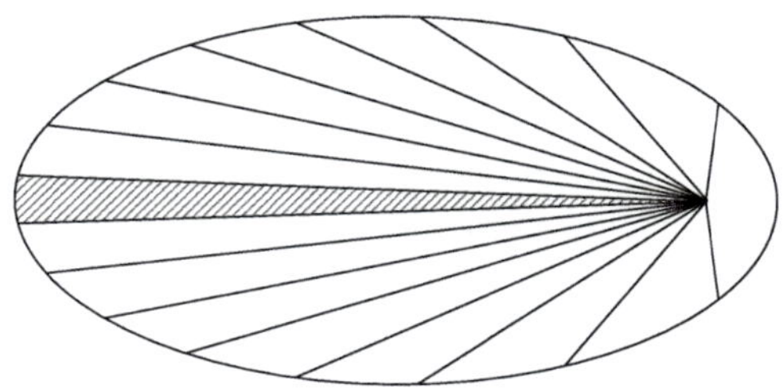

Fig. 1.2 Schematic illustration of Kepler's second law stating that the areal velocity is constant

Kepler's First Law *The orbit of each planet is elliptic with the Sun at one of its foci.*

Kepler's Second Law *The radius vector connecting a planet and the Sun sweeps out equal areas of the ellipse during equal intervals of time.*

Now we recall how Kepler discovered these laws. He knew that the orbital period of Mars is 687 days, therefore, after this time period Mars returns to the same place, while the Earth has orbited the Sun almost twice. This actually defines two different directions in which Mars is projected on the celestial sphere and Kepler could therefore determine its location in the orbital plane (see Fig. 1.1). Repeating this procedure for different time instants, with the help of coordinates of Mars measured by Brahe, Kepler could draw the entire orbit of Mars and find that its trajectory is elliptic (see Kepler's first law).

When Kepler ascribed relevant time data to the respective positions of Mars, he discovered the second law (see Fig. 1.2).

Let $a \geq b$ stand for the lengths of the semiaxes of an elliptic orbit of the planet. For simplicity, we will also denote these semiaxes by the same symbols a and b. Here a is called the *semimajor axis* and b the *semiminor axis* provided $a > b$.

The distance ε of the focus of the ellipse to its center is called the *linear eccentricity* and is defined by the relation (see Fig. 1.3)

$$\varepsilon = \sqrt{a^2 - b^2}.$$

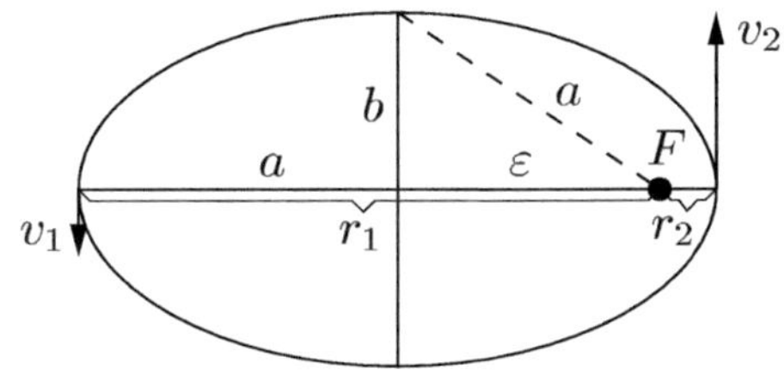

Fig. 1.3 Keplerian trajectory

Similarly

$$e = \frac{\varepsilon}{a} \tag{1.1}$$

is called the *numerical* or *orbital eccentricity*. For simplicity, we will call this dimensionless number only the *eccentricity*. We see that

$$b = a\sqrt{1 - e^2}. \tag{1.2}$$

The case $e = 0$ corresponds to a circle. The eccentricity of an ellipse is always less than one.

Kepler was actually very lucky that he concentrated his attention mainly to Mars, because its orbit has a relatively large eccentricity $e = 0.0935 \approx 0.1$ (see Fig. 1.1). A consequence of Kepler's first and second law is (see [155]):

Kepler's Third Law *The square of the orbital period of a planet is proportional to the cube of the semimajor axis of its orbit.*

Kepler discovered the third law due to his consistent and systematic work. Essentially, he found it empirically, when he left Prague. Kepler's third law can be formulated as

$$\boxed{T^2 = Ca^3.} \tag{1.3}$$

Here, T is the orbital period of the planets and $C > 0$ is a constant. At that time the multiplication of numbers was accomplished by means of the sum of logarithms and after using the inverse procedure, Kepler obtained the product. He noticed a simple dependence

$$2 \log_{10} T - 3 \log_{10} a = \text{const.}$$

between the decimal logarithms of the measured value of the orbital period T and the length of the semimajor axis a. It was then a short step to the formulation of the third "harmonic law", which Kepler published in 1619 in his work *Harmonices mundi libri V*. In 1621, Kepler found that his new law also applies to the four large moons of Jupiter.

Remark 1.1 From (1.3) we may calculate e.g. distances of all the planets from the Sun knowing their orbital periods and the Earth-Sun distance.

Now we will introduce some further useful relations for elliptic planetary orbits. Assume that the mass of a planet is negligible relative to the mass of the Sun. The point of its furthest (closest) approach to the Sun is called the *aphelion* (*perihelion*). If r_1 and r_2 are the respective distances of the planet at aphelion and perihelion from the focus F, where the Sun is located, then $r_1 = a + \varepsilon$ and $r_2 = a - \varepsilon$ (see Fig. 1.3).

From the relations

$$2a = r_1 + r_2, \quad 2\varepsilon = r_1 - r_2, \tag{1.4}$$

and $b = \sqrt{a^2 - \varepsilon^2}$ we see that the length of the semimajor axis a is equal to the arithmetic mean of distances r_1 and r_2, i.e.,

$$a = \frac{r_1 + r_2}{2}, \tag{1.5}$$

while the length of the semiminor axis b is equal to their geometric mean, i.e.

$$b = \sqrt{\left(\frac{r_1 + r_2}{2}\right)^2 - \left(\frac{r_1 - r_2}{2}\right)^2} = \sqrt{r_1 r_2}. \tag{1.6}$$

Denote by v_1 and v_2 the respective velocities of a planet at aphelion and perihelion. By the law of conservation of angular momentum (mrv =const.) it follows that

$$r_1 v_1 = r_2 v_2.$$

From this, (1.4), and (1.1) we obtain

$$\frac{v_2}{v_1} = \frac{r_1}{r_2} = \frac{a + \varepsilon}{a - \varepsilon} = \frac{1 + e}{1 - e}. \tag{1.7}$$

For a fixed e, the ratio v_2/v_1 is thus a constant independent of the size of the elliptic orbit. For the eccentricity $e = 0.0935$ of Mars' orbit we obtain a fairly high ratio $v_2/v_1 = 1.206$, which actually helped Kepler to discover his second law. For the eccentricity $e = 0.2056$ of Mercury's orbit (see Table 1.1) we get a much higher ratio $v_2/v_1 > 1.2/0.8 = 1.5$. Moreover, according to (1.7), we find that

$$e = \frac{v_2 - v_1}{v_2 + v_1} = \frac{r_1 - r_2}{r_1 + r_2}, \tag{1.8}$$

where $r_1 = a(1 + e)$ and $r_2 = a(1 - e)$.

Each planet is generally moving along its orbit nonuniformly. It is therefore important to be able to determine its coordinates at a given time instant in the plane

Table 1.1 The symbol i denotes the ith planet, m_i its mass in kg, a_i is the semimajor axis of its orbit in meters, T_i is the orbital period in years, v_i is the mean orbital velocity of the planet around the Sun in km/s, R_i is the mean radius of the planet in km, and e_i is the orbital eccentricity

i	Planet	$m_i/10^{24}$	$a_i/10^9$	T_i	v_i	R_i	e_i
1	Mercury	0.33022	57.9	0.241	47.9	2440	0.2056
2	Venus	4.8676	108.2	0.615	35.0	6052	0.0068
3	Earth	5.97219	149.6	1	29.8	6371	0.0167
4	Mars	0.64185	229.97	1.881	24.0	3390	0.0935
5	Jupiter	1898.6	778.4	11.861	13.1	69911	0.0487
6	Saturn	568.46	1427.0	29.457	9.6	58232	0.0520
7	Uranus	86.81	2869.6	87.011	6.8	25362	0.0472
8	Neptune	102.43	4496.6	164.79	5.4	24624	0.0086

of its elliptic path, and those are what is determined by the Kepler equation, see e.g. [159, 184] for details. In this case, the sum of kinetic and potential energy is constant.

1.3 Newton's Gravitational Law

Newton's law of universal gravitation played a crucial role in the development of physics. It significantly contributed to understanding the structure and evolution of the universe. Sometimes it is called the discovery of the last millennium, as will be indeed seen from the following paragraphs. It was first formulated by Sir Isaac Newton (1642–1726) in his treatise *Principia* from 1687. In particular, he was inspired by Kepler's third law.

According to Newton's law of gravitation, the size of the gravitational force between two point masses is equal to

$$\boxed{F = G\frac{mM}{r^2},} \tag{1.9}$$

where m and M are their masses, r is their distance, and

$$G = 6.674 \cdot 10^{-11}\,\mathrm{m^3kg^{-1}s^{-2}}$$

is the *gravitational constant* (cf. Remark 10.1).

Remark 1.2 If m is the mass of a planet and $M = M_\odot$ the mass of the Sun, then for a circular planetary orbit of radius $r = a$ by Newton's law of action and reaction, the gravitational force (1.9) is equal to the centripetal force, i.e.,

$$G\frac{mM}{a^2} = \frac{mv^2}{a},$$

where v is the orbital velocity of the planet. From this for $v = 2\pi a/T$ we obtain as a result Kepler's third law in the form

$$\frac{a^3}{T^2} = \frac{GM}{4\pi^2} \ (= C), \tag{1.10}$$

where it is implicitly assumed that $m \ll M$.

Remark 1.3 Observe that relation (1.10) connects three important quantities with the physical dimensions of length, mass, and time. In the system of units SI, they are expressed in

m, kg, s.

Knowing two of them, we can calculate the third one. Therein lies the beauty of Kepler's third law.

In some textbooks it is inaccurately stated that F in (1.9) is the size of gravitational force between two bodies. In doing so, the reader will not know how to define precisely their distance. For example, what is the distance r between a homogeneous tiny ring (almost a circle) and a point mass at its center? That is: ⊙

1. If the distance r is equal to the radius of the ring, we do not get the correct answer by (1.9), since the resulting total force (and thus also its size) is zero.
2. However, for $F = 0$ relation (1.9) yields $r = \infty$, which certainly has nothing to do with reality.
3. If r is the distance between the centers of gravity of both bodies, then we surprisingly obtain $F = \infty$, because the denominator in (1.9) is $r = 0$. Again, we get into trouble.

Thus we see that a mechanical use of the law of gravity can lead to unexpected paradoxes. In what follows we shall, therefore, often restrict ourselves only to idealized point masses, which in the physical world do not exist. However, relation (1.9) remains unchanged if instead of point masses we consider balls with special density distributions as follows from the next theorem.

Theorem 1.1 (Newton's First Theorem) *If the density distribution of a ball of mass M is spherically symmetric, then the size of the force between the ball and a point mass m that lies outside the interior of the ball, is given by (1.9), where r is the distance between the point and the center of the ball.*

The proof is based on the specific shape of the gravitational potential of mass uniformly distributed on a spherical surface. The resulting equation (1.9) is then obtained by integration, see e.g. [15, p. 149]. Theorem 1.1 can evidently be generalized to the interaction between two balls with spherically symmetric mass density distribution (see Fig. 1.4).

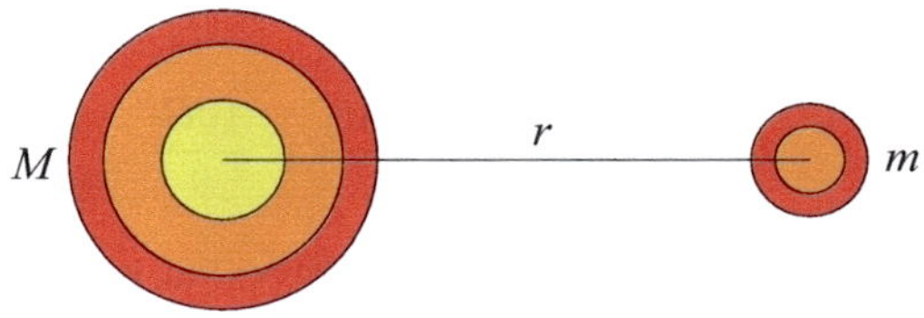

Fig. 1.4 Illustration of Newton's first theorem for two spherically symmetric bodies

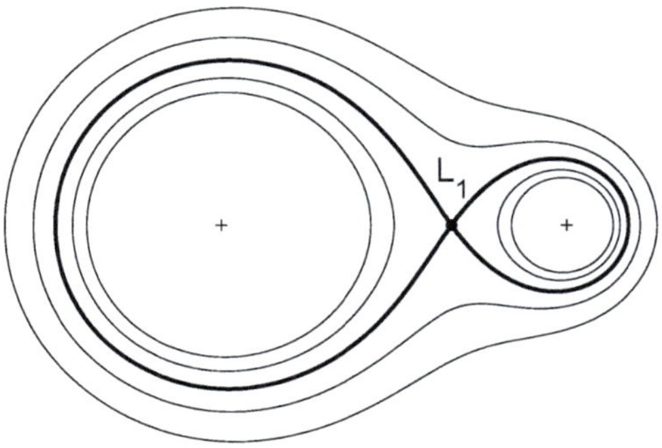

Fig. 1.5 Roche lobe of two closely orbiting bodies

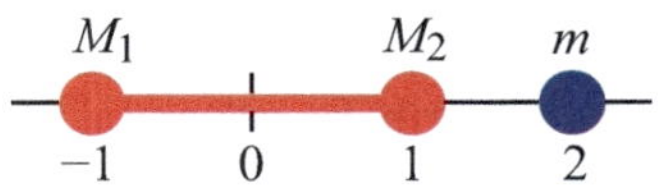

Fig. 1.6 The force between two bodies cannot generally be substituted by the force between two point masses located at their centers of gravity

Remark 1.4 Celestial bodies with diameter of over $\approx$1000 km are automatically formed by gravity in an approximate spherical shape and their density distribution also becomes almost spherically symmetric. This process is called *gravitational differentiation*. However, two large closely orbiting bodies are not spherically symmetric due to tidal bulges. In Fig. 1.5 we see the intersection of contour lines of the gravitational potential at the Lagrangian point L_1 of the corresponding Roche lobe (cf. [28, p. 350]).

Example 1.1 In general, bodies that do not have a spherically symmetric mass density distribution cannot be replaced by point masses at their centers of gravity (e.g. the elongated asteroid Ida which is orbited by the satellite Dactyl). To see this, it is enough to consider how a dumbbell-shaped body with mass $M = M_1 + M_2$ acts on the second body, which is a homogeneous ball of mass m. The mass of the straight middle part of the dumbbell is neglected. Set $M_1 = M_2 = m = 1$ kg. The scale on the horizontal axis in Fig. 1.6 is indicated in meters. Then the magnitude of the total force between the two bodies is equal to

$$F = G\frac{mM_1}{3^2} + G\frac{mM_2}{1^2} = \frac{10G}{9}.$$

However, if we concentrate the mass of the dumbbell at its center of gravity at 0, then the resulting force $\tilde{F}$ will be essentially different from F, i.e.

$$\tilde{F} = G\frac{m(M_1 + M_2)}{2^2} = \frac{G}{2} \quad \text{and} \quad 2\tilde{F} < F.$$

The quadratic nonlinearity r^2 in the denominator of Eq. (1.9) thus causes that the force F is more than two times bigger than the force $\tilde{F}$ corresponding to the mass of the first body concentrated at its center of gravity.

Example 1.2 Is it possible to measure the mean mass density of the Sun using a protractor? To solve this seemingly absurd problem assume for simplicity that the Earth's orbit is circular. From (1.9), Newton's second and third law (of action and reaction), we get as in Remark 1.2 that

$$G\frac{M_\odot m}{r^2} = \frac{mv^2}{r}, \tag{1.11}$$

where $M_\odot$ is the mass of the Sun, m is the mass of the Earth, r is their mutual distance, and v is the speed of the Earth. By a "careful" use of the protractor, we find that the viewing angle of the Sun is approximately $\delta = 32'$. Hence, $R_\odot = r \sin \frac{1}{2}\delta$ is the radius of the Sun, see Fig. 1.7. Clearly,

$$v = \frac{2\pi r}{T}, \tag{1.12}$$

where $T = 31{,}558{,}149.54$ s $= 365.25636$ days is the Earth's orbital period (the sidereal year). Denoting by $V = \frac{4}{3}\pi R_\odot^3$ the volume of the Sun and substituting $M_\odot$

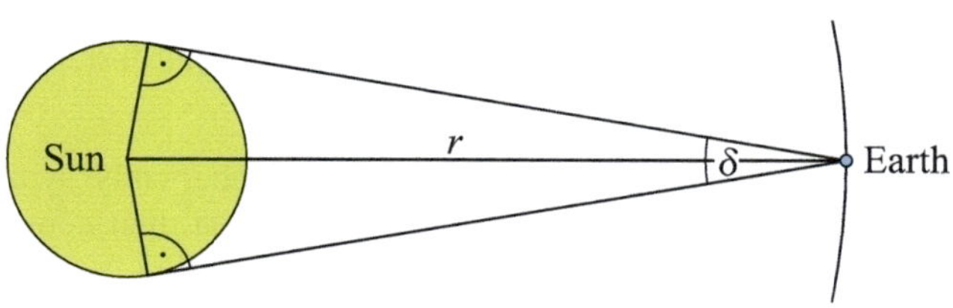

Fig. 1.7 The mean mass density of the Sun can be determined from its viewing angle δ and the Earth's orbital period (see (1.13))

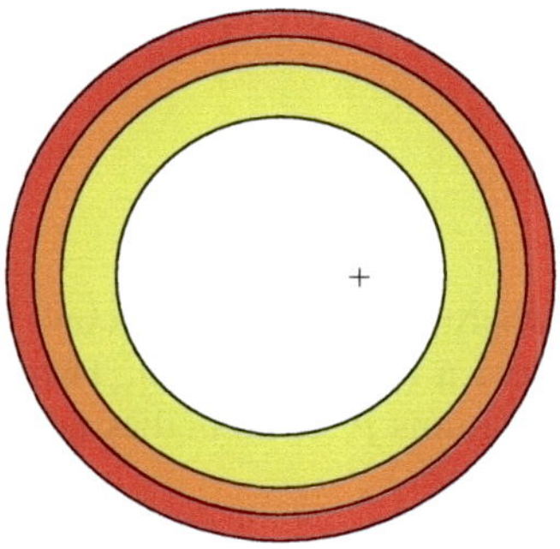

Fig. 1.8 Illustration of Newton's Second Theorem 1.2 in a cross-section of the spherical layer. The force acting on a point mass inside the cavity marked by + is zero for a spherically symmetric mass density distribution. The gravitational potential inside the cavity is constant

from (1.11), we get by (1.12) that the mean mass density is

$$\rho = \frac{M_\odot}{V} = \frac{v^2 r}{GV} = \frac{(2\pi r)^2 \cdot r}{T^2 G \cdot \frac{4}{3}\pi (r \sin \frac{1}{2}\delta)^3} = \frac{3\pi}{T^2 G \sin^3 \frac{1}{2}\delta} = 1409 \text{ kg/m}^3, \tag{1.13}$$

i.e., it is about 40 % higher than the density of water. Moreover, using Kepler's third law (1.10), we find that the mass of the Sun is about

$$M_\odot = 2 \cdot 10^{30} \text{ kg}.$$

Another important theorem for a spherical layer states (see Fig. 1.8):

Theorem 1.2 (Newton's Second Theorem = Shell Theorem) *A spherical layer with spherically symmetric mass density distribution exerts no force on a point mass located inside.*

Remark 1.5 Newtonian celestial mechanics is very elegant and perfectly models reality on short time scales. It is the simplest theory of gravitation. Claudius Ptolemy (ca. 100–170 AD) described motion of planets by very complicated epicycles. On the other hand, the General Theory of Relativity (and its extensions) is too complicated to describe the evolution of the Solar system as we shall see in Sect. 5.3.

1.4 The N-Body Problem

For an integer $N \geq 2$ consider an isolated system of N point-mass bodies that mutually interact only gravitationally. Let $r_i = r_i(t)$, $i \in \{1, \ldots, N\}$, be the so-called radius-vector in Euclidean space $\mathbb{E}^3$ describing the position of the ith body with mass m_i in time $t \geq 0$. According to Newton's second law of motion, the force between them is linearly proportional to the acceleration. By (1.9) the classical N-body problem is described (see [1, 156, 184, 234]) by the following nonlinear system of ordinary differential equations for the unknown trajectories r_i

$$\ddot{r}_i = G \sum_{i \neq j}^{N} \frac{m_j (r_j - r_i)}{|r_j - r_i|^3}, \tag{1.14}$$

where $\ddot{r}_i$ denotes the second time derivative of r_i. However, this system does not have a unique solution. Therefore, at time $t = 0$ we impose *initial conditions* on positions p_i of all bodies and their velocities v_i,

$$r_i(0) = p_i, \quad \dot{r}_i(0) = v_i \quad \text{for } i = 1, \ldots, N \tag{1.15}$$

which, of course, have an essential influence on the resulting motion of the bodies. Note that Eq. (1.14) are called *autonomous*, since their right-hand side does not explicitly depend on time. Hence, without loss of generality, we can take the initial conditions at $t = 0$.

From the theory of ordinary differential equations, we know that the solution of problem (1.14) and (1.15) exists uniquely on an arbitrarily long time interval, where all mutual distances $|r_i - r_j|$ for $i < j$ are nonzero, i.e., the bodies do not collide. Note that the probability that two or more point masses collide within a fixed time interval is equal to zero, even though this case may theoretically occur. Although there are numerous collisions in the physical world, we should not forget that in the classical multiple body problem only point masses are considered, while actual bodies always have positive sizes.

The N-body problem has many real-world technical applications. As an example let us recall the complex gravitational ping-pong of the journey of Voyager 2 to Jupiter, Saturn, Uranus, and Neptune, when they all were on the same side of the Sun. The idea of accelerating probes by planetary gravitational fields was developed around 1960 during numerical simulations of the multiple body problem. It is a concept of the American mathematician Michael Andrew Minovitch. The actual trajectory is obtained by the numerical solution of a system of differential equations (1.14). Each planet greatly accelerated the probe by means of its gravitational field and also changed its direction with respect to the Sun. The probe actually fell into the gravitational potential well of each planet. Sometimes this orbit is referred as a *gravitational slingshot*. Higher kinetic energy is gained by the probe at the expense of the total energy of the planet. The speed of the probe after a close flyby about a planet decreases, of course.

Remark 1.6 Newton's gravitational law also enabled us to discover e.g. the planet Neptune and white dwarf Sirius B, to land six Apollo missions on the Moon, to obtain unique photos of planets and their moons, to calculate trajectories of asteroids threatening the Earth, to launch meteorological and telecommunication satellites (including GPS), to place telescopes in space, and to establish various physical parameters in the Solar system (cf. Example 1.2) and distant universe.

Remark 1.7 System (1.14) does not satisfy the well-known Carathéodory sufficient conditions for the existence of the solution, since the right-hand side of (1.14) is not continuous when $r_j \to r_i$. Despite this, the solution exists locally as long as the point-bodies do not collide. The system (1.14) is nonlinear and has a large number of unrealistic solutions. For instance, a simple two-body problem on one line admits arbitrarily large (superluminal) velocities, see [184, p. 69]. In [313] five bodies are considered such that all of them reach infinity in a finite time for appropriate initial conditions. The Newtonian theory of gravitation thus may describe completely absurd situations. Papers [65, 252] present also a solution of (1.14) when three equally massive particles orbit along a trajectory that looks like ∞ which has not yet been observed in the universe. In this special case, the total orbital momentum is zero and the trajectory is stable with respect to small long-term perturbations.

Remark 1.8 The system of ordinary differential equations (1.14) is sometimes also used to simulate the evolution of galaxies or even galaxy clusters when galaxies are replaced by point masses. Note that galaxies have huge speeds reaching up to a few percent of the speed of light. However, system (1.14) does not describe relativistic phenomena nor effects of tides, rotation of galaxies, the finite speed of propagation of the gravitational interaction, and so on. Hence, it is only an approximate model. The N-body problem should not be applied to strong gravitational fields either.

1.5 General Computational Scheme

The general computational scheme for solving problems of mathematical physics is sketched in Fig. 1.9. Since no equation describes physical reality absolutely exactly, we always make a *modeling error* e_0. The true solution is usually searched in suitable function spaces whose dimension is infinite. However, for computer implementation we have to approximate them by finite dimensional problems, which yields a *discretization error* e_1. This error may also include the error of numerical integration, error of approximation of the boundary of the examined region, interpolation of input data, etc. Finally, a *computational error* e_2 arises during the computation of the discrete model itself on a computer. It contains, of course, rounding errors, but may include other errors (like iteration error, etc.).

The errors e_0, e_1, and e_2 should be handled with care. They are generally not canceled. If just one is large, then the total error $e_3 = e_0 + e_1 + e_2$ is also large. Thus there is no point in trying to make another error very small. Therefore, it is computationally advantageous that all these errors are about the same size if possible.

Remark 1.9 Some authors include a physical model between the physical reality and its mathematical model in Fig. 1.9. The corresponding error analysis is similar.

Without any exception each equation of mathematical physics has some limitations on the size of the investigated objects. For instance, the standard heat

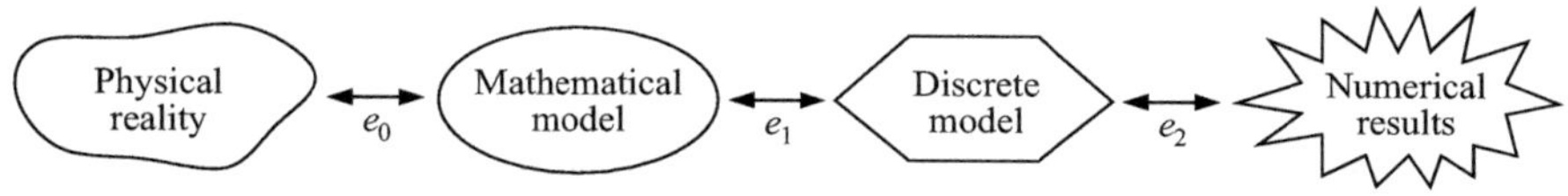

Fig. 1.9 The modeling error e_0 is the difference between physical reality and its mathematical description. A discrete finite dimensional model differs from the mathematical model by the discretization error e_1. Finally, in e_2 are included computational errors (rounding errors, iteration errors, etc.)

conduction equation (stationary or nonstationary) approximates very well the true temperature in solids of size about 1 meter, which can be verified by direct measurements [191]. However, in applying the heat equation on the atomic level in the cube with edge 10^{-10} m, we get nonsensical numbers, as well as in the cube with edge 10^{10} m, which would immediately collapse into a black hole (see Example 7.3 below). Analogously, we should not apply Keplerian laws on scales of 10^{-10} m or the Schrödinger equation on objects that have the size of a cat. Similar arguments apply also for time scales. For instance, the standard predator-prey equation or the N-body problem on the interval of 10^{10} years do not describe the reality well. Therefore, in any calculation we have to take care of the modeling error which is small only under some restrictive space-time limits.

In spite of that, when deriving the standard cosmological Friedmann equation, Einstein's equations are applied to the entire universe. This is considered as a platitude and almost nobody deals with the question, whether it is justified to perform such a fearless cosmological extrapolation without any observational support, since general relativity has only been "tested" on much smaller scales like the Solar system (bending of light in the gravitational field of the Sun, measuring the curvature of spacetime near the rotating Earth by means of the Lense-Thirring precession effect, perihelion advance of Mercury's orbit, Shapiro effect, etc.). Note that galaxies have a diameter of the order of 10^{10} astronomical units and the universe has at least five orders of magnitude larger diameter.

Fritz Zwicky [385] and Vera Rubin [307] used too simple models and insufficient data involving diverse selection effects which led to the prediction of dark matter [183]. For instance, at the end of the last century astronomers believed that only 3% of all stars are red dwarfs [29, p. 93]. However, at present from the Gaia data we know that at least 76% of all stars are red dwarfs. Therefore, the proposed ratio 27 : 5 between the amount of dark matter and baryonic matter is almost certainly overestimated. According to [196, p. 442], the (normalized) Friedmann equation was derived under an incorrect cosmological extrapolation ignoring the modeling error. Of course, Alexander Friedmann when he published in 1922 his famous paper [108] had no idea about the size of our universe, since galaxies were discovered only in 1924 by Edwin Powell Hubble. Thus, it seems that cosmologists solve incorrectly extrapolated equations and then present their results very exactly (up to three/four significant digits).

1.6 Estimation of Modeling and Numerical Errors

The difference between the true solution of a mathematical model and its numerical approximation is usually estimated using various a priori or a posteriori error estimates. This step is very important, since real-life problems are often nonlinear, large scale, ill-conditioned or unstable. They may have multiple solutions, singularities, etc. If we do not perform a reliable error analysis, we do not know, in fact, how close the numerical solution is to the exact one. First we will again deal with the N-body problem (1.14).

Can we believe numerical simulations of the evolution of the Solar system based on (1.14) for billions of years in the past or future?

The answer is NO for several reasons. Newtonian mechanics does not allow any delay given by the finite speed of gravitational interaction. System (1.14) is only a vector ordinary differential equation whose solution on the interval $[0, T]$ depends only on positions and velocities at the point $t_0 = 0$ and not on the history. An extremely small modeling error $\varepsilon > 0$ during one year may yield an error of order $\varepsilon 10^9$ after 1 Gyr that may be quite large [161]. Also various nongravitational forces (Yarkovsky effect, YORP effect, electromagnetic forces, etc.) are not included in (1.14). The corresponding calculations over time intervals of several Gyr thus resemble computations of weather forecast for many months in advance. The reason is that system (1.14) is not Lyapunov stable in time, i.e., extremely small changes of initial conditions or other perturbations may cause very large changes in the final state provided the time interval is long enough. For $N > 2$ Isaac Newton stated:

An exact solution exceeds, if I am not mistaken, the force of any human mind.

Henri Poincaré knew that the N-body problem for $N > 2$ has an analytical solution only in a few special cases, but there is no explicit formula for its general solution (see Remark 4.8 below). Therefore, this problem has to be solved numerically by means of Runge-Kutta methods, multi-step methods, symplectic methods, etc., see e.g. [1, 184, 234]. For instance, we cannot calculate a reliable trajectory of our Moon in the Solar system on intervals longer than 10,000 years due to Lyapunov instability (see e.g. references in [340]).

If an integration step h gives almost the same numerical results as $h/2$ the integration of the system (1.14) is usually stopped. However, this heuristic criterion does not need to work properly. Here we present another way how to check whether our numerical results are reliable. It is based on backward integration of (1.14) as sketched on Fig. 1.10. Let $r = (r_1, \ldots, r_N)$ denote a vector function with $3N$ entries, i.e., each component r_i of r is a three-dimensional vector function. Let $f = f(r)$ stand for the right-hand side of (1.14) in a three-dimensional space. Then we can formulate the following theorem [162].

Theorem 1.3 *Let $r = r(t)$ be the unique solution of the system*

$$\ddot{r} = f(r) \tag{1.16}$$

on the interval $[0, T]$ *with given initial conditions*

$$r(0) = r_0 \quad \textit{and} \quad \dot{r}(0) = v_0. \tag{1.17}$$

Then the function $s = s(t)$ *defined by*

$$s(t) = r(T - t) \tag{1.18}$$

solves the same system (1.16) with initial conditions $s(0) = r(T)$, $\dot{s}(0) = -\dot{r}(T)$ *and we have*

$$s(T) = r_0 \quad \textit{and} \quad \dot{s}(T) = -v_0. \tag{1.19}$$

Proof According to (1.18) and (1.16), we obtain

$$\ddot{s}(t) = (-\dot{r}(T - t))^{\cdot} = \ddot{r}(T - t) = f(r(T - t)) = f(s(t)).$$

We see that s satisfies the same system (1.16) like r. For the final conditions by (1.18) and (1.17) we get relations (1.19),

$$s(T) = r(0) = r_0 \quad \text{and} \quad \dot{s}(T) = -\dot{r}(T - T) = -\dot{r}(0) = -v_0.$$

This ends the proof. □

Example 1.3 Since the N-body problem is exponentially unstable, the above theorem can be applied follows. Denote by r^* and s^* a numerical solution of the system (1.14) with initial conditions (1.18) and

$$s^*(0) = r^*(T) \quad \text{and} \quad \dot{s}^*(0) = -\dot{r}^*(T),$$

respectively. If $\delta > 0$ is a given tolerance and

$$|s^*(T) - r_0| + |\dot{s}^*(T) + v_0| \gg \delta,$$

then we cannot claim that $|r(T) - r^*(T)| + |\dot{r}(T) - \dot{r}^*(T)| < \delta$, where r is the exact solution, i.e., the numerical error of the original problem (1.14) and (1.18) would be also large, as schematically depicted in Fig. 1.10.

Finally, let us point out that the previous Theorem 1.3 and computational strategy can be easily generalized to evolution partial differential equations.

From (1.14) we observe that when two bodies are close to each other ($r_j \approx r_i$) which is an important case e.g. in space aeronautics, then we subtract in the denominator two numbers of almost the same size. This may lead to a catastrophic loss of accuracy (see e.g. [45, 162, 255] for several shocking examples).

Moreover, numerical error generally grows exponentially with time. This fact has serious consequences. For instance, on April 13, 2029, the distance of the asteroid

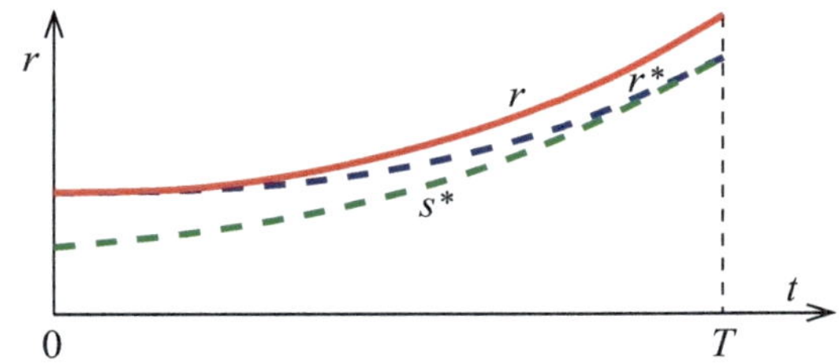

Fig. 1.10 Application of Theorem 1.3 in numerical solution of the N body problem. The symbol r stands for the true solution, r^* is the numerical solution, and s^* is the control backward solution

Apophis from the Earth will be about 30,000 km. However, for the next approach in 2036 the distance cannot be reliably calculated, since the system (1.14) is unstable.

Remark 1.10 There is a general belief in the current astrophysical community that Einstein's equations best describe gravity [167]. However, their extreme complexity prevents one from verifying whether they model e.g. the Solar system better than Newtonian N-body simulations with $N > 1$. Hence, in this case Einstein's equations are hardly testable in their general form, since their analytical solution is not known and they are, in fact, non-computable by present computer facilities, as we shall see in Sects. 5.1 and 5.3. Moreover, it will be obvious why we are unable to calculate reliably trajectories of planets by means of Einstein's equations, e.g. in the Jupiter-Sun system. The reason is that the mass of Jupiter is not negligible with respect to the Sun's mass.

On the other hand, such trajectories can be calculated numerically very precisely on short time intervals by the Newtonian N-body simulations (e.g. with 8 planets) even though their analytical solution is not known. Thus the modeling error e_0, which is the difference between observed trajectories and the analytical solution, is also not known. However, the modeling error e_0 (see Fig. 1.9) can be easily estimated by the triangle inequality

$$|e_0| \leq |e_1| + |e_2| + |e_3|,$$

where e_3 is the total error which is the difference between observed trajectories and numerically calculated trajectories. The sizes of e_1 and e_2 can be established by various a priori and a posteriori error estimates frequently used in numerical analysis.

Remark 1.11 A *light cone*, i.e. a future cone, is the path that a flash of light, emanating from a single event (localized to a single point in space and a single moment in time) and traveling in all directions, would take through spacetime (cf. Fig. 1.11). Thus the N-body problem (1.14) does not satisfy the *principle of causality:*

All real events necessarily have a cause.

The reason is that in Newtonian formulation an infinite speed of gravitational interaction is assumed which is not physical. Hence, any information about changing

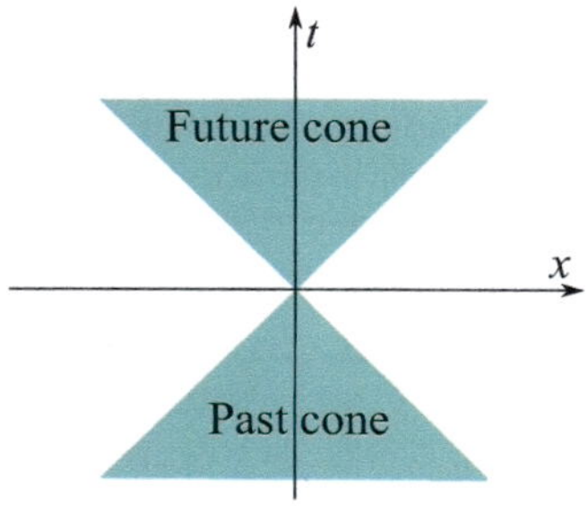

Fig. 1.11 Newton's theory of gravitation does not satisfy the principle of causality, since it assumes that the speed of gravitational interaction is infinite. Thus if any body changes its position, then information about this occurrence immediately travels outside the corresponding light cone along the horizontal axis which represents the spatial coordinates (x, y, z), where two spatial dimensions are suppressed

position gets immediately horizontally out the corresponding light cone. The modeling error of the speed of gravitational interaction is thus infinite.

Remark 1.12 Newtonian celestial mechanics is designed in such a way that the laws of conservation of energy and angular momentum hold absolutely exactly. However, in Sects. 9.2, 9.3, and 10.4, etc., we shall see that this is not true for the physical universe. Moreover, on different scales we observe qualitatively completely different hierarchical structures (stars, planetary systems, globular clusters, spiral galaxies, galaxy clusters, galaxy filaments, cosmic web) that are stable for billions of years. This indicates that physical gravity is not scale invariant.

Chapter 2
Paradoxes in the Special Theory of Relativity

2.1 Short Introduction to the Special Theory of Relativity

This chapter is based on our papers [168–171]. In Theorem 2.1 and Examples 2.2 and 2.5 we show that a finite speed of light and the Doppler effect can produce more dominant and entirely opposite effects for relativistic speeds than those predicted by the Special Theory of Relativity (STR), in particular, the clock paradox, time dilation, and length contraction. This important fact is not emphasized in most of the books on STR.

For instance, by STR a moving clock ticks slower than at rest. However, an observer will measure a higher frequency of an approaching clock than the same clock has at rest due to the Doppler effect. We also prove that under certain conditions an approaching bar on a photo may seem to have a larger length for a relativistic speed than at rest, even though by STR lengths of moving objects are contracted.

According to *Newton's first law of inertia*, a body will remain at rest or in uniform motion in a straight line unless acted upon by an external force. This fundamental physical principle serves to introduce inertial systems in STR, see [107, p. 211]. Consider a fixed coordinate system S with orthogonal axes x, y, z containing a fixed system of hypothetical synchronized clocks that define the time coordinate $t \in (-\infty, \infty)$ of a uniformly flowing time. The coordinate system S is called *inertial* if it obeys Newton's first law of motion.

Let S' be another coordinate system with orthogonal axes x', y', z' which are for simplicity parallel with x, y, z and have the same scale at rest [384]. The time $t' \in (-\infty, \infty)$ in S' is introduced similarly using a fixed system of synchronized clocks in S' having also the same time scale at rest. Let the origin of S' move along the x axis at a constant speed $v \in (-c, c)$, where c is the speed of light in a vacuum, see Fig. 2.1.

M. Křížek, L. Somer, *Mathematical Aspects of Paradoxes in Cosmology*,
https://doi.org/10.1007/978-3-031-31768-2_2

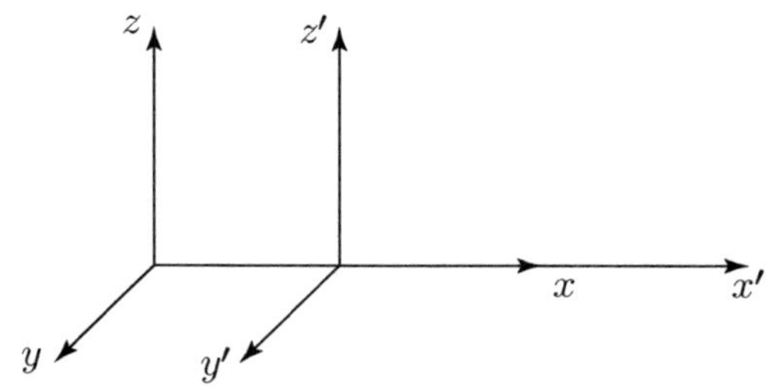

Fig. 2.1 The inertial system S' is moving with speed $v \in (-c, c)$ with respect to the system S

Remark 2.1 The basic postulate of STR that the speed of light c has the same size in all inertial systems was verified experimentally on the Earth by the well-known Michelson experiments [244]. The inertial systems S and S' move equally with each other.

Lorentz's transformation (see [227] and also [293]) is a fundamental tool of STR. The parameter defined by

$$\gamma_v = \frac{1}{\sqrt{1 - \frac{v^2}{c^2}}} \geq 1 \tag{2.1}$$

is called the *Lorentz factor*. Notice that its reciprocal value fulfills the equation for the unit circle

$$\left(\frac{v}{c}\right)^2 + \left(\frac{1}{\gamma_v}\right)^2 = 1.$$

Points of the four-dimensional spacetime are called *events*. This space is sometimes referred as flat, since it is not curved. Unless otherwise stated, we will restrict ourselves to one pair of the above described inertial systems, where the encounter of the origins of S and S' determines the beginning of time counting in the first and in the second inertial system, respectively, i.e., $t = 0$ in S and $t' = 0$ in S'. In this special case the *Lorentz transformation* has the form $\mathcal{L}_v : (x, y, z, t) \mapsto (x', y', z', t')$,

$$x' = \gamma_v(x - vt), \tag{2.2}$$

$$y' = y,$$

$$z' = z,$$

$$t' = \gamma_v\left(t - \frac{v}{c^2}x\right), \tag{2.3}$$

where $x, y, z, t \in (-\infty, \infty)$ and the last equality expresses how to transform a uniformly flowing proper time during the transition from S to S'. Events which are simultaneous in S are given by the identity $t \equiv t_0$, where t_0 is a fixed constant.

By (2.3) we see that the time t' depends not only on t but also on the position x, i.e., t' is not constant and thus the corresponding events do not have to be simultaneous in S' for $v \neq 0$.

Remark 2.2 Albert Einstein uses the transformation (2.2) and (2.3) in his pioneering paper [89, p. 902] from 1905, but does not use the term inertial. He also does not cite Lorentz's paper [227, p. 185] from 1892 nor does he mention Hendrik Lorentz himself. Einstein probably knew Lorentz's work [227], since the titles of their two papers are very similar. Moreover, Lorentz was very famous after receiving the Nobel Prize in 1902.

Let us emphasize that any two different events which are simultaneous in S are not causally connected. An observer can verify that they were really simultaneous, e.g., when he is in the middle and their future light cones just intersected.

Notice that the right-hand sides of relations (2.2) and (2.3) are linear functions in variables x and t for any fixed v. Thus, for $\boldsymbol{x} = (ct, x, y, z)$ and $\boldsymbol{x}' = (ct', x', y', z')$ the Lorentz transformation can be rewritten into the matrix form

$$\boldsymbol{x}' = \boldsymbol{L}_v \boldsymbol{x},$$

where

$$\boldsymbol{L}_v = \begin{pmatrix} \gamma_v & -\frac{v}{c}\gamma_v & 0 & 0 \\ -\frac{v}{c}\gamma_v & \gamma_v & 0 & 0 \\ 0 & 0 & 1 & 0 \\ 0 & 0 & 0 & 1 \end{pmatrix} \tag{2.4}$$

is a block diagonal symmetric and positive definite matrix. Note that the physical dimension of all entries of the vectors $\boldsymbol{x}$ and $\boldsymbol{x}'$ is one meter.

The inverse matrix $\boldsymbol{L}_v^{-1}$ has a similar form as $\boldsymbol{L}_v$, only the two minus signs in (2.4) have to be replaced by plus. Therefore, the Lorentz transformation $\mathcal{L}_v$ is a one-to-one mapping from $\mathbb{E}^4$ onto $\mathbb{E}^4$ for $v \in (-c, c)$.

Let us point out that in the limit case $|v| = c$, the matrix (2.4) becomes singular, since its first two rows are linearly dependent. It this case, the inverse of $\mathcal{L}_v$ does not exist. Consequently, the Lorentz transformation should not be applied to the surface of the light cone.

2.2 Time Dilation

The relation (2.3) is to be understood as the time which we would record at the moment when the two clocks in S and S' are closely passing each other at one single x-coordinate (e.g. at the origin). So we can compare only time data t and t' of local clocks, because the concept of the present is relative. By definition, all clocks in each inertial system at rest show the same time in the entire infinite three-dimensional space (e.g. at the beginning and at the end of a motionless bar). Thus, when we are exactly in the middle between any two fixed clocks, they will show us the same time.

Consider a fixed time interval

$$\Delta t' = t_2' - t_1',$$

where t_i' are space-independent coordinates in S'. For an arbitrary fixed point x in S we determine the corresponding t_2 and t_1 from formulae (cf. (2.3))

$$t_2' = \gamma_v\Big(t_2 - \frac{v}{c^2}x\Big), \qquad t_1' = \gamma_v\Big(t_1 - \frac{v}{c^2}x\Big),$$

and we set $\Delta t = t_2 - t_1$. From this we get the well-known *time dilation* (see e.g. [133, p. 430])

$$\Delta t' = \gamma_v(t_2 - \frac{v}{c^2}x - t_1 + \frac{v}{c^2}x) = \gamma_v \Delta t. \tag{2.5}$$

By (2.1) we see that $\Delta t' > \Delta t$ for any $v \neq 0$ independently of the sign of v. The relation (2.5) actually expresses that the time, measured by a clock located at a fixed coordinate x' in a moving system S', runs slower than the time measured by a clock that is at rest with respect to S. Hence:

The clock at rest is fastest.

The time dilation is usually theoretically justified as follows: A photon launched from the origin of the systems S and S' at time $t = t' = 0$ in the z direction flies obliquely in S' with speed c, see Fig. 2.2. Therefore, in terms of an observer in S' this photon needs longer time to reach the plane $z' = z = 1$ than in terms of an observer in the system S.

Example 2.1 The experimental verification of time dilation can be demonstrated by means of particles called muons whose mean half-life time at rest is $\tau = 2.2 \cdot 10^{-6}$ s. From observations of cosmic rays we know that if muons move linearly at almost the speed of light, they will travel on average a much longer distance than $c\tau = 660$ m. However, it should be emphasized that in the inertial system associated with muons their decay will not slow down.

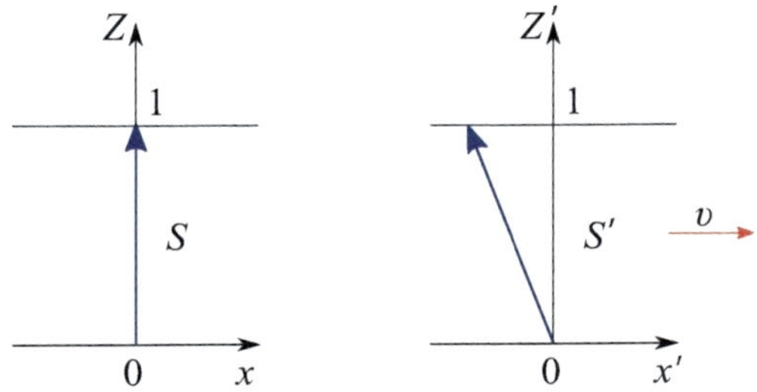

Fig. 2.2 Two (blue) trajectories of the same photon in the inertial systems S and S'

Remark 2.3 The time dilation is also verified by means of the transverse relativistic Doppler effect, see e.g. [38]. The transverse relativistic Doppler effect was first measured by Ives [138] already in 1938. In classical mechanics, this transverse effect does not occur, because it is given by time dilation (2.5) only. Note that the Hafele-Keating experiment [123] with two atomic clocks in airplanes and one on the Earth is not too credible, since none of the corresponding three systems was inertial.

Remark 2.4 Olaf Rømer (in Journal des Sçavans, 1676) suggested an elegant method to measure the speed of light. When the Earth moved toward Jupiter, the time interval between successive eclipses of Jupiter's moon Io became steadily shorter with respect to the terrestrial time. When the Earth moved away from Jupiter these eclipses became steadily longer, i.e., they were behind the expected values. Rømer thus actually found a phenomenon, which was later named after Christian Doppler [184, p. 28].

The non-relativistic longitudinal Doppler effect [83] is described by the relation

$$f_v = \frac{c}{c - v} f, \tag{2.6}$$

where f is the source frequency at rest, v is the speed of the source approaching an observer along the axis x, f_v is the frequency measured by the observer, and c is the speed of signal. For relativistic speeds this relationship needs to be corrected by time dilation [107]. All physical processes including clock speed in S' will run by (2.3) slower when observed from S. Thus by (2.6), the new relation will be of the form

$$f_v = \frac{c}{c - v} f', \tag{2.7}$$

where c is the speed of light and

$$f' = \gamma_v^{-1} f \tag{2.8}$$

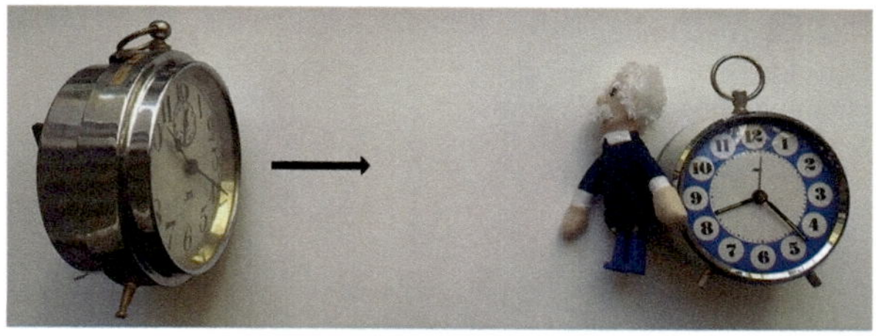

Fig. 2.3 The observer on the right is at rest with his clock. He will see a higher frequency of ticks on the clock on the left approaching him at relativistic speed. This is because the Doppler effect is greater than time dilation

corresponds to the lower frequency calculated from (2.5). By (2.7), (2.8), and (2.1) we obtain a relativistic Doppler relation for the frequency detected in S (see [89]),

$$f_v = \frac{1}{1 - \frac{v}{c}} f' = \frac{\gamma_v^{-1}}{1 - \frac{v}{c}} f = \frac{\sqrt{1 - \left(\frac{v}{c}\right)^2}}{1 - \frac{v}{c}} f = \sqrt{\frac{c + v}{c - v}} f. \tag{2.9}$$

From this and (2.8) we immediately get the following theorem.

Theorem 2.1 *For any* $v \in (0, c)$ *we have that* $f_v/f' > f_v/f > 1$. *Moreover,* $f_v/f \to \infty$ *as* $v \to c$.

Consequently, the Doppler effect manifests more than the time dilation itself, whenever the clock approaches the observer, see Fig. 2.3. Hence, the higher the speed v, the greater the Doppler effect. Special relativity effects for large v are of higher order in v/c than those arising from the Doppler effect.

It is therefore very important to distinguish consistently between reconstructions (calculations by means of the Lorentz transformation) and observations (measurements, detections, photos, videos, and recording in general). The notion "observer" in STR is somewhat confusing. It should not be a person who only applies relations (2.2) and (2.3). The observer performs real observations and measurements including all effects together as it is usually understood, i.e., the observer measures incoming frequencies.

Example 2.2 Suppose that a clock is approaching the origin of S at relativistic speed $v = 0.8c$. Its proper time passes slower than on clocks fixed in the system S, since by (2.5) we have

$$\gamma_v = \frac{1}{\sqrt{1 - 0.64}} = \frac{5}{3}$$

and

$$\Delta t' = \frac{5}{3}\Delta t.$$

However, substituting $v = 0.8c$ into (2.9) and (2.8), we find that

$$f_v = 3f \quad \text{and} \quad f_v = 5f', \tag{2.10}$$

i.e., the observer at the origin of S will detect a $3\times$ higher (blue-shifted) frequency f_v than the same clock has at rest in the system S, and this frequency f_v is even $5\times$ higher than the frequency f' which the time dilation predicts (see (2.8)). This may seem to be paradoxical, since due to the Doppler effect **the observer sees an opposite effect than STR claims.**

For a clock receding the origin by the speed $(-v)$, the observer will detect by (2.9) a $3\times$ lower (red-shifted) frequency than f. So there is a jump in these constant frequencies at the origin and only at this single point the observer can theoretically detect the proper frequency f'. So the Doppler effect plays an essential role.

Remark 2.5 The observer usually does not have a possibility to measure directly the speed v of some distant object so that he could immediately use the Lorentz transformation. However, he can measure the observing frequency f_v (blue-shifted or red-shifted) of some characteristic spectral line of a certain chemical compound and establish the corresponding rest frequency f, i.e. in the corresponding rest frame. From this and (2.9) he can establish the speed v (or its radial component in a general case). Then he can determine the factor γ_v^{-1} and find how significant are the corresponding relativistic effects (2.2)–(2.3).

2.3 The Lorentz Transformation Does not Allow Superluminal Velocities

First we recall Einstein's formula [89] for a relativistic addition of velocities, see also [270, Chapt. I.6].

Theorem 2.2 (Einstein) *Let $u \in (-c, c)$ and $w \in (-c, c)$ be constant velocities of a point-like object in the system S and S', respectively, in the direction of the horizontal axis. Then*

$$u = \frac{v + w}{1 + \dfrac{vw}{c^2}}, \tag{2.11}$$

where $v \in (-c, c)$ is a constant speed of S' with respect to S.

Proof Velocities u and w are constant in S and S', respectively. Therefore,

$$u = \frac{\mathrm{d}x}{\mathrm{d}t} \quad \text{a} \quad w = \frac{\mathrm{d}x'}{\mathrm{d}t'}. \tag{2.12}$$

By (2.3) we get

$$\frac{\mathrm{d}t'}{\mathrm{d}t} = \gamma_v\Big(1 - \frac{v}{c^2}\frac{\mathrm{d}x}{\mathrm{d}t}\Big) = \gamma_v\Big(1 - \frac{uv}{c^2}\Big),$$

where the difference in parenthesis is obviously positive. Using this equality, (2.12), and (2.2), we find that

$$w = \frac{\mathrm{d}x'}{\mathrm{d}t'} = \frac{\mathrm{d}x'}{\mathrm{d}t}\frac{\mathrm{d}t}{\mathrm{d}t'} = \gamma_v\Big(\frac{\mathrm{d}x}{\mathrm{d}t} - v\Big)\gamma_v^{-1}\Big(1 - \frac{uv}{c^2}\Big)^{-1} = \frac{u - v}{1 - \dfrac{uv}{c^2}}.$$

From this it follows that

$$u - v = w - \frac{uvw}{c^2}.$$

Now it is enough to evaluate u and we obtain (2.11). □

For example, by Theorem 2.2 we see that for $v = w = \frac{2}{3}c$ the speed $u = \frac{12}{13}c$ is less than c. In the next theorem we prove that from (2.11) we can never get the speed of light or faster-than-light speed u, even if $|v|$ and $|w|$ are arbitrarily close to c.

Let us recall that a *group* G is a set equipped with an associative binary operation $\circ : G \times G \to G$ and with the neutral element e such that for any $g \in G$ there exists exactly one inverse element $g^{-1} \in G$ for which

$$g \circ g^{-1} = e = g^{-1} \circ g.$$

The composition $\circ$ of two Lorentz transformations $\mathcal{L}_v$ and $\mathcal{L}_w$ given by relations (2.2) and (2.3) for $v, w \in (-c, c)$ is defined as follows

$$\mathcal{L}_u = \mathcal{L}_v \circ \mathcal{L}_w, \tag{2.13}$$

where u satisfies Einstein's formula (2.11).

Theorem 2.3 *Lorentz transformations $\mathcal{L}_v$ for all $v \in (-c, c)$ defined by (2.2)–(2.3) form an Abelian group.*

Proof If $v, w \in (-c, c)$ then obviously

$$\Big(1 + \frac{v}{c}\Big)\Big(1 + \frac{w}{c}\Big) > 0 \qquad \text{and} \qquad \Big(1 - \frac{v}{c}\Big)\Big(1 - \frac{w}{c}\Big) > 0.$$

From this we find that

$$-\left(1+\frac{vw}{c^2}\right) < \frac{v+w}{c} < 1+\frac{vw}{c^2},$$

and thus

$$-c < \frac{v+w}{1+\dfrac{vw}{c^2}} < c.$$

Comparing with Einstein's formula (2.11), we see that $u \in (-c, c)$, i.e., $|u|$ is always less than c.

Using (2.1) for $v = 0$, we find that $\gamma_0 = 1$ and the corresponding transformation $\mathcal{L}_0$ is the identity, i.e. the neutral element.

From (2.11) and (2.13) we immediately get that

$$\mathcal{L}_v \circ \mathcal{L}_{-v} = \mathcal{L}_0 = \mathcal{L}_{-v} \circ \mathcal{L}_v,$$

where $\mathcal{L}_{-v}$ is the inverse transformation, i.e., $x = \gamma_v(x' + vt')$, $t = \gamma_v(t' + vx'/c^2)$.

The composition $\circ$ is commutative, since the special block diagonal matrices $\boldsymbol{L}_v$ and $\boldsymbol{L}_w$ defined by (2.4) are commutative, i.e.

$$\boldsymbol{L}_v \boldsymbol{L}_w = \begin{pmatrix} \gamma_v & -\frac{v}{c}\gamma_v & 0 & 0 \\ -\frac{v}{c}\gamma_v & \gamma_v & 0 & 0 \\ 0 & 0 & 1 & 0 \\ 0 & 0 & 0 & 1 \end{pmatrix} \begin{pmatrix} \gamma_w & -\frac{w}{c}\gamma_w & 0 & 0 \\ -\frac{w}{c}\gamma_w & \gamma_w & 0 & 0 \\ 0 & 0 & 1 & 0 \\ 0 & 0 & 0 & 1 \end{pmatrix}$$

$$= \begin{pmatrix} \gamma_v\gamma_w + \frac{vw}{c^2}\gamma_v\gamma_w & -\frac{v+w}{c}\gamma_v\gamma_w & 0 & 0 \\ -\frac{v+w}{c}\gamma_v\gamma_w & \gamma_v\gamma_w + \frac{vw}{c^2}\gamma_v\gamma_w & 0 & 0 \\ 0 & 0 & 1 & 0 \\ 0 & 0 & 0 & 1 \end{pmatrix} = \boldsymbol{L}_w \boldsymbol{L}_v$$

for all $v, w \in (-c, c)$. The associativity of the operation $\circ$ is due to the fact that matrix multiplication is associative. □

Remark 2.6 According to [248], astronomers observe superluminal velocities of jets produced by black holes in our Galaxy. However, in [158, 184] we present a detailed mathematical justification of this paradox. We show that this is only an illusion and that the actual velocity of jets is less than c.

Remark 2.7 The classical geocentric model with motionless Earth in the center contradicts the Theory of Relativity. To see this consider the planet Neptune which is 4 light hours far from the Earth. Hence, the length of its orbit is $2\pi \cdot 4 > 24$ light hours. So Neptune would need a superluminal speed to orbit the Earth in one day which is impossible.

2.4 Length Contraction

Lorentz's length contraction is an immediate consequence of the Lorentz transformation. On the horizontal axis x' consider a fixed bar which is at rest in the system S'. Denote its length by

$$\Delta x' = x_2' - x_1', \tag{2.14}$$

where x_i' are fixed time independent coordinates of its ends in S'. For an arbitrary fixed time instant t in S we determine the corresponding x_2 and x_1 from formulae (cf. (2.2))

$$x_2' = \gamma_v(x_2 - vt), \qquad x_1' = \gamma_v(x_1 - vt),$$

and we set $\Delta x = x_2 - x_1$. Substituting this into (2.14), we get (cf. (2.5))

$$\Delta x' = \gamma_v(x_2 - vt - x_1 + vt) = \gamma_v \Delta x. \tag{2.15}$$

Denoting $\ell_0 = \Delta x'$ and $\ell = \Delta x$, we get by (2.1) the well-known *length contraction*

$$\ell = \ell_0 \sqrt{1 - \frac{v^2}{c^2}}. \tag{2.16}$$

Hence:

The bar at rest has the greatest length.

Remark 2.8 For the time being there is no direct experimental evidence of length contraction (2.16). Anyway, we can verify it indirectly by means of muons mentioned in Example 2.1. In the system S associated with these muons at rest we will observe their usual mean half-life time $\tau = 2.2 \cdot 10^{-6}$ s. However, in S' connected with Earth's atmosphere they will travel a longer distance than $c\tau = 660$ m. This demonstrates the length contraction in S' as seen from S.

In 1959, Roger Penrose published a paper [274] (see also [277, 337, p. 431]) describing why a quickly flying non-rotating ball in a photo should again look like a stationary ball. In the same year, his thoughts were elaborated in more detail by James Terrell [342] using light aberration. Here is a specific example showing the substantial effect of light aberration for relativistic speeds.

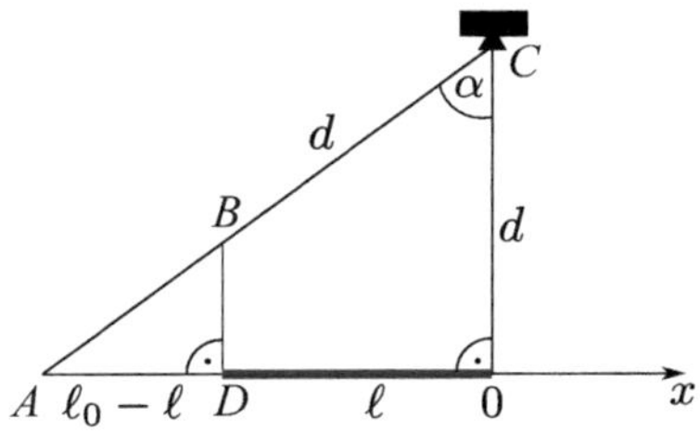

Fig. 2.4 The length of the legs of the larger (or smaller) right triangle is 1 and 0.75 (or 0.4 and 0.3) m. The ratio between the lengths of sides of both the triangles is 5 : 4 : 3. Due to light aberration the flying bar from the left to the right at the speed $0.8c$ has the same length in the photo as the same bar at rest. A photon emitted to the camera C from the rear end of the moving bar will have the same x coordinate as this rear end in this special case

Example 2.3 Consider a bar with length $\ell_0 = 1$ m. Assume that it moves from the left to the right along the axis x by the constant speed $v = 0.8c$ and that its front end just reached the origin of the coordinate system S. By (2.16) the bar as seen from S is shortened to

$$\ell = \ell_0\sqrt{1 - 0.64} = 0.6 \text{ m},$$

and thus the length of the straight line segment AD in Fig. 2.4 is $|AD| = 0.4$ m. We will photograph this bar from the axis z by a fixed nonrotating camera C which is placed at the distance

$$d = 0.75 \text{ m} \tag{2.17}$$

from the origin. Using the similarity of right triangles from Fig. 2.4, we find that $|BD| = |AD|d/\ell_0 = 0.3$ m. From this we have

$$|AB| = \sqrt{0.4^2 + 0.3^2} = 0.5$$

measured in meters. The segment on the hypotenuse from B to the camera C has the same length in meters as d in (2.17),

$$\sqrt{1^2 + 0.75^2} - |AB| = 1.25 - 0.5 = d. \tag{2.18}$$

To avoid blurred photos, we assume that our idealized camera can take pictures within 1 picosecond. During this time period, the light will only fly 0.3 mm and a possible blurring will not play a significant role. For simplicity, we shall analyze only that photo, in which the front end of the bar just reached the coordinate origin of S. However, the rear end of the bar will be on the photo farther than ℓ, since the light from the front end flies along a shorter distance d than the light from the rear end (see Fig. 2.4). That is why there will be recorded photons on the photo from the rear end of the bar that were emitted earlier than those from the front end. During the time period, when the rear end of the bar moves from A to D, a photon pointing from A to the camera will travel the distance $|AB|$, since $v/c = |AD|/|AB| = 0.8$. Hence, thanks to light aberration and (2.18) the moving bar will have on the photo the same length as the fixed one meter long bar.

Let us point out that a photon will travel the distance d from the origin to the camera C during the time period $\Delta t = d/c$. During this period, the bar will shift about $v\Delta t = 0.8d = 0.6$ m, i.e., it will be placed entirely to the right of point 0.

Example 2.4 Let again $\ell_0 = 1$ m and $v = 0.8c$. Hence, $\ell = 0.6$ m. This time, however, we place the camera closer to the axis x, i.e. $d < 0.75$ m. We shall again analyze the image, where the right end of the bar is at the origin. The left end of the bar will shift from the point $A = (-a, 0)$ to the point $(-\ell, 0)$ during the time period $\Delta t = (a - \ell)/v$. During this period, a photon will travel the distance $c\Delta t$ from the point A to the camera C. From the relation $a^2 + d^2 = ((a - \ell)c/v + d)^2$ we can derive the following inverse formula

$$d = \frac{a^2\left(\frac{v}{c} - \frac{c}{v}\right) + 2a\ell\frac{c}{v} - \ell^2\frac{c}{v}}{2(a - \ell)}.$$

For instance, when $a = 2$ m we obtain $d = \frac{15}{56} = 0.26\ldots$ m. So if we place the camera on the axis z at a distance of 26 cm from the origin, the one meter flying bar will appear extended in the photo as two meters long. Similarly for $v = 0.9$ and $d = 4$ cm we even get $a = 4$ m. The main cause for these surprising phenomena is that photons, which simultaneously passed through the lens, were not emitted simultaneously in S.

For $d > 0.75$ m we shall see the bar in the photo as shorter than 1 meter. The same bar will also be shorter than ℓ, if we photograph it such that its left end is at the origin. If it is placed exactly symmetrically with respect to the origin, its length on the photo will be just ℓ, but nonlinearly deformed.

Example 2.5 Now let us consider a 2 m long moving bar ($\ell_0 = 2$ m) again with speed $v = 0.8c$, and let $d = \frac{15}{56} = 0.26\ldots$ m as in Example 2.4. When the bar is exactly symmetric with respect to the origin (cf. Fig. 2.5), the length of its left part

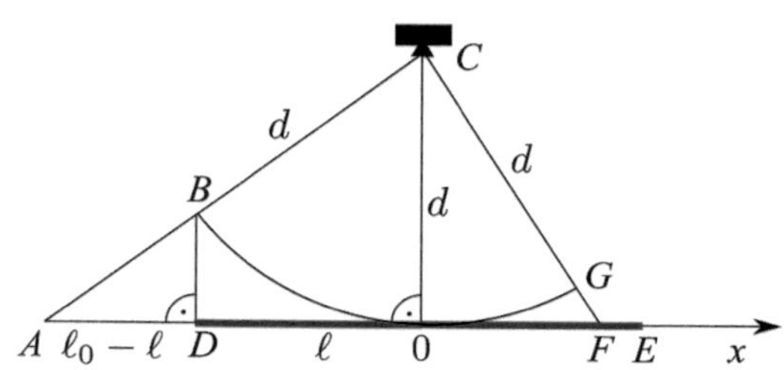

Fig. 2.5 Left: the front end of the (blue) bar reaches the origin. Right: the rear end of the (red) bar reaches the origin, where $|BC| = |CG| = d$, $|DE| = 2\ell$, and $v|FG| = c|EF|$

will be 2 m on the photo due to Example 2.4. Since its right part has positive length, **we will see that the bar is longer than 2 m on the photo.**

Approaching objects show a blue shift (i.e. shortening the wave length). However, due to aberration they may seem to be prolonged, which is paradoxical. On the other hand, receding objects that are manifested by red shift may seem to be shortened. Due to Examples 2.4 and 2.5 some pictures from the popularization book [112, Chapt. 1] are confusing. In the article [373], Weisskopf describes an apparent deformation of a quickly flying cube on a photo.

2.5 The Twin Paradox

In a series of publications [103, 112], [249, p. 167], [250, 253], ... the twin paradox (also called the clock paradox) is described by means of time dilation as follows:

There is a set of twins, Adam and Bob. Adam stays on Earth, while Bob flies in a rocket e.g. to the star Sirius about 8 light years away with constant relativistic speed v. Bob's time runs slower with respect to Adam's clock due to time dilation, and his distance to Sirius is less than 8 light years due to length contraction. When Bob returns with velocity $(-v)$ back on Earth, he finds out that his brother Adam is much older than him.

Now let us look at the twin paradox in more detail. First we present several false arguments sometimes circulating among the general public and on internet.

Sometimes it is incorrectly claimed that Bob will return as old as Adam. Bob seems to be at rest in his rocket and thus it is Adam who moves. Hence, Adam's time runs more slowly than Bob's time due to time dilation (2.5). This symmetry holds for all inertial systems. However, we have to realize that Bob has changed his inertial system when he was near Sirius, and thus this situation is not symmetric with respect to the two twins. Therefore, the above reasoning is not correct.

To present another incorrect idea, we first present the following definition. The *redshift* z of an object resulting from its radial motions is defined by

$$z := \frac{\lambda_1}{\lambda_0} - 1 = \frac{\nu_0}{\nu_1} - 1,$$

where λ_1 and ν_1 are the observed wavelength and frequency, λ_0 and ν_0 are the emitted wavelength and frequency in the rest frame, and $\lambda_i \nu_i = c$ for $i = 0, 1$. If $z < 0$, then z is called the *blueshift*. Note that the coordinate z appearing in (2.2) has a different meaning. However, no ambiguity can arise.

Some people argue as follows: Adam first observes the redshift of the receding twin Bob due to the Doppler effect, i.e., Adam sees that Bob's clock ticks slower. After the turnover at Sirius, Adam will watch the blueshift of Bob's clock. These phenomena are thus balanced and the twins will be the same age. This reasoning is also not correct, since the time dilation (2.5) was not taken into account.

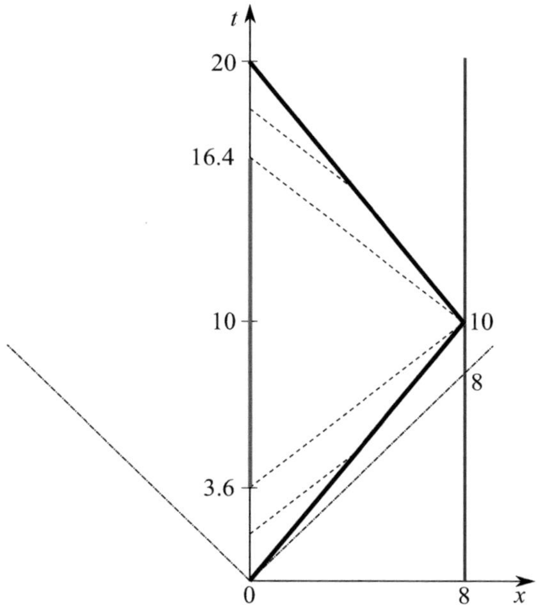

Fig. 2.6 The vertical axis t shows the time in years and the horizontal axis x shows the distance in light years. The left vertical line is the world line of Adam who stays on Earth. The right vertical line corresponds to the world line of Sirius. The world line of flying Bob is marked with a thick piecewise linear line given by equations $t = \frac{5}{4}x$ and $t = 20 - \frac{5}{4}x$. The future light cone $t = |x|$ is marked by the dot-and-dash line and the dashed lines stand for events with simultaneous times in S' and S'' such that t' and t'' are constant

Another false argument is that when Bob changes the inertial system, he feels an infinite acceleration during the turnover at Sirius and that this is the main cause of the twin paradox. This is also not true. In more detail, Bob's proper time t' (and also t'') was not taken into account correctly as we shall see in Example 2.6. From Theorem 2.1 and Example 2.2 we know that the Doppler effect plays an important role in STR. Bob feels an infinite deceleration[1] during his turnover at Sirius. The reader is confused that this is the main reason for the twin paradox, since Bob changes its inertial system. However, this cannot be true. When Bob reaches Sirius, he can just transmit information (e.g. a photo) about his actual age in the system S' to another traveler who is passing by Sirius and is fixed in the system S'' associated with speed $(-v)$. The same applies for Bob's departure and arrival back to the Earth when he also feels an infinite acceleration or deceleration.

To be more specific, assume that $v = 0.8c$. Sometimes it is also said that Bob will observe a large jump in time on Earth, since no event from the red interval (3.6, 16.4) on the vertical axis in Fig. 2.6 is simultaneous with Bob. In Example 2.6, we will show that Bob will see only a jump in frequencies and no jump in time in S during his turnover (see Fig. 2.7).

[1] Roger Penrose [277, p. 421] rounds the corresponding world lines on a small neighborhood around the three critical points to avoid an infinite acceleration or deceleration.

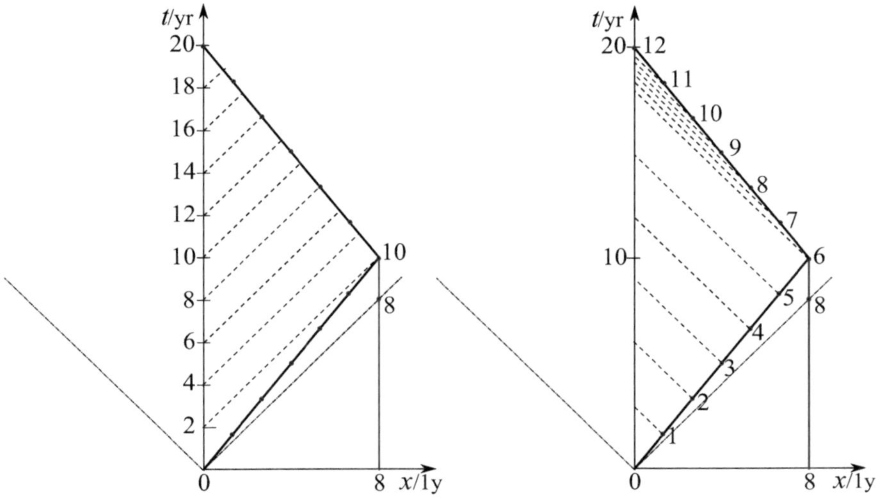

Fig. 2.7 The notation is the same as in Fig. 2.6 except for dashed lines. Here dashed lines indicate trajectories of photons launched every year by Adam and Bob. Left: Adam sends a periodic signal to Bob. Right: Bob sends a period signal to Adam. There is no jump in time—only a jump in received frequencies

2.6 Correct Solution of the Twin Paradox

To solve the twin paradox we first prove the following theorem.

Theorem 2.4 *The difference $(ct)^2 - x^2$ is invariant with respect to the Lorentz transformation.*

Proof From (2.2) and (2.3) we see that

$$\begin{aligned}(ct')^2 - x'^2 &= (ct' - x')(ct' + x')\\ &= \gamma_v\Big(ct - \frac{vx}{c} - x + vt\Big)\gamma_v\Big(ct - \frac{vx}{c} + x - vt\Big)\\ &= \gamma_v^2\Big(1 + \frac{v}{c}\Big)(ct - x)\Big(1 - \frac{v}{c}\Big)(ct + x) = (ct)^2 - x^2. \qquad (2.19)\end{aligned}$$

Hence, this difference of squares is invariant with respect to the Lorentz transformation. □

Example 2.6 Let again $v = 0.8c$. (For general $v \in (-c, c)$ we may proceed analogously.) Since Bob is at rest in the system S', we have that $x' = 0$. Thus from (2.19) we get for $\Delta t = 10$ yr and $\Delta x = 8$ ly

$$c\Delta t' = \sqrt{(c\Delta t)^2 - (\Delta x)^2} = \sqrt{100 - 64}\ \text{ly} = 6\ \text{ly}.$$

Consequently, Bob will spend 6 years flying to Sirius according to his proper time (see the right part of Fig. 2.7). Thus his clock at Sirius will show 6 years. After the turnover, Bob will fly the same distance back to Earth at speed $(-v)$ for another 6 years with respect to his own proper time. On arrival, he will be $20 - 2 \cdot 6 = 8$ years younger than his twin Adam.

Finally, let us see how the frequency of received signals changes. The Doppler effect, which is not included in the relations (2.2) and (2.3), cannot be removed during observations. Adam and Bob have the same clocks and let f_0 be their frequency at rest. When Bob moves away from Adam at speed v, then according to the relativistic Doppler effect Adam will detect the following frequency in S

$$f_v = \sqrt{\frac{c - v}{c + v}}\, f_0. \tag{2.20}$$

Substituting $v = 0.8c$, we find that Adam will observe $3\times$ lower frequency from Bob's clock, since

$$\frac{f_{-v}}{f_0} = \sqrt{\frac{c - v}{c + v}} = \frac{1}{3}.$$

Similarly we find that after the turnover of Bob, Adam will observe the frequency $f_v = 3f_0$, see (2.10) and Fig. 2.7. Adam will receive the same frequencies from Bob due to the relativity principle. Bob will see all instants on Earth even though with another frequency, i.e., no time interval will be skipped.

Example 2.7 For simplicity assume that Niel Armstrong was flying to the Moon and back by the constant speed $v = 10$ km/s and that $\Delta x = 384{,}000$ km is the mean distance to the Moon. *When he returned from the Moon, how many seconds younger was he?*

His one way flight lasted

$$\Delta t = \frac{\Delta x}{v} = 38{,}400 \text{ s} \tag{2.21}$$

with respect to Earth's clock. Then by (2.20) we get

$$\Delta t' = \sqrt{(\Delta t)^2 - \frac{(\Delta x)^2}{c^2}} = \sqrt{38{,}400^2 - 1.28^2 \text{ s}} = \sqrt{1{,}474{,}559{,}998.3616} \text{ s}$$
$$= 38{,}399.999{,}978{,}\overline{666} \text{ s}.$$

From this and (2.21) we obtain

$$\Delta t - \Delta t' = 0.000021344 \text{ s}.$$

For the return trip we get the same value. Hence, according to STR, Armstrong has returned from the Moon 43 μs younger than if he had remained on Earth.

Example 2.8 Now let us take into account the Earth's gravitational field and, for simplicity, neglect the Moon's gravitational field. According to General Theory of Relativity (GTR), any clock on the Earth's surface ticks slower than on the Moon. Let us ask the same question as in Example 2.7: *How many seconds younger was Armstrong upon his return from the Moon?*

For the emitted frequency $f_{\rm em}$ in the gravitational field and the detected frequency f in zero gravitation the following relativistic relation holds (see e.g. [327, p. 261])

$$f = f_{\rm em}\sqrt{1 - \frac{r}{R}}, \tag{2.22}$$

where we substitute for $R = 6373$ km the mean radius of the Earth and for $r = 2GM/c^2 = 8.87$ mm the Schwarzschild radius of the Earth (cf. (3.10) below), where G is the gravitational constant and $M = 5.972 \cdot 10^{24}$ kg is Earth's mass. From this we get

$$\frac{f}{f_{\rm em}} = 1 - 6.5 \cdot 10^{-10}.$$

During the time interval $2\Delta t = 76\,800$ s (see (2.21)), the clock on Earth is delayed by

$$2\Delta t\Big(1 - \frac{f}{f_{\rm em}}\Big) = 50\ \mu\text{s}.$$

By comparison with the previous Example 2.7, we find that Armstrong returned even older than if he had remained on Earth. In this case, the effects of Special and General Relativity are subtracted from each other (similarly to the precise GPS setting).

Example 2.9 Albert Einstein asked: *Is a clock at a pole faster than one at the Earth equator?* This example can be solved similarly as Example 2.8. Therefore, we will concentrate only on the main points. The equatorial radius of the Earth is $R_1 = 6{,}378{,}135$ m and polar $R_2 = 6{,}356{,}750$ m. Assume that the clock on a pole has a frequency $f = 1\ \text{s}^{-1}$. The clock on the equator has the speed $v = 2\pi R_1/T = 465.1$ m/s, where $T = 86{,}164.1$ s is the length of the sidereal day. We may suppose that they are in an inertial system on a very short time interval. According to (2.8), the frequency f' satisfies $f' = \gamma_v^{-1} f$, and thus for the difference of these two frequencies we find from (2.1) that

$$f - f' = \Big(1 - \sqrt{1 - \frac{v^2}{c^2}}\Big) f = 1.2 \cdot 10^{-12}\ \text{s}^{-1}.$$

By (2.22) the frequency $\tilde{f}$ of a non-moving clock on the equator with radius R_1 is

$$\tilde{f} - f = 2.3 \cdot 10^{-12} \text{ s}^{-1}$$

higher than is the frequency f of the clock at the pole whose distance from the Earth gravity center is R_2, In this case, GTR produces a larger effect than time dilation proposed by STR.

Finally, let us ask the question: *For what speed v is the time dilation equal to the slower motion of a clock in the gravitational field of some body, e.g. the Earth, Mars, Jupiter?* The answer is given by the next theorem.

Theorem 2.5 *Assume that an observer is far from all gravitational sources. Time on the surface of a spherically symmetric body B with mass M runs as fast as the observer's own time moving relative to the body B at the escape velocity* $v = \sqrt{2GM/R}$.

Proof According to time dilation (2.5), the ratio of the corresponding frequencies satisfies the equality $f'/f = \gamma^{-1}$. Thus by (2.1) and (2.22) we require that

$$\sqrt{1 - \frac{v^2}{c^2}} = \sqrt{1 - \frac{r}{R}}.$$

Hence, we have

$$v^2 = \frac{rc^2}{R} = \frac{2GM}{R}.$$

The balancing of both the times therefore occurs for the speed

$$v = \sqrt{\frac{2GM}{R}}.$$

This relation is the well-known escape velocity which is necessary to leave the gravitational field of a given body from the radius R. □

In the monograph [341, Sect. 3.5], one can find what the escape velocity means in General Theory of Relativity.

2.7 Conclusions

The Special Theory of Relativity has a number of unexpected claims that contradict our intuition. According to STR, no experiment can be made to decide whether the body is at rest or moving. All inertial systems for describing physical phenomena are equivalent and there is no preferred inertial system. However, at present we know that the cosmic microwave background radiation (CMB) actually determines a certain kind of a fixed reference system in our neighborhood. Thus there arise speculations whether the principle of relativity in the physical universe holds.

Remark 2.9 In the limiting case $w = \pm c$, Einstein's formula (2.11) for $v \in (-c, c)$ gives $u = c$. Hence, every photon always has the speed of light in any inertial system.

It is often said that the Lorentz transformation for low speeds $|v| \ll c$ changes into the Galileo transformation

$$x' = x - vt,$$
$$y' = y,$$
$$z' = z,$$
$$t' = t.$$

This is not true (see [21, 355]), since for an arbitrarily small fixed $v > 0$ we can always find x such that the term vx/c^2 in (2.3) will dominate significantly over t. Nevertheless, from (2.2) and (2.3) it follows that the Lorentz transformation changes into the Galileo transformation for a fixed v, if we treat c as a parameter and assume that $c \to \infty$. In this case the term vx/c^2 becomes negligible. However, for an infinite speed of light there would be no Doppler effect nor aberration of light.

Remark 2.10 For $\vec{x} = (x, y, z)$ and a constant velocity vector $\vec{v} \in \mathbb{E}^3$ with length $|\vec{v}| \in (0, c)$ the *general Lorentz transformation* is of the form (see e.g. [133, p. 434])

$$\vec{x}\,' = \vec{x} + \Big(\frac{\gamma - 1}{|\vec{v}|^2}\vec{v} \cdot \vec{x} - \gamma t\Big)\vec{v}, \tag{2.23}$$

$$t' = \gamma\Big(t - \frac{\vec{v} \cdot \vec{x}}{c^2}\Big). \tag{2.24}$$

Here the Lorentz factor γ is defined similarly as in (2.1), only v^2 needs to be rewritten as $|\vec{v}|^2$. It is easy to find that for nonzero $\vec{v} = (v, 0, 0)$, where $v \in (-c, c)$, relations (2.23)–(2.24) change to (2.2)–(2.3). By Wikipedia [389] (see also [12]), the Einstein addition of velocities is neither commutative nor associative, in general.

We conclude by stating that the longitudinal Doppler effect and aberration of light may cause that we observe completely opposite phenomena than those predicted by the Special Theory of Relativity by means of (2.2) and (2.3), see Examples 2.2 and 2.5. Note that relations (2.2) and (2.3) represent only a transformation of spacetime coordinates of points from one inertial systems into spacetime coordinates of the second inertial system. We saw that some other effects than time dilation and length contraction can manifest itself in a stronger manner and these effects cannot be shielded in any way. For a visualization of several further accompanying effects (like nonlinear distortion) we refer to [388].

Chapter 3
Einstein's Equations

3.1 Historical Facts of Importance

Karl Schwarzschild was probably the first scientist who ever realized that at any fixed time instant, our universe might be non-Euclidean and that it could be modeled by the three-dimensional sphere $\mathbb{S}^3$ (called also a hypersphere) or the three-dimensional hyperbolic pseudosphere $\mathbb{H}^3$, see his paper [320] from 1900. Around 1915, he became famous for calculating the first nontrivial solution of Einstein's vacuum equations (cf. (3.6) below).

Albert Einstein replaced the Newtonian gravitational force (1.9) by a curved spacetime, where all bodies move along *geodesics*—the shortest paths between two nearby points. This was a genial concept of how gravitational forces work (that resembles the well-known Fermat principle from optics). On November 18, 1915, Einstein submitted his famous paper [90] about Mercury's perihelion shift. Here the gravitational field is described by the following equations

$$\Gamma^{\varkappa}_{\mu\nu,\varkappa} - \Gamma^{\lambda}_{\mu\varkappa}\Gamma^{\varkappa}_{\lambda\nu} = 0, \quad \varkappa, \lambda, \mu, \nu = 0, 1, 2, 3, \tag{3.1}$$

where the present standard notation and *Einstein's summation convention* over all the values of repeated indices are used,

$$\Gamma^{\varkappa}_{\mu\nu} := \frac{1}{2} g^{\varkappa\lambda}(g_{\nu\lambda,\mu} + g_{\lambda\mu,\nu} - g_{\mu\nu,\lambda}) \tag{3.2}$$

are the *Christoffel symbols of the second kind* (sometimes also called the *connection coefficients*) and $g_{\mu\nu} = g_{\mu\nu}(x^0, x^1, x^2, x^3)$ are dimensionless components of the unknown 4×4 twice differentiable symmetric *metric tensor* (represented by a nonsingular matrix) of the spacetime of one time variable $x^0 = ct$ and three space

M. Křížek, L. Somer, *Mathematical Aspects of Paradoxes in Cosmology*,
https://doi.org/10.1007/978-3-031-31768-2_3

variables x^1, x^2, x^3, and

$$\det(g_{\mu\nu}) = -1. \tag{3.3}$$

Furthermore, $g^{\mu\nu}$ is the 4×4 symmetric inverse tensor to $g_{\mu\nu}$. For brevity, the first classical derivatives of a function $f = f(x^0, x^1, x^2, x^3)$ are denoted as follows $f_{,\varkappa} := \partial f / \partial x^\varkappa$. For simplicity, the dependence of all functions on the spacetime coordinates will often not be indicated. An infinitesimally small spacetime distance $\mathrm{d}s$ is usually expressed in the literature on General Relativity as follows

$$\mathrm{d}s^2 = g_{\mu\nu}\mathrm{d}x^\mu \mathrm{d}x^\nu = \begin{pmatrix} \mathrm{d}x^0 \\ \mathrm{d}x^1 \\ \mathrm{d}x^2 \\ \mathrm{d}x^3 \end{pmatrix}^{\top} \begin{pmatrix} g_{00} & g_{01} & g_{02} & g_{03} \\ g_{10} & g_{11} & g_{12} & g_{13} \\ g_{20} & g_{21} & g_{22} & g_{23} \\ g_{30} & g_{31} & g_{32} & g_{33} \end{pmatrix} \begin{pmatrix} \mathrm{d}x^0 \\ \mathrm{d}x^1 \\ \mathrm{d}x^2 \\ \mathrm{d}x^3 \end{pmatrix}.$$

Note that $\mathrm{d}s^2$ means $(\mathrm{d}s)^2$ and not $\mathrm{d}(s^2)$. Let us also point out that the number of the Christoffel symbols appearing in (3.1) is $10(4 + 4 \cdot 4) = 200$. For the fundamentals of tensor calculus, we refer e.g. to [107, 249].

On November 20, 1915, David Hilbert submitted the paper [129] which was published on March 31, 1916. He did not require the validity of the restrictive algebraic constraint (3.3). Using a variational principle, he derived the following complete equations (see Fig. 3.1)

$$R_{\mu\nu} := \Gamma^{\varkappa}_{\mu\nu,\varkappa} \underline{\underline{- \Gamma^{\varkappa}_{\mu\varkappa,\nu} + \Gamma^{\lambda}_{\mu\nu}\Gamma^{\varkappa}_{\lambda\varkappa}}} - \Gamma^{\lambda}_{\mu\varkappa}\Gamma^{\varkappa}_{\lambda\nu} = 0 \tag{3.4}$$

for the gravitational field, where the doubly underlined terms do not appear in (3.1), see Remark 3.2 below. Note that $-\left\{{\mu\nu \atop \varkappa}\right\}$ stands for the Christoffel symbol $\Gamma^{\varkappa}_{\mu\nu}$

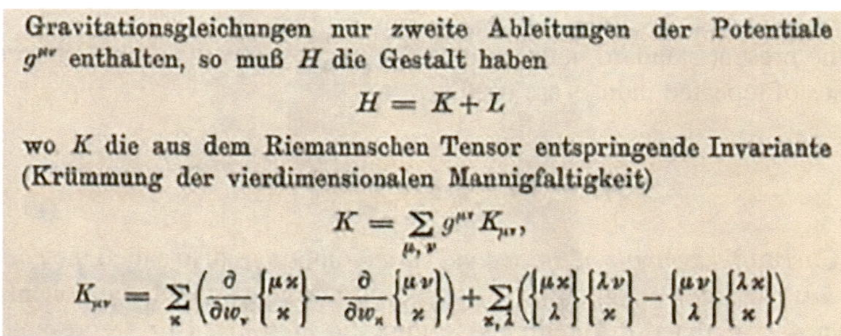
Gravitationsgleichungen nur zweite Ableitungen der Potentiale $g^{\mu\nu}$ enthalten, so muß H die Gestalt haben

$$H = K + L$$

wo K die aus dem Riemannschen Tensor entspringende Invariante (Krümmung der vierdimensionalen Mannigfaltigkeit)

$$K = \sum_{\mu,\nu} g^{\mu\nu} K_{\mu\nu},$$

$$K_{\mu\nu} = \sum_{\varkappa}\left(\frac{\partial}{\partial w_\nu}\left\{{\mu\varkappa \atop \varkappa}\right\} - \frac{\partial}{\partial w_\varkappa}\left\{{\mu\nu \atop \varkappa}\right\}\right) + \sum_{\varkappa,\lambda}\left(\left\{{\mu\varkappa \atop \lambda}\right\}\left\{{\lambda\nu \atop \varkappa}\right\} - \left\{{\mu\nu \atop \lambda}\right\}\left\{{\lambda\varkappa \atop \varkappa}\right\}\right)$$

Fig. 3.1 Hilbert's original paper (see [129, p. 402]) contains the Ricci tensor $R_{\mu\nu}$ from (3.4) which is denoted here by $K_{\mu\nu}$ a K is the Ricci scalar. Hilbert's paper was submitted for publication 5 days earlier than Einstein's note [91]

in Fig. 3.1. Hilbert's main results are formulated in mathematical theorems with precise assumptions and statements.

From (3.2) we observe that (3.4) is a system of nonlinear second order partial differential equations. Its left-hand side $R_{\mu\nu}$ is at present called the *Ricci tensor* and the equations $R_{\mu\nu} = 0$ are called *Einstein's vacuum equations*. By Sauer [317, p. 569], Hilbert's knowledge and understanding of the calculus of variation and of invariant theory readily put him into a position to fully grasp Einstein's gravitational theory.

On November 25, 1915, Albert Einstein submitted a three and a half page note [91] which contains the same expression of the Ricci tensor as in (3.4), but the condition (3.3) is kept. The way in which Einstein's equations (see (3.33) below) were derived is not described there. Einstein's short note [91] was published on December 2, 1915, i.e. within one week.

Now we show that the classical Minkowski metric tensor defined by

$$g_{\mu\nu} = \mathrm{diag}(-1, 1, 1, 1), \tag{3.5}$$

where all non-diagonal entries are zeros, is a trivial solution of Einstein's vacuum equations.

Theorem 3.1 *The Minkowski metric tensor (3.5) is a solution of Einstein's vacuum equations (3.4).*

Proof Since all Christoffel symbols (3.2) are zeros, the theorem immediately follows. □

In the same way we can prove that any other 4×4 constant symmetric tensor is also a solution of Einstein's equations (3.4).

3.2 Exterior Schwarzschild Solution

On January 13, 1916, Karl Schwarzschild submitted the paper [321] containing the first nontrivial static spherically symmetric solution of Einstein's vacuum equations. His solution is at present written as the following diagonal tensor

$$g_{\mu\nu} = \mathrm{diag}\Big(-\frac{r-S}{r}, \frac{r}{r-S}, r^2, r^2\sin^2\theta\Big), \tag{3.6}$$

where $S \geq 0$ is a fixed real constant and (r, θ, ϕ) are the standard spherical coordinates such that

$$x^1 = r\cos\phi\cos\theta, \quad x^2 = r\sin\phi\cos\theta, \quad x^3 = r\sin\theta,$$

$r > S$, $\phi \in [0, 2\pi)$, and $\theta \in [0, \pi]$. The time independent metric tensor (3.6) is called the *exterior Schwarzschild solution* of Eq. (3.4) and we have

$$c^2 d\tau^2 = -c^2(1 - S/r)dt^2 + (1 - S/r)^{-1}dr^2 + r^2 d\theta^2 + r^2 \sin^2\theta \, d\phi^2.$$

We see that for $S = 0$ the formula (3.6) reduces to the Minkowski metric tensor in spherical coordinates. In this case, (3.6) is called the *isotropic metric tensor*.

Remark 3.1 On December 22, 1915, Karl Schwarzschild had already written a letter to Einstein announcing that he was reading Einstein's paper on Mercury and found a solution of the field equations (see [90] for the English translation of this letter). Schwarzschild's solution looks like (3.6). However, the meaning of the radial coordinate r is slightly different in his original paper [321] than in Theorem 3.2 below.

On March 20, 1916, Einstein submitted his fundamental work [92], where General Theory of Relativity was established. One year later, Einstein in [94] added the cosmological term $\Lambda g_{\mu\nu}$ to the right-hand side of his field equations (see (3.52) below) and in 1917 Willem de Sitter [80] found their first vacuum cosmological solution which describes the behavior of the expansion function on a closed universe with no mass. In 1932, Einstein and de Sitter [96] published another cosmological solution with positive mass.

Classical relativistic tests [90, 237, 249, 324, 377] (such as bending of light, Mercury's perihelion shift, gravitational redshift, Shapiro's fourth test of General Relativity) are based on verification of very simple algebraic formulae derived by various simplifications and approximations of the Schwarzschild solution (3.6) which is very special and corresponds only to one spherically symmetric nonrotating body. For instance, the bending angle $\phi = 1.75''$ at the distance $R = R_\odot$ is obtained from the relation

$$\phi = \frac{4GM}{c^2 R} \tag{3.7}$$

which is derived from (3.6) under several simplifications, see [332] (cf. also [222, p. 11]). Relations (4.2) and (11.10) below represent further simple formulae that are tested. Therefore, in this chapter we will first present several theorems on the exterior Schwarzschild solution.

Einstein and also Schwarzschild assumed that the gravitational field has the following properties:

1. It is static, i.e., all components $g_{\mu\nu}$ are independent of the time variable x^0.
2. It is spherically symmetric with respect to the coordinate origin, i.e., the same solutions is obtained after a linear orthogonal transformation.
3. The equations $g_{\rho 0} = g_{0\rho} = 0$ hold for any $\rho = 1, 2, 3$.
4. The spacetime is asymptotically flat, i.e., the metric tensor $g_{\mu\nu}$ tends to the Minkowski metric tensor at infinity.

Theorem 3.2 (Schwarzschild) *The metric tensor (3.6) with $S > 0$ satisfies Eq. (3.4), but does not satisfy (3.1).*

Proof We will proceed in three steps:

1. At first, we introduce the following definition of the *Christoffel symbols of the first kind*

$$\Gamma_{\lambda\mu\nu} := \frac{1}{2}(g_{\lambda\mu,\nu} + g_{\nu\lambda,\mu} - g_{\mu\nu,\lambda}), \qquad \lambda, \mu, \nu = 0, 1, 2, 3. \tag{3.8}$$

From this and the relation $g_{\mu\nu} = g_{\nu\mu}$ we find that the Christoffel symbols are, in general, symmetric with respect to the second and third subscript

$$\Gamma_{\lambda\mu\nu} = \frac{1}{2}(g_{\lambda\nu,\mu} + g_{\mu\lambda,\nu} - g_{\nu\mu,\lambda}) = \Gamma_{\lambda\nu\mu}. \tag{3.9}$$

Hence, their number is not $4 \times 4 \times 4 = 64$ but only $4 \times (1 + 2 + 3 + 4) = 40$. Using (3.8) and (3.6), we obtain

$$\begin{aligned}
\Gamma_{001} &= \tfrac{1}{2} g_{00,1} = -\frac{r - (r - S)}{2r^2} = -\frac{S}{2r^2},\\
\Gamma_{100} &= -\tfrac{1}{2} g_{00,1} = \frac{S}{2r^2},\\
\Gamma_{111} &= \tfrac{1}{2} g_{11,1} = -\frac{S}{2(r - S)^2},\\
\Gamma_{122} &= -\tfrac{1}{2} g_{22,1} = -r,\\
\Gamma_{133} &= -\tfrac{1}{2} g_{33,1} = -r \sin^2\theta,\\
\Gamma_{221} &= \tfrac{1}{2} g_{22,1} = r,\\
\Gamma_{233} &= \tfrac{1}{2} g_{33,2} = r^2 \sin\theta \cos\theta,\\
\Gamma_{331} &= \tfrac{1}{2} g_{33,1} = r \sin^2\theta,\\
\Gamma_{332} &= \tfrac{1}{2} g_{33,2} = r^2 \sin\theta \cos\theta,
\end{aligned}$$

and the other components $\Gamma_{\lambda\mu\nu} = \Gamma_{\lambda\nu\mu}$ are zeros.

2. Similarly to (3.9) we can find that the Christoffel symbols of the second kind are symmetric with respect to the second and third (lower) indices and by (3.2) and (3.8) we have

$$\Gamma^{\varkappa}_{\mu\nu} = g^{\varkappa\lambda}\Gamma_{\lambda\mu\nu}.$$

Since the metric tensor (3.6) is diagonal, its inverse reads

$$g^{\mu\nu} = \mathrm{diag}\Big(-\frac{r}{r-S}, \frac{r-S}{r}, \frac{1}{r^2}, \frac{1}{r^2 \sin^2\theta} \Big)$$

and all non-diagonal entries are zeros. From this and Step 1 we get

$$\begin{aligned}
\Gamma^0_{\ 01} &= g^{00}\Gamma_{001} = \frac{S}{2r(r-S)}\,, \\
\Gamma^1_{\ 00} &= g^{11}\Gamma_{100} = \frac{S(r-S)}{2r^3}\,, \\
\Gamma^1_{\ 11} &= g^{11}\Gamma_{111} = -\frac{S}{2r(r-S)}\,, \\
\Gamma^1_{\ 22} &= g^{11}\Gamma_{122} = -(r-S), \\
\Gamma^1_{\ 33} &= g^{11}\Gamma_{133} = -(r-S)\sin^2\theta\,, \\
\Gamma^2_{\ 12} &= g^{22}\Gamma_{221} = \frac{1}{r^2}r = \frac{1}{r}\,, \\
\Gamma^2_{\ 33} &= g^{22}\Gamma_{233} = \sin\theta\cos\theta\,, \\
\Gamma^3_{\ 31} &= g^{33}\Gamma_{331} = \frac{1}{r^2\sin^2\theta} r\sin^2\theta = \frac{1}{r}\,, \\
\Gamma^3_{\ 32} &= g^{33}\Gamma_{332} = \frac{1}{r^2\sin^2\theta} r^2\sin\theta\cos\theta = \mathrm{cotan}\,\theta,
\end{aligned}$$

and the other components $\Gamma^{\varkappa}_{\ \mu\nu} = \Gamma^{\varkappa}_{\ \nu\mu}$ are zeros.

3. Finally, we will evaluate all entries of the Ricci tensor defined in (3.4). In particular, using formula for the inverse $g^{\mu\nu}$ in Step 2, we obtain

$$\begin{aligned}
R_{00} &= \Gamma^{\varkappa}_{\ 00,\varkappa} \underline{\underline{-\Gamma^{\varkappa}_{\ 0\varkappa,0} + \Gamma^{\lambda}_{\ 00}\Gamma^{\varkappa}_{\ \lambda\varkappa}}} - \Gamma^{\lambda}_{\ 0\varkappa}\Gamma^{\varkappa}_{\ \lambda 0} \\
&= \Gamma^1_{\ 00,1} \underline{\underline{+\Gamma^1_{\ 00}(\Gamma^2_{\ 12} + \Gamma^3_{\ 13})}} - 2\Gamma^1_{\ 00}\Gamma^0_{\ 01} \\
&= \frac{S2r^3 - S(r-S)6r^2}{4r^6} \underline{\underline{+\frac{S(r-S)}{2r^3}\Big(\frac{1}{r} + \frac{1}{r}\Big)}} - 2\frac{S(r-S)}{2r^3}\frac{S}{2r(r-S)} \\
&= \frac{rS}{2r^4} - \frac{3S(r-S)}{2r^4} \underline{\underline{+\frac{S(r-S)}{2r^3}\frac{2}{r}}} - \frac{S^2}{2r^4} = 0,
\end{aligned}$$

where the doubly underlined terms correspond to the doubly underlined terms in (3.4). From this it is obvious that the Schwarzschild solution (3.6) does not satisfy (3.1), since $S(r-S)/r^4 \neq 0$ for $S > 0$.

Similarly, we find that

$$R_{11} = \Gamma^{\varkappa}_{11,\varkappa} \underline{\underline{- \Gamma^{\varkappa}_{1\varkappa,1} + \Gamma^{\lambda}_{11}\Gamma^{\varkappa}_{\lambda\varkappa}}} - \Gamma^{\lambda}_{1\varkappa}\Gamma^{\varkappa}_{\lambda 1}$$

$$= \Gamma^{1}_{11,1} \underline{\underline{- \Gamma^{0}_{10,1} - \Gamma^{1}_{11,1} - \Gamma^{2}_{12,1} - \Gamma^{3}_{13,1} + \Gamma^{1}_{11}(\Gamma^{0}_{01} + \Gamma^{2}_{12} + \Gamma^{3}_{13})}}$$

$$- \Gamma^{0}_{01}\Gamma^{0}_{01} - \Gamma^{2}_{12}\Gamma^{2}_{12} - \Gamma^{3}_{13}\Gamma^{3}_{13}$$

$$= \underline{\underline{\frac{S(2r-S)}{2r^2(r-S)^2} + \frac{2}{r} - \frac{S}{2r(r-S)}\left(\frac{S}{2r(r-S)} + \frac{2}{r}\right)}} - \frac{S^2}{4r^2(r-S)^2} - \frac{2}{r^2} = 0,$$

$$R_{22} = \Gamma^{\varkappa}_{22,\varkappa} \underline{\underline{- \Gamma^{\varkappa}_{2\varkappa,2} + \Gamma^{\lambda}_{22}\Gamma^{\varkappa}_{\lambda\varkappa}}} - \Gamma^{\lambda}_{2\varkappa}\Gamma^{\varkappa}_{\lambda 2}$$

$$= \Gamma^{1}_{22,1} \underline{\underline{- \Gamma^{3}_{23,2} + \Gamma^{1}_{22}(\Gamma^{2}_{12} + \Gamma^{3}_{13})}} - 2\Gamma^{2}_{21}\Gamma^{1}_{22} - \Gamma^{3}_{23}\Gamma^{3}_{23}$$

$$= -1 + \underline{\underline{\frac{1}{\sin^2\theta} - (r-S)\frac{2}{r}}} + \frac{2}{r}(r-S) - \mathrm{cotan}^2\theta = 0,$$

$$R_{33} = \Gamma^{\varkappa}_{33,\varkappa} \underline{\underline{- \Gamma^{\varkappa}_{3\varkappa,3} + \Gamma^{\lambda}_{33}\Gamma^{\varkappa}_{\lambda\varkappa}}} - \Gamma^{\lambda}_{3\varkappa}\Gamma^{\varkappa}_{\lambda 3}$$

$$= \underline{\underline{\Gamma^{1}_{33}\Gamma^{3}_{13} + \Gamma^{2}_{33}\Gamma^{3}_{23}}} - \Gamma^{3}_{31}\Gamma^{1}_{33} - \Gamma^{3}_{32}\Gamma^{2}_{33} = 0,$$

and the other non-diagonal entries of $R_{\mu\nu}$ are also zeros. Therefore, (3.4) is satisfied.

□

The above theorem nicely demonstrates Schwarzschild's ingenuity to find a nontrivial solution of a very complicated system of partial differential equations (3.4) for an arbitrary constant $S > 0$ (see also [107, 167, 249, 369]). For a spherically symmetric nonrotating object with mass $M > 0$, the constant

$$S = \frac{2MG}{c^2} \tag{3.10}$$

is called its *Schwarzschild gravitational radius*, where G is the gravitational constant and c is the speed of light in a vacuum. Applying the Newtonian formula for the escape velocity $v = \sqrt{2MG/r}$ from a circular orbit with radius $r = S$, we find that $v = c$.

Remark 3.2 Denote by

$$S_{\mu\nu} = -\Gamma^{\varkappa}_{\ \mu\varkappa,\nu} + \Gamma^{\lambda}_{\ \mu\nu}\Gamma^{\varkappa}_{\ \lambda\varkappa}$$

the doubly underlined terms in (3.4). By Einstein [91] one additional algebraic condition (3.3) on volume surprisingly guarantees that the 10 differential operators $S_{\mu\nu} = S_{\nu\mu}$ simultaneously vanish when (3.4) is valid, see [107, p. 73]. The reason is that $\Gamma^{\varkappa}_{\ \lambda\varkappa} = (\ln\sqrt{-g})_{,\lambda}$, where $g = \det(g_{\mu\nu}) = -1$. Since $\ln 1 = 0$, we find that $\Gamma^{\varkappa}_{\ \lambda\varkappa} = 0$, and thus (3.1) holds, i.e., the doubly underlined terms in (3.4) will disappear when (3.3) is valid.

For positive numbers $R < Q$ consider the spherical shell

$$\{(x^1, x^2, x^3) \in \mathbb{E}^3 \mid R^2 \le (x^1)^2 + (x^2)^2 + (x^3)^2 \le Q^2\}$$

with interior radius R and exterior radius Q (see Fig. 3.2). Its volume in Euclidean space $\mathbb{E}^3$ is equal to

$$V = \frac{4}{3}\pi(Q^3 - R^3). \tag{3.11}$$

Now we will derive a formula for the proper volume $\tilde{V}$ of the spherical shell with coordinate radii $R < Q$ in a curved space around a nonrotating ball with mass M and coordinate radius R. Here the tilde indicates a curved space. It is really curved, even though it is empty (modeling a vacuum). By (3.6) we find that the exterior

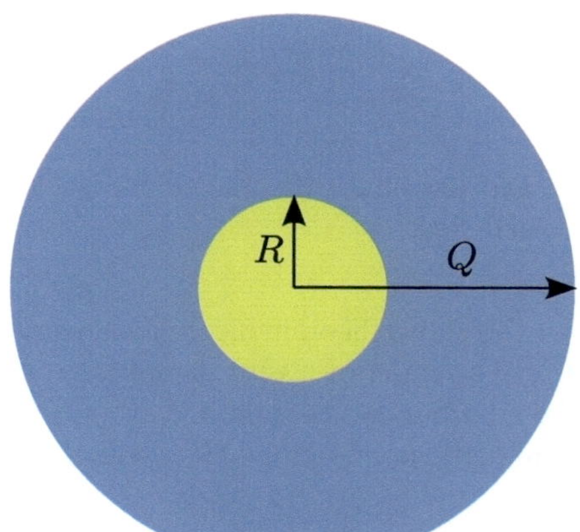

Fig. 3.2 Spherical shell between two concentric spheres

space volume element is equal to

$$\mathrm{d}\tilde{V} = \sqrt{\frac{r}{r-S}}\,\mathrm{d}r \cdot (r \sin\theta\,\mathrm{d}\phi) \cdot (r\,\mathrm{d}\theta).$$

Therefore, by Fubini's Theorem [296, p. 581] the *proper (relativistic) volume* is defined as

$$\tilde{V} = \int_R^Q r^2 \sqrt{\frac{r}{r-S}}\,\mathrm{d}r \cdot \int_0^\pi \Big(\int_0^{2\pi} \sin\theta\,\mathrm{d}\phi\Big)\mathrm{d}\theta = 4\pi \int_R^Q r^2 \sqrt{\frac{r}{r-S}}\,\mathrm{d}r. \tag{3.12}$$

In [182] we prove the following astonishing theorem.

Theorem 3.3 *Let $M > 0$ and $R > S$ be any fixed numbers and let (3.10)–(3.12) hold. Then*

$$\tilde{V} - V \to \infty \quad \textit{as } Q \to \infty.$$

Proof The primitive function (i.e. antiderivative) to (3.12) is given e.g. in [120, p. 97] and by differentiation we can verify that

$$\int r^2 \sqrt{\frac{r}{r-S}}\,\mathrm{d}r = \Big(\frac{r^2}{3} + \frac{5Sr}{12} + \frac{5S^2}{8}\Big)\sqrt{r(r-S)} + \frac{5S^3}{16}\ln(2\sqrt{r(r-S)} + 2r - S).$$

From this, (3.12), and (3.11) we get

$$\tilde{V} - V = 4\pi \int_R^Q r^2 \sqrt{\frac{r}{r-S}}\,\mathrm{d}r - \frac{4}{3}\pi(Q^3 - R^3) =$$

$$= \frac{4\pi}{3}\Big[\Big(Q^2 + \frac{5SQ}{4} + \frac{15S^2}{8}\Big)\sqrt{Q(Q-S)} + \frac{15S^3}{16}\ln(2\sqrt{Q(Q-S)} + 2Q - S)$$

$$- \Big(R^2 + \frac{5SR}{4} + \frac{15S^2}{8}\Big)\sqrt{R(R-S)} - \frac{15S^3}{16}\ln(2\sqrt{R(R-S)} + 2R - S)$$

$$- Q^3 + R^3\Big]. \tag{3.13}$$

We see that the relation (3.13) was derived from the Schwarzschild solution (3.6) exactly without any approximations. The difference of the two terms containing ln in (3.13) is positive, because

$$Q > R > S$$

and the logarithmic function is increasing, (Since ln $a-$ ln b $=\ln(a/b)$, the argument in parenthesis is dimensionless.) Thus from the inequality

$$\sqrt{Q(Q-S)} > Q-S \tag{3.14}$$

we get the following lower bound

$$\tilde{V}-V > \Big(Q^2+\frac{5SQ}{4}+\frac{15S^2}{8}\Big)(Q-S)-Q^3+C=\frac{SQ^2}{4}+\frac{5S^2Q}{8}+\overline{C},$$

where C contains all remaining terms not depending on Q and where $\overline{C}=C-15S^3/8$. Letting $Q\to\infty$, we obtain the statement of the theorem. □

Consequently, a natural question arises: How large can Q be so that the relativistic relation (3.12) approximates reality well. According to astronomical observations, space is locally almost flat at any fixed time, and thus to consider a very large Q makes good sense.

From Theorem 3.3 we see that the difference of volumes $\tilde{V}-V$ increases beyond all limits for $Q\to\infty$, which is a quite surprising property. Namely, this theorem can be applied for instance to a billiard ball or a small steel ball bearing (see Example 3.1 below) or an imperceptible pinhead, since the fixed mass $M>0$ can be arbitrarily small. Consequently, a natural question arises whether the exterior Schwarzschild solution (3.6) approximates reality well on large scales (see also [110]).

Example 3.1 Setting $M=0.033$ kg, $R=0.01$ m, and $Q=5\cdot10^{20}$ m, which is the radius of our Galaxy, we find from (3.10) that $S=5\cdot10^{-29}$ m and by (3.13) the difference

$$\tilde{V}-V\approx 10\,000\ \text{km}^3.$$

This is about 10^{19} times more than the volume of the ball itself. From this we see that the application of Einstein's equations to galactic distances is questionable. Their application to cosmological distances may yield, by Theorem 3.3, surprising results, which may have nothing to do with reality (see also [198]).

Remark 3.3 If $M>0$ and $Q>S$ are any fixed numbers such that (3.10) holds, then by (3.13) we get $\tilde{V}<\infty$ for $R\to S^+$, i.e., there is no singularity in volume near the ball with Schwarzschild radius.

Remark 3.4 According to (3.6), the curved Schwarzschild space is in the radial direction described by the relation

$$\mathrm{d}\rho=\sqrt{\frac{r}{r-S}}\mathrm{d}r.$$

By differentiation, we can check that

$$\int \sqrt{\frac{r}{r-S}}\,\mathrm{d}r = \sqrt{r(r-S)} + S\ln(\sqrt{r}+\sqrt{r-S}).$$

Replacing the volume V by another expression $\hat{V} = \frac{4}{3}\pi((\rho(Q))^3 - (\rho(R))^3)$ in Theorem 3.3, we can similarly derive that $\hat{V} - \tilde{V} \to \infty$ as $Q \to \infty$, see [182].

3.3 Interior Schwarzschild Solution

If o is the circumference of a nonrotating mass ball, then its apparent radius

$$R = \frac{o}{2\pi} \tag{3.15}$$

is called the *coordinate radius*.

In 1916, Karl Schwarzschild [322] found the first static nonvacuum solution of Einstein's equations, cf. (3.16) below. He assumed that the ball with coordinate radius $R > 0$ is formed by an ideal incompressible nonrotating fluid with constant density to avoid a possible internal mechanical stress in the solid that may have a nonnegligible influence on the resulting gravitational field. Then by Ellis [98] (see also [105, p. 529], [332, p. 213], [390]) the corresponding metric tensor is given by

$$g_{\mu\nu} = \mathrm{diag}\left(-\frac{1}{4}\left(3\sqrt{1-\frac{S}{R}} - \sqrt{1-\frac{Sr^2}{R^3}}\right)^2, \frac{R^3}{R^3 - Sr^2}, r^2, r^2\sin^2\theta\right), \tag{3.16}$$

where $r \in [0, R]$. The tensor (3.16) is called the *interior Schwarzschild solution*, see Stephani [332, p. 213]. It is again a static solution, meaning that it is independent of time. To avoid division by zero in the component g_{11}, we require

$$R^3 - Sr^2 > 0 \quad \text{for all } r \in [0, R]$$

which leads to the inequality

$$\boxed{R > S.}$$

We observe that for $r = R$ the solution (3.16) for the interior of the ball continuously matches the Schwarzschild solution (3.6) defined outside the ball. We shall return to this important issue in Sect. 3.4.

From (3.16) we see that the corresponding time independent metric tensor (i.e. when $\mathrm{d}t = 0$) satisfies

$$\mathrm{d}l^2 = \frac{R^3}{R^3 - Sr^2}\,\mathrm{d}r^2 + r^2\mathrm{d}\theta^2 + r^2\sin^2\theta\,\mathrm{d}\phi^2$$

$$= \begin{pmatrix}\mathrm{d}r\\ \mathrm{d}\theta\\ \mathrm{d}\phi\end{pmatrix}^{\top} \begin{pmatrix}\frac{R^3}{R^3 - Sr^2} & 0 & 0\\ 0 & r^2 & 0\\ 0 & 0 & r^2\sin^2\theta\end{pmatrix} \begin{pmatrix}\mathrm{d}r\\ \mathrm{d}\theta\\ \mathrm{d}\phi\end{pmatrix}, \tag{3.17}$$

where $r \in [0, R]$. Now we will calculate the proper radius of the nonrotating ball defined by (3.22) below. For $\mathrm{d}\phi = 0$ and $\mathrm{d}\theta = 0$ the equality (3.17) obviously reduces to

$$\mathrm{d}l^2 = \frac{\mathrm{d}r^2}{1 - r^2 s^2}, \tag{3.18}$$

where

$$s := \sqrt{\frac{S}{R^3}}. \tag{3.19}$$

For $r \in [0, s^{-1})$ we can easily verify that

$$F(r) = \frac{1}{s}\arcsin(rs) \tag{3.20}$$

is a primitive function of

$$f(r) = \frac{1}{\sqrt{1 - r^2 s^2}}. \tag{3.21}$$

From the inequality $R > S$ and (3.19) we see that

$$R < R\sqrt{\frac{R}{S}} = s^{-1}.$$

Hence, the primitive function F can be considered on the shorter interval $[0, R] \subset [0, s^{-1})$.

Using (3.18), (3.20), and (3.21), we now define the *proper (relativistic) radius* of the homogeneous nonrotating mass ball as

$$\tilde{R} = \int_0^R \frac{\mathrm{d}r}{\sqrt{1 - r^2 s^2}}.$$

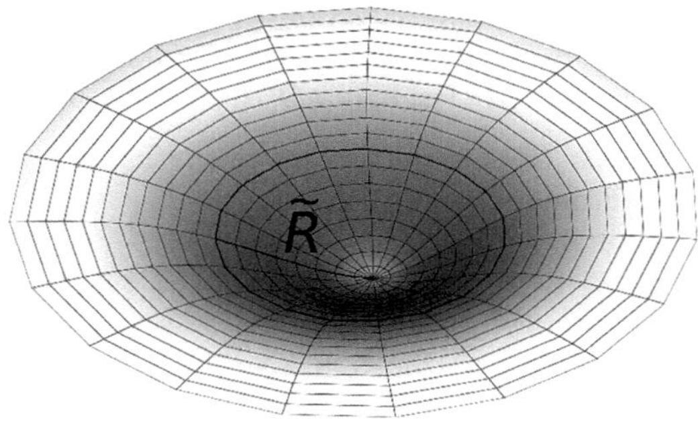

Fig. 3.3 Relativistic radius $\tilde{R}$ is greater than the coordinate radius $R = o/(2\pi)$

From (3.20) we find that (see Fig. 3.3)

$$\tilde{R} = \frac{1}{s} \arcsin(Rs) > R, \tag{3.22}$$

since $\arcsin x > x$ for $x \in (0, 1]$.

Remark 3.5 The proper radius $\tilde{R}$ represents a geodesic in a curved spacetime inside the nonrotating mass ball. It is the shortest path between any point on the surface of the ball and its center $(0, 0, 0)$, cf. Fig. 3.3. As usual, the length of this geodesic is denoted by the same symbol $\tilde{R}$. For $0 \leq r_1 < r_2 \leq R$ we immediately get by (3.21) the inequality

$$\int_{r_1}^{r_2} \frac{1}{\sqrt{1 - r^2 s^2}} \mathrm{d}r > \int_{r_1}^{r_2} \mathrm{d}r = r_2 - r_1.$$

From this we observe that the length element in the radial direction along $\tilde{R}$ is longer than the length element in the corresponding Euclidean space.

Theorem 3.4 *For a given fixed $R > 0$ we have*

$$\tilde{R} \to R \quad \textit{for } s \to 0.$$

Proof From (3.22) we find by l'Hospital's rule [296, p. 388] that

$$\lim_{s \to 0} \frac{\arcsin(Rs)}{s} = \lim_{s \to 0} \frac{R}{\sqrt{1 - (Rs)^2}} = R.$$

The proof is completed. □

From (3.10) and (3.19) we see that $s \to 0$ if and only if $M \to 0$. By Theorem 3.4 the relativistic radius thus converges to the coordinate radius when $M \to 0$.

The coordinate radius of a given fixed ball with circumference o is uniquely defined by (3.15), whereas its proper radius $\tilde{R}$ is not uniquely determined. It depends on the mass M. The larger M is, the larger $\tilde{R}$ is.

Example 3.2 Consider a homogeneous ball with mass $M_\odot = 2 \cdot 10^{30}$ kg and coordinate radius

$$R_\odot = 695{,}700 \text{ km} \tag{3.23}$$

corresponding to the Sun. (Note that by the B3 resolution of the International Astronomical Union in 2015 (Hawaii) the radius of the Sun is given only to the above 4 significant digits.) Then by (3.22) we have $\tilde{R}_\odot - R_\odot = 492$ m. Hence, the associated Euclidean circumference is about 3 km shorter than $2\pi \tilde{R}_\odot$, where $\tilde{R}_\odot$ is the proper radius [163]. This relativistic effect is thus negligible when observing the Sun.

Remark 3.6 In 1932, Lev D. Landau briefly mentions stars which could have the density of an atomic nucleus, see [212, p. 228]. Two years later Walter Baade and Fritz Zwicky postulated the existence of neutron stars as remainders of supernova explosions and stated the hypothesis that this could be a source of cosmic rays (see [19, 20]). On August 6, 1967, Jocelyn Bell [127] observed periodically repeating signals at the Mullard Radio Astronomy Observatory in Cambridge (the radiotelescope was constructed by Antony Hewish). Then F. Pacini [267] and T. Gold [117] found that these signals are generated by rapidly rotating neutron stars. As of 2022 the fastest-spinning neutron star known is PSR J1748-2446ad, rotating at a rate of 716 times a second. A natural question arises:

How far does its strong magnetic field act?

Note that at the distance 70 km from the rotational axis the speed v of magnetic field lines in the equatorial plane would be

$$v = 716\,\text{s}^{-1} \cdot 2\pi \cdot 70\,\text{km} = 3.14 \cdot 10^5 \text{ km/s},$$

which is more than the speed of light! Since magnetic field is a mass object, it may act only in very narrow relativistic jets along the rotational axis. Similar jets are observed also at black holes (see Fig. 12.4) and they could have the same origin. For the time being there is no generally accepted theory that explains the generation and collimation of these jets. However, it is not excluded that the kernel of any physical rotating black hole also produces strong magnetic fields, even though the theory admits only electric fields.

Note that the current astrophysical hypothesis claims that the magnetic field is attributed not to the black hole itself. Rather, it is somehow generated from the plasma surrounding and falling into the black hole.

Remark 3.7 Let us emphasize that one should carefully distinguish between an observed black hole and a theoretical black hole on a piece of paper. The interior of a physical black hole can contain a compact object with neutron star density (see [166] and Fig. 12.2 below) instead of the proclaimed point singularity with infinite mass density. Infinity is only an abstract mathematical term. Moreover, it is important to realize that an arbitrarily large number is always in a close neighborhood of the origin on the real axis for a suitably chosen scale.

At present, it is assumed that the core of a neutron star consists of cold quark-gluon plasma and that its mean density could be 2–3 times higher than the density of a neutron itself.

Example 3.3 Consider an idealized homogeneous nonrotating neutron star with mass $M_\odot$ and radius $R = 10$ km, which is in good agreement with [331]. Then by (3.10), (3.19), and (3.22) we obtain that $\tilde{R} = 10.583$ km (see Fig. 3.3).

Theorem 3.5 *If $M > 0$ is fixed, then the function $R \mapsto \tilde{R} - R$ is decreasing and strictly convex on the interval (S, ∞).*

The proof follows easily from (3.19) and (3.20) under the condition $R > S$ (cf. Fig. 3.4).

Theorem 3.6 *The proper volume of a homogeneous mass ball with coordinate radius $R > S > 0$ is*

$$\tilde{V} = \frac{2\pi}{s^2}(\tilde{R} - R\sqrt{1 - R^2 s^2}),$$

where $\tilde{R}$ is given by (3.22) and s by (3.19).

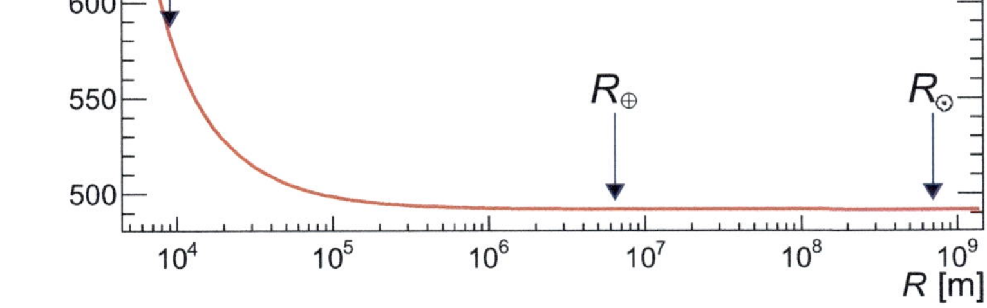

Fig. 3.4 Dependence of the difference $\tilde{R} - R$ on the coordinate radius R for the fixed mass $M = M_\odot = 2 \cdot 10^{30}$ kg (see Theorem 3.5). Here $R_\odot = 695\,700$ km is the coordinate radius of the Sun. The radius $R_\oplus = 6378$ km corresponds approximately to a white dwarf and $R_\circ \approx 10$ km to a hypothetical neutron star

Proof By (3.17) and (3.18) the volume element reads

$$\mathrm{d}\tilde{V} = \frac{\mathrm{d}r}{\sqrt{1 - r^2 s^2}} \cdot (r\,\mathrm{d}\theta) \cdot (r \sin\theta\,\mathrm{d}\phi).$$

By differentiation we can easily check that

$$H(r) = \frac{1}{2s^3} \arcsin(rs) - \frac{r}{2s^2}\sqrt{1 - r^2 s^2}$$

is a primitive function of

$$h(r) = \frac{r^2}{\sqrt{1 - r^2 s^2}}$$

on the interval $[0, s^{-1})$, cf. [272, p. 298].

By Fubini's Theorem [296, p. 581] the proper (relativistic) volume of the homogeneous fixed mass ball can be expressed as follows

$$\tilde{V} = \int_0^R \frac{r^2\,\mathrm{d}r}{\sqrt{1 - r^2 s^2}} \cdot \int_0^\pi \sin\theta\,\mathrm{d}\theta \cdot \int_0^{2\pi} \mathrm{d}\phi = H(R) \cdot 2 \cdot 2\pi$$
$$= \frac{2\pi}{s^2}\Big(\frac{\arcsin(Rs)}{s} - R\sqrt{1 - R^2 s^2}\Big). \quad (3.24)$$

Hence, by (3.22) we find that the theorem is valid. □

Theorem 3.7 *Under the assumptions of Theorem 3.6 we have*

$$\tilde{V} \to V := \frac{4}{3}\pi R^3 \quad \textit{for } s \to 0. \quad (3.25)$$

Proof From Theorem 3.6, formula (3.22), and l'Hospital's rule [296, p. 388] we obtain

$$\lim_{s\to 0} \frac{\arcsin(Rs) - Rs\sqrt{1 - R^2 s^2}}{s^3} = \lim_{s\to 0} \frac{\dfrac{R - (R - R^3 s^2) + R^3 s^2}{\sqrt{1 - R^2 s^2}}}{3s^2} = \frac{2}{3} R^3.$$

This together with (3.24) yields (3.25). □

Example 3.4 Consider a homogeneous ball with mass $M = M_\oplus = 5.97219 \cdot 10^{24}$ kg and coordinate radius $R = R_\oplus = 6378$ km corresponding to the Earth. Then by (3.24) we get a surprisingly large relativistic effect, namely,

$$\tilde{V}_\oplus - V_\oplus = 457 \text{ km}^3,$$

where

$$V_{\oplus} \approx 10^{12} \text{ km}^3. \tag{3.26}$$

Theorem 3.8 *If $M > 0$ is fixed, then the function $R \mapsto \tilde{V}/V$ is decreasing and strictly convex on the interval (S, ∞).*

The proof directly follows from relations (3.19) and (3.24), and the inequality $R > S$.

Remark 3.8 The convergence in (3.25) is monotonically decreasing. Hence, the proper volume $\tilde{V}$ is larger than the coordinate volume V for $M > 0$, namely,

$$\tilde{V} > V.$$

The matter responds to gravitational compression in such a way that it increases its proper volume $\tilde{V}$ when R is fixed and M increases. In other words, the larger M is, the larger $\tilde{V}$ is for a fixed circumference and fixed V. Note that the mineral zeolite has a similar property due to its special crystal lattice. Whenever it is compressed, it increases its volume. Its atoms are arranged such that they have the maximum possible density.

Example 3.5 For the Sun with coordinate radius $R_{\odot} = 695\,700$ km we obtain by (3.24) that (cf. (3.26))

$$\tilde{V}_{\odot} - V_{\odot} = 1.796 \cdot 10^{12} \text{ km}^3, \quad \overline{V}_{\odot} - \tilde{V}_{\odot} = 1.197 \cdot 10^{12} \text{ km}^3,$$

where $V_{\odot} = \frac{4}{3}\pi R_{\odot}^3$ and $\overline{V}_{\odot} = \frac{4}{3}\pi \tilde{R}_{\odot}^3$ (cf. (3.26) and Example 3.2). For comparison, by integration of the vacuum solution we get by (3.13) that

$$\tilde{V} - V = 4.49 \cdot 10^{12} \text{ km}^3$$

for $Q = R_{\odot}$ and $R \searrow S_{\odot}$, where

$$S_{\odot} = 2.953 \text{ km} \tag{3.27}$$

is the Schwarzschild radius of the Sun.

Example 3.6 We will now show that there is a big difference between classical Euclidean geometry and the geometry of a curved spacetime inside high density objects. Consider an idealized neutron star described by a homogeneous ball with mass $M = M_{\circ} = 3M_{\odot} = 6 \cdot 10^{30}$ kg and coordinate radius $R = R_{\circ} = 10$ km (cf. Fig. 3.4). Then by (3.24) and (3.25) we get quite a large relativistic effect, namely, the ratio between the relativistic volume and Euclidean volume equals

$$\frac{\tilde{V}}{V} = \frac{3}{2S}(\tilde{R} - R\sqrt{1 - S/R}) = 1.666. \tag{3.28}$$

Consequently, the relativistic volume $\tilde{V}$ is 67% larger than the Euclidean volume V defined in (3.25). This means that the mean mass density $\tilde{\rho} = M/\tilde{V}$ is smaller than $\rho = M/V$ and $\tilde{\rho}/\rho = 0.6$.

Since the density gradient of a physical neutron star is not known, formula (3.28) represents only an approximate lower estimate. Moreover, the relativistic volume extension could be sufficient to prevent density increase above the neutron or quark densities. The definition of mass density under strong gravitational field conditions is not an evident quantity, because the volume is influenced by the curved geometry of space under gravity. For derivations of density expressions under such conditions we refer to [104] (see also [100]).

3.4 Composite Schwarzschild Metric Tensor

For $r_0 = R$ define the *composite Schwarzschild metric tensor* $g_{\mu\nu}$ by (3.6) for $r > r_0$ and by (3.16) for $r \in [0, r_0]$. It is easy to check that $g_{\mu\nu}$ is continuous everywhere. However, we show that the first classical derivatives of the composite metric tensor have jumps. These derivatives appear in Einstein's equations which are of the second order.

Theorem 3.9 *The composite Schwarzschild metric tensor $g_{\mu\nu}$ is not differentiable for $r = r_0$.*

Proof We will show that the component g_{11} is not differentiable. From (3.6) and (3.16) we find that its first classical derivative does not exist at $r = r_0$ (see Fig. 3.5). Namely, the component $g_{11}(r)$ of the interior solution (3.16) is an increasing function on $[0, r_0]$, whereas from (3.6) we see that the one-sided limit of

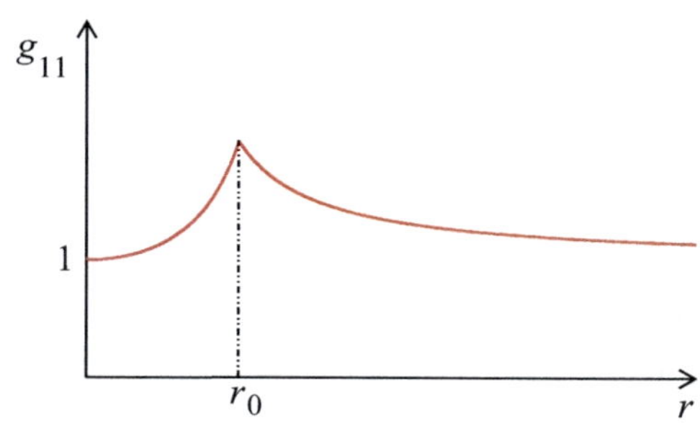

Fig. 3.5 The behavior of the non-differentiable component $g_{11} = g_{11}(r)$ of the composite metric tensor from (3.6) and (3.16). The first classical derivative $(\partial g_{11}/\partial r)(r_0)$ is not defined. The piecewise rational function g_{11} cannot be smoothed near r_0, since then Einstein's equations would not be valid in a close neighborhood of r_0

the first derivative of the component $g_{11}(r) = r/(r - S)$ of the exterior solution is negative

$$\lim_{r \to r_0^+} \frac{\partial g_{11}}{\partial r}(r) < 0.$$

The proof is completed. □

To show why the composite Schwarzschild metric tensor (3.6)+(3.16) is not a global solution of Einstein's equation, we first present the following trivial example.

Example 3.7 Consider the function $u(x) = |x|$ on the interval $(-1, 1)$. It solves the second order differential equation

$$\ddot{u}(x) = 0$$

locally on the intervals $(-1, 0)$ and $(0, 1)$. However, $u = u(x)$ is not a global solution on the interval $(-1, 1)$, since its first derivative $\dot{u}$ has a jump at 0.

Thus from Theorem 3.9 we observe that the composite Schwarzschild metric tensor (3.6)+(3.16) is not a global solution of an idealized spherically symmetric star and its close neighborhood (see Fig. 3.2). All Riemannian spacetime manifolds have to be locally Euclidean which is not true in this case. Recall that an *n-dimensional manifold* with $n \in \mathbb{N} = \{1, 2, 3, \dots\}$ is a set of points for which there exists an open neighborhood that can be continuously mapped onto an open set in $\mathbb{E}^n$ such that the inverse is continuous, too. An example of a manifold is the graph of a parabola, a hyperboloid of two sheets, the surface of a torus, and so on. On the other hand, the union of the hyperplane $x_1 = 0$ and axis x_1 in $\mathbb{E}^n$ is not a manifold for $n > 1$. Also the set of rational numbers is not a manifold.

Example 3.8 For comparison, note that the first order classical derivatives of the Newton potential u of a similar problem are continuous. This problem is described by the *Poisson equation*

$$\Delta u = 4\pi G\rho,$$

where ρ stands for the mass density. Let the right-hand side $f = 4\pi G\rho$ be spherically symmetric and such that

$$f(r) = \begin{cases} 1 & \text{if } r \in [0, 1], \\ 0 & \text{otherwise.} \end{cases}$$

The Laplace operator in spherical coordinates reads (see [296, p. 236])

$$\Delta u = \frac{\partial^2 u}{\partial r^2} + \frac{2}{r}\frac{\partial u}{\partial r} + \frac{1}{r^2}\Big(\frac{\partial^2 u}{\partial \theta^2} + \operatorname{cotan}\theta \frac{\partial u}{\partial \theta} + \frac{1}{\sin^2\theta}\frac{\partial^2 u}{\partial \phi^2}\Big).$$

The term in parenthesis on the right-hand side is zero for the spherically symmetric case. By the well-known method of variations of constants, we find the following solution of the above Poisson equation

$$u(r,\theta,\phi) = \begin{cases} \frac{1}{6}r^2 - \frac{1}{2} & \text{for } r \in [0,1], \\ -\frac{1}{3r} & \text{otherwise.} \end{cases}$$

Hence, we see that both u and $\partial u/\partial r$ are continuous at $r_0 = 1$. The second derivatives are square integrable in a neighborhood of $r_0 = 1$.

3.5 Einstein's Equations with General Right-Hand Side

Einstein's field equations (cf. [92]) without the cosmological constant are at present usually written in the following general form

$$R_{\mu\nu} - \frac{1}{2}Rg_{\mu\nu} = \frac{8\pi G}{c^4}T_{\mu\nu}, \quad \mu,\nu = 0,1,2,3. \tag{3.29}$$

This is the system of 10 equations for $10 = 1+2+3+4$ components of the unknown twice differentiable symmetric metric tensor $g_{\mu\nu}$. Now we describe the individual terms. The symbols $R_{\mu\nu}$ stands for the symmetric Ricci tensor defined by (3.4), its contraction

$$R = g^{\mu\nu}R_{\mu\nu} \tag{3.30}$$

is called the *Ricci scalar* (i.e. the scalar curvature), and $T_{\mu\nu}$ $(= T^{\varkappa\lambda}g_{\varkappa\mu}g_{\lambda\nu})$ is the symmetric *tensor of density of energy and momentum.* The left-hand side of (3.29)

$$G_{\mu\nu} := R_{\mu\nu} - \frac{1}{2}Rg_{\mu\nu}$$

is called the *Einstein tensor.* Contrary to the conventions in mathematical texts, the dependence of $R_{\mu\nu}$ and R on $g_{\mu\nu}$ is not indicated in the GTR.

Remark 3.9 Let us emphasize that Einstein's 10 equations (3.29) are not independent, since the covariant divergence of the right-hand side is supposed to be zero (see e.g. [249, p. 261]), i.e.,

$$T^{\mu\nu}_{\ \ ;\nu} := T^{\mu\nu}_{\ \ ,\nu} + \Gamma^{\mu}_{\lambda\nu}T^{\lambda\nu} + \Gamma^{\nu}_{\lambda\nu}T^{\mu\lambda} = 0, \tag{3.31}$$

where $T^{\mu\nu} = g^{\mu\varkappa}g^{\lambda\nu}T_{\varkappa\lambda}$ and where all Christoffel symbols vanish when the spacetime is not curved. Note that the covariant divergence of the Ricci tensor $R^{\mu\nu}_{\ \ ;\nu}$

is nonzero, in general. However, by the Bianchi identities (see e.g. [369, p. 147]) the covariant divergence of the whole left-hand side of (3.29) (i.e. the Einstein tensor) is zero automatically for $g_{\mu\nu}$ smooth enough, e.g., if the third derivatives of $g_{\mu\nu}$ are continuous (which is not the case in Fig. 3.5). Therefore, we have only 6 independent equations in (3.29). The number of independent components of the metric tensor is also 6, since one has to choose 4 coordinates. Relation (3.31) is called the *law of local energy-momentum conservation*, see [249, p. 386].

Recall (see [107, p. 4]) that the *principle of general covariance* requires invariance of the form of physical laws under arbitrary diffeomorphic changes of coordinates.

Finally note that the contravariant symmetric 4×4 metric tensor $g^{\mu\nu}$ in (3.29) which is inverse to the covariant metric tensor $g_{\mu\nu}$ satisfies

$$g^{\mu\nu} = \frac{g^*_{\mu\nu}}{\det(g_{\mu\nu})}, \qquad \det(g_{\mu\nu}) := \sum_{\pi \in S_4} (-1)^{\operatorname{sgn}\pi} g_{0\nu_0} g_{1\nu_1} g_{2\nu_2} g_{3\nu_3}, \tag{3.32}$$

where the entries $g^*_{\mu\nu}$ form the 4×4 matrix of 3×3 algebraic adjoints of $g_{\mu\nu}$, S_4 is the symmetric group of 24 permutations π of indices $(\nu_0, \nu_1, \nu_2, \nu_3)$, $\operatorname{sgn}\pi = 0$ for an even permutation and $\operatorname{sgn}\pi = 1$ for an odd permutation.

Remark 3.10 Notice that the constant on the right-hand side of (3.29) is extremely small in the SI base units. Namely,

$$\frac{8\pi G}{c^4} \approx \frac{8\pi \cdot 6.674 \cdot 10^{-11}}{3^4 \cdot 10^{32}} \approx 2 \cdot 10^{-43} \ (\mathrm{kg}^{-1}\mathrm{m}^{-1}\mathrm{s}^2),$$

implying that the Ricci curvature tensor $R_{\mu\nu}$ is also very small (if components of $T_{\mu\nu}$ are not too large).

Einstein in [91] presented the field equations (3.29) in an equivalent form that is presently written as follows:

$$R_{\mu\nu} = \frac{8\pi G}{c^4}\left(T_{\mu\nu} - \frac{1}{2} T g_{\mu\nu}\right), \tag{3.33}$$

where $T := T_{\mu\nu} g^{\mu\nu} = T^{\mu}_{\mu}$ denotes the trace of $T_{\mu\nu}$. To see that (3.33) is equivalent with (3.29), we multiply (3.29) by $g^{\mu\nu}$. Then the traces of the corresponding tensors satisfy

$$-R = R - 2R = \frac{8\pi G}{c^4} T.$$

Substituting this into (3.29), we get (3.33).

Now we present a quite surprising nonuniqueness property of Einstein's equations with nonzero right-hand side, in general (compare with sign conventions in [249, p. 0]).

Theorem 3.10 *If $g_{\mu\nu}$ is a solution of (3.29), then $(-g_{\mu\nu})$ also solves (3.29).*

Proof From (3.2), we find that the Christoffel symbols of the second kind remain the same if we replace $g_{\mu\nu}$ by $(-g_{\mu\nu})$, namely,

$$\Gamma^{\varkappa}{}_{\mu\nu} = \frac{1}{2}(-g^{\varkappa\lambda})(-g_{\nu\lambda,\mu} - g_{\lambda\mu,\nu} + g_{\mu\nu,\lambda}).$$

Using (3.4), we find that the Ricci tensor $R_{\mu\nu}$ in (3.29) also does not change. Concerning the second term on the left-hand side of (3.29), we observe from (3.30) that $\left(-\frac{1}{2}Rg_{\mu\nu}\right)$ also remains unchanged if we replace $g_{\mu\nu}$ by $(-g_{\mu\nu})$. □

Now we will derive the relation (3.4) from (3.29).

Theorem 3.11 *If $T_{\mu\nu} = 0$ then Einstein's equations (3.29) imply that $R_{\mu\nu} = 0$.*

Proof Multiplying (3.29) by $g^{\mu\nu}$ and summing over all μ and ν, we obtain by (3.30) that

$$0 = g^{\mu\nu}R_{\mu\nu} - \frac{1}{2}Rg^{\mu\nu}g_{\mu\nu} = R - \frac{1}{2}R\delta^{\mu}{}_{\mu} = R - \frac{1}{2}4R,$$

where $\delta^{\mu}{}_{\nu}$ is the *Kronecker delta* (i.e. $\delta^0{}_0 = \delta^1{}_1 = \delta^2{}_2 = \delta^3{}_3 = 1$ and $\delta^{\mu}{}_{\nu} = 0$ for $\mu \neq \nu$) and $\delta^{\mu}{}_{\mu} = 1+1+1+1 = 4$. Thus, $R = 0$ and Einstein's vacuum equations can be rewritten in the equivalent form

$$R_{\mu\nu} = 0, \tag{3.34}$$

where the unknown metric tensor $g_{\mu\nu}$ is not indicated. □

Concerning nonuniqueness, we observe from (3.2) and (3.4) that we can add any constant to any component $g_{\mu\nu} = g_{\nu\mu}$ and Einstein's equations $R_{\mu\nu} = 0$ will still be valid.

In 1923, George David Birkhoff proved the following theorem (see [30]). Note that a similar statement was introduced by Jørg Tofte Jebsen two years earlier [142].

Theorem 3.12 (Birkhoff) *Any spherically symmetric solution of (3.34) has to be static and asymptotically flat.*

This means that the exterior solution of vacuum Einstein's equations (i.e. the spacetime outside of a spherically symmetric and nonrotating gravitating body) must be given by the exterior Schwarzschild solution (3.6).

Remark 3.11 In 1963, Roy P. Kerr found an axially symmetric solution of Einstein's vacuum equations, see [144]. His solution (called the *Kerr metric tensor*) is a generalization of the Schwarzschild metric tensor when the central body has a nonzero angular momentum. Kerr's solution is described by two parameters: the mass M and the distance parameter a, where Mca is the angular momentum,

$$c^2 \mathrm{d}\tau^2 = -\frac{\Delta}{\rho^2}(c\mathrm{d}t - a\sin^2\theta \mathrm{d}\phi)^2 + \frac{\rho^2}{\Delta}\mathrm{d}r^2 + \rho^2 \mathrm{d}\theta^2 + \frac{\sin^2\theta}{\rho^2}[(r^2+a^2)\mathrm{d}\phi - ac\mathrm{d}t]^2, \tag{3.35}$$

where r, θ, ϕ are the standard spherical coordinates and

$$\Delta = r^2 - 2MGr/c^2 + a^2, \qquad \rho^2 = r^2 + a^2\cos^2\theta.$$

If a is positive, the metric tensor will be nondiagonal. For $a = 0$ and $\rho \geq 0$ we see that $\rho = r$ and relation (3.35) reduces to

$$c^2 \mathrm{d}\tau^2 = \Big(-1 + \frac{2MG}{c^2 r}\Big)\mathrm{d}t^2 + \Big(1 - \frac{2MG}{c^2 r}\Big)^{-1}\mathrm{d}r^2 + r^2\mathrm{d}\theta^2 + r^2\sin^2\theta \mathrm{d}\phi^2.$$

Using (3.10), we easily find that this leads to the Schwarzschild exterior solution (3.6), i.e., if the rotation in the Kerr solution is put to zero, the Schwarzschild solution is retrieved.

If $0 \leq a \leq S/2$, then the radius $U = (S + \sqrt{S^2 - 4a^2})/2$ has a similar property as the Schwarzschild radius S given by (3.10). There exists a special surface defined by $V = (S + \sqrt{S^2 - 4a^2\cos^2\theta})/2$, where $\theta \in [0, \pi/2]$ is the angle from the rotational axis. The region between U and V is called an *ergosphere*. The behavior of geodesics lying in the equatorial plane is examined in [41].

From a mathematical point of view, it is confusing to call (3.35) a "metric". The reason is that a function $\rho : M \times M \to \mathbb{E}^1$ is called a *metric* on a manifold M if:

1. $\rho(A, B) \geq 0 \quad \forall A, B \in M$,
2. $\rho(A, B) = 0 \iff A = B$,
3. $\rho(A, B) = \rho(B, A) \quad \forall A, B \in M$,
4. $\rho(A, B) \leq \rho(A, C) + \rho(C, B) \quad \forall A, B, C \in M$ (triangle inequality).

3.6 Black Holes

A *black hole* is a region of spacetime where gravity is so strong that nothing (no particles or even electromagnetic radiation) can escape from it. A spherical surface with Schwarzschild radius (3.10) is called an *event horizon*. Its topology is $\mathbb{S}^2$. The curvature above the event horizon is usually described by the Schwarzschild metric tensor, or the Kerr metric tensor if the black hole is rotating, see (3.6) and Remark 3.11.

Stellar black holes are formed when very massive stars collapse at the end of their life cycle. For the origin of intermediate and supermassive black holes we refer to the paper [208] by Pavel Kroupa et at. Indirect evidence of their existence can be obtained from fast motions of orbiting stars, see e.g. [9, 319]. Anyway, for the time being no mechanism is known that some black holes grew to be a billion times the mass of the Sun just 1 Gyr after the Big Bang, even though the universe had much higher mean mass density than at present. These giant black holes could also be the remnants of the last Big Crunch.

Remark 3.12 An English natural philosopher John Michell (1724–1793) was the first person who predicted the existence of black holes. He called them *dark stars*. In his 1784 paper [243], he writes: "*..., if the semi-diameter of a sphære of the same density with the sun were to exceed that of the sun in the proportion of 500 to 1, a body falling from an infinite height towards it, would have acquired at its surface a greater velocity than that of the light, and consequently, supposing light to be attracted by the same force in proportion to its vis inertiæ, with other bodies, all light emitted from such a body would be made to return towards it, by its own proper gravity.*"

Example 3.9 Let us verify the above Michell's original statement. The mass of a ball 500 times larger than the Sun with the same density is $500^3 M_\odot = 1.25 \cdot 10^8 M_\odot$. By (3.10) and (3.27) its Schwarzschild radius is $S = 3.75 \cdot 10^8$ km. We see that this number is by (3.23) indeed comparable with radius of Michell's ball $r = 500 \cdot 695{,}700$ km $\approx 3.5 \cdot 10^8$ km.

Remark 3.13 In 1799, Pierre Simon Laplace [215] came to a similar conclusion as Michell: "*...the attractive force of a heavenly body could be so large, that light could not flow out of it.*" In fact, Laplace also gave some details of his statement. In particular, from (3.10) we see that

$$c^2 = \frac{2MG}{S}. \tag{3.36}$$

This equality resembles the Newtonian relation for the escape velocity of an object with mass m from a ball with radius S and mass M, where $0 < m \ll M$, see e.g. [184, p. 52].

Remark 3.14 The origin of the term *black hole* is usually attributed to John Archibald Wheeler, see [375] which was published in 1968. However, already in 1963 this term circulated during corridor conversations in the Texas Symposium on Relativistic Astrophysics in Dallas, see the historical survey [126]. Moreover, in 1964 Ann Ewing published the paper [99]: "Black holes" in space.

Finally, note that the entire universe should not be considered as a black hole as is sometimes claimed, since it has a different topology ($\mathbb{S}^3$ or $\mathbb{E}^3$ or $\mathbb{H}^3$) than a three-dimensional ball in a curved spacetime (cf. Sect. 6.2).

3.7 Einstein's Equations for Euclidean Space

Throughout this section we shall assume that the spatial part of the 4×4 metric tensor is the following constant 3×3 diagonal tensor associated to the three-dimensional Euclidean space

$$g^{\alpha\beta} = g_{\alpha\beta} = \delta_{\alpha\beta} \quad \text{and let} \quad g^{\alpha 0} = g^{0\beta} = g_{\alpha 0} = g_{0\beta} = 0 \quad \text{for all } \alpha, \beta = 1, 2, 3. \tag{3.37}$$

That is,

$$g_{\mu\nu} = \begin{pmatrix} -g_{00} & 0 & 0 & 0 \\ 0 & 1 & 0 & 0 \\ 0 & 0 & 1 & 0 \\ 0 & 0 & 0 & 1 \end{pmatrix}. \tag{3.38}$$

However, $g_{00} = g_{00}(x^0, x^1, x^2, x^3)$ may depend on all four coordinates x^0, x^1, x^2, x^3, in general, and for the determinant g of $g_{\mu\nu}$ we have

$$g := -g_{00}. \tag{3.39}$$

Note that the above notations of the tensor $g_{\mu\nu}$ used in General Relativity differs from the standard notations in linear algebra, where $g_{\mu\nu}$ denotes only one particular entry. The next theorem was inspired by the paper [190].

Theorem 3.13 *Assume that the metric tensor* $g_{\mu\nu} = g_{\mu\nu}(x^0, x^1, x^2, x^3)$ *satisfying Einstein's equations (3.34) is of the form (3.38). Then*

$$g_{00} = g_{00}(x^0),$$

i.e., it is independent of x^1, x^2, x^3.

Proof First, we prove that the Christoffel symbols of the first kind will be zero for the case (3.37) if at most one index is zero. Since all nondiagonal entries of the metric tensor (3.38) are zeros and the time derivative of a constant is zero as well, we obtain that (see [173]):

1. There are $6 = 1 + 2 + 3$ independent components of $\Gamma_{0\alpha\beta}$ due to the symmetry (3.9) and we have

$$\Gamma_{0\alpha\beta} = \frac{1}{2}(g_{0\alpha,\beta} + g_{\beta 0,\alpha} - g_{\alpha\beta,0}) = 0 \quad \text{for } \alpha, \beta = 1, 2, 3.$$

2. There are $9 = 3 \times 3$ independent components of $\Gamma_{\alpha 0\beta}$,

$$\Gamma_{\alpha 0\beta} = \frac{1}{2}(g_{\alpha 0,\beta} + g_{\beta\alpha,0} - g_{0\beta,\alpha}) = 0 \quad \text{for } \alpha, \beta = 1, 2, 3. \tag{3.40}$$

3. There are $18 = 3 \times 6$ independent components of $\Gamma_{\alpha\beta\gamma}$,

$$\Gamma_{\alpha\beta\gamma} = \frac{1}{2}(g_{\alpha\beta,\gamma} + g_{\gamma\alpha,\beta} - g_{\beta\gamma,\alpha}) = 0 \quad \text{for } \alpha, \beta, \gamma = 1, 2, 3. \tag{3.41}$$

Second, we prove that the Christoffel symbols will be, in general, nonzero for the case (3.37) if at least two indices are zero. Hence, by the symmetry $g_{\mu\nu} = g_{\nu\mu}$, (3.8), (3.37), (3.38), (3.40), and (3.41) we have only the following nontrivial cases:

4. By (3.39) there are 4 component such that

$$\Gamma_{0\varkappa 0} = \frac{1}{2}(g_{0\varkappa,0} + g_{00,\varkappa} - g_{\varkappa 0,0}) = \frac{1}{2}g_{00,\varkappa} = -\frac{1}{2}g_{,\varkappa} \quad \text{for } \varkappa = 0, 1, 2, 3. \tag{3.42}$$

5. Finally, for the remaining 3 components we get

$$\Gamma_{\alpha 00} = \frac{1}{2}(g_{\alpha 0,0} + g_{0\alpha,0} - g_{00,\alpha}) = \frac{1}{2}(-g_{00,\alpha}) = \frac{1}{2}g_{,\alpha} \quad \text{for } \alpha = 1, 2, 3. \tag{3.43}$$

So altogether we have evaluated

$$6 + 9 + 18 + 4 + 3 = 40$$

symbols $\Gamma_{\varkappa\mu\nu}$, but only 7 of them are nonzero, in general.

The Christoffel symbols of the second kind are defined by (3.2). However, since in our case the metric tensor is diagonal (3.38), we find that

$$\Gamma^{\varkappa}{}_{\mu\nu} = g^{\varkappa\varkappa}\Gamma_{\varkappa\mu\nu}, \qquad \varkappa, \mu, \nu = 0, 1, 2, 3. \tag{3.44}$$

Here we do not use Einstein's summation convention. Thus, by (3.2), (3.37), (3.39), (3.42), and (3.43) we get

$$\Gamma^{0}{}_{\mu 0} = -\frac{g_{,\mu}}{2g} \quad \text{for } \mu = 0, 1, 2, 3, \tag{3.45}$$

where g does not vanish, and

$$\Gamma^{\alpha}{}_{00} = -\frac{1}{2}g_{,\alpha} \quad \text{for } \alpha = 1, 2, 3. \tag{3.46}$$

By (3.9) and (3.2) we obtain the symmetry in the last two indices of the Christoffel symbols of the second kind,

$$\Gamma^{\varkappa}{}_{\mu\nu} = \Gamma^{\varkappa}{}_{\nu\mu}\,. \tag{3.47}$$

Further, we shall calculate the Ricci tensor. From (3.4) for the entry R_{00} we get

$$R_{00} = \Gamma^{\varkappa}_{\ 00,\varkappa} - \Gamma^{\varkappa}_{\ 0\varkappa,0} + \Gamma^{\lambda}_{\ 00}\Gamma^{\varkappa}_{\ \lambda\varkappa} - \Gamma^{\lambda}_{\ 0\varkappa}\Gamma^{\varkappa}_{\ 0\lambda}. \tag{3.48}$$

Below we shall not use Einstein's summation convention for the index $\alpha = 1, 2, 3$. For the first part of (3.48) with derivatives we obtain by (3.46) that

$$\Gamma^{\varkappa}_{\ 00,\varkappa} - \Gamma^{\varkappa}_{\ 0\varkappa,0} = \left(\Gamma^{0}_{\ 00,0} - \Gamma^{0}_{\ 00,0}\right) + \sum_{\alpha=1}^{3} \Gamma^{\alpha}_{\ 00,\alpha} = -\frac{1}{2}\sum_{\alpha=1}^{3} g_{,\alpha\alpha}\,, \tag{3.49}$$

since by (3.40), (3.41), and (3.44) we have

$$\Gamma^{\alpha}_{\ \lambda\alpha} = 0 \quad \text{for } \alpha = 1, 2, 3, \quad \lambda = 0, 1, 2, 3.$$

From this for the second part of (3.4) with nonlinearities we have by (3.39) and (3.45)–(3.47) that

$$\Gamma^{\lambda}_{\ 00}\Gamma^{\varkappa}_{\ \lambda\varkappa} - \Gamma^{\lambda}_{\ 0\varkappa}\Gamma^{\varkappa}_{\ 0\lambda} = \Gamma^{\lambda}_{\ 00}\Gamma^{0}_{\ \lambda 0} - \Gamma^{\lambda}_{\ 00}\Gamma^{0}_{\ 0\lambda} - \sum_{\alpha=1}^{3}\Gamma^{0}_{\ 0\alpha}\Gamma^{\alpha}_{\ 00} = \frac{1}{4}\sum_{\alpha=1}^{3}\frac{g_{,\alpha}}{g}g_{,\alpha}.$$

Substituting this and (3.49) into (3.48), we obtain by (3.34) the following second order partial differential equation for the unknown function, see (3.39),

$$g = g(x^0, x^1, x^2, x^3),$$

namely,

$$R_{00} = \sum_{\alpha=1}^{3}\left(g_{,\alpha\alpha} - \frac{(g_{,\alpha})^2}{2g}\right) = 0 \tag{3.50}$$

with the standard Laplace operator and the squared gradient.

Now, we show that for the metric tensor (3.38) the entries $R_{0\beta}$ of the corresponding Ricci tensor are identically zeros without using (3.34). This means that we do not get any further condition on g. Using (3.4), (3.47), and (3.45), and the equality $\Gamma^{\varkappa}_{\ \mu\nu} = 0$ if at most one index is zero, we obtain for $\beta = 1, 2, 3$ that

$$\begin{aligned} R_{0\beta} &= \Gamma^{\varkappa}_{\ 0\beta,\varkappa} - \Gamma^{\varkappa}_{\ 0\varkappa,\beta} + \Gamma^{\lambda}_{\ 0\beta}\Gamma^{\varkappa}_{\ \lambda\varkappa} - \Gamma^{\lambda}_{\ 0\varkappa}\Gamma^{\varkappa}_{\ \beta\lambda} \\ &= \Gamma^{0}_{\ 0\beta,0} - \Gamma^{0}_{\ 00,\beta} + \Gamma^{0}_{\ 0\beta}\Gamma^{0}_{\ 00} - \Gamma^{0}_{\ 00}\Gamma^{0}_{\ \beta 0} = \Gamma^{0}_{\ 0\beta,0} - \Gamma^{0}_{\ 00,\beta} \\ &= -\frac{1}{2}\left(\frac{g_{,\beta}}{g}\right)_{,0} + \frac{1}{2}\left(\frac{g_{,0}}{g}\right)_{,\beta} = -\frac{1}{2}\frac{g_{,\beta 0}\,g - g_{,\beta}\,g_{,0} - g_{,0\beta}\,g + g_{,0}\,g_{,\beta}}{g^2} = 0. \end{aligned}$$

Further, we find that

$$\begin{aligned}
R_{\alpha\beta} &= \Gamma^{\varkappa}_{\alpha\beta,\varkappa} - \Gamma^{\varkappa}_{\alpha\varkappa,\beta} + \Gamma^{\lambda}_{\alpha\beta}\Gamma^{\varkappa}_{\lambda\varkappa} - \Gamma^{\lambda}_{\alpha\varkappa}\Gamma^{\varkappa}_{\beta\lambda} \\
&= \Gamma^{0}_{\alpha\beta,0} - \Gamma^{0}_{\alpha 0,\beta} + \Gamma^{0}_{\alpha\beta}\Gamma^{0}_{00} - \Gamma^{0}_{\alpha 0}\Gamma^{0}_{\beta 0} = -\Gamma^{0}_{\alpha 0,\beta} - \Gamma^{0}_{\alpha 0}\Gamma^{0}_{\beta 0} \\
&= \left(\frac{g_{,\alpha}}{2g}\right)_{,\beta} - \frac{1}{4}\frac{g_{,\alpha}}{g}\frac{g_{,\beta}}{g} = \frac{g_{,\alpha\beta}\, g - g_{,\alpha}\, g_{,\beta}}{2g^2} - \frac{g_{,\alpha}\, g_{,\beta}}{4g^2} \\
&= \frac{g_{,\alpha\beta}}{2g} - \frac{3g_{,\alpha}\, g_{,\beta}}{4g^2}, \qquad \alpha, \beta = 1, 2, 3.
\end{aligned}$$

From this and (3.34) we obtain 6 partial differential equations different from those in (3.50), namely,

$$g_{,\alpha\beta} - \frac{3g_{,\alpha}\, g_{,\beta}}{2g} = 0, \qquad \alpha, \beta = 1, 2, 3. \tag{3.51}$$

Finally, we show that a general solution of the system (3.50) and (3.51) is independent of the space variables x^1, x^2, x^3. Setting $\beta := \alpha$ and summing (3.51) over all α, we obtain

$$\sum_{\alpha=1}^{3}\left(g_{,\alpha\alpha} - \frac{3(g_{,\alpha})^2}{2g}\right) = 0.$$

Subtracting this from (3.50) and multiplying by g gives

$$\sum_{\alpha=1}^{3}(g_{,\alpha})^2 = 0$$

which yields

$$g_{,1} = g_{,2} = g_{,3} = 0.$$

Hence, a general solution is of the form (see (3.39))

$$g = g(x^0), \text{ i.e. } g_{00} = g_{00}(x^0).$$

The proof is completed. □

The class of all solutions of (3.50) and (3.51) is therefore not too rich when compared with Sects. 3.2 and 3.3. Theorem 3.13 may have applications in cosmology for a flat space with zero cosmological constant.

3.8 Cosmological Constant

In 1917, Albert Einstein introduced a new form of his equations (see [94]) which are at present usually written as follows

$$R_{\mu\nu} - \frac{1}{2} R g_{\mu\nu} + \Lambda g_{\mu\nu} = \frac{8\pi G}{c^4} T_{\mu\nu} \tag{3.52}$$

with non-zero *cosmological constant* $\Lambda \in \mathbb{E}^1$. A positive value of this constant produces a repulsive force which does not allow a gravitational collapse of the entire universe which follows from (3.29). Einstein wanted to keep his original idea of a static universe. We will derive a formula for the total mass of Einstein's static universe in Remark 7.22.

The left-hand side of (3.52) is divergence-free, since the Einstein tensor is divergence free $G^{\mu\nu}{}_{;\nu} = 0$ and the metric tensor is also divergence-free $g^{\mu\nu}{}_{;\nu} = 0$, see e.g. [107, p. 80] for the proof. Hence, the law of energy-momentum conservation remains upheld.

Remark 3.15 It is said that the gravitational constant G is the worst established constant of all fundamental physical constants, since its value is known only to 3 (or 4) significant digits. However, we do not know any significant digit of the cosmological constant Λ, yet. We even do not know whether it is positive or negative, if it exists, although there are hundreds of papers about this idea of Einstein. Some of these papers claim that the GTR is valid with precision 99 %. Nevertheless, the solution of Einstein's equation essentially depends on Λ, see e.g. Fig. 6.7 below. The current standard cosmological model only assumes that $\Lambda \approx 10^{-52}$ m^{-2}, i.e., an extremely small number in the SI base units that can play an essential role only on cosmological scales. This value can be obtained from (6.31), (6.29), and Remark 6.14.

Theorem 3.14 *Let $g_{\mu\nu}$ be a solution of (3.52). Then $(-g_{\mu\nu})$ solves (3.52) if Λ is replaced by $(-\Lambda)$.*

Proof We will proceed similarly as in the proof of Theorem 3.10. Using (3.2), we observe that the Christoffel's symbols of the second kind remain the same if $g_{\mu\nu}$ is replaced by $(-g_{\mu\nu})$,

$$\Gamma^{\mu}_{\ \varkappa\sigma} = \frac{1}{2}(-g^{\varkappa\lambda})(-g_{\nu\lambda,\mu} - g_{\lambda\mu,\nu} + g_{\mu\nu,\lambda}).$$

By (3.4) we find that the Ricci tensor $R_{\mu\nu}$ in (3.52) does not change as well. Concerning the second term on the left-hand side of (3.52), we see from (3.30) that $\left(-\frac{1}{2} R g_{\mu\nu}\right)$ remains also unchanged if $g_{\mu\nu}$ is replaced by $(-g_{\mu\nu})$. Finally, for the third term we get $\Lambda g_{\mu\nu} = -\Lambda(-g_{\mu\nu})$. □

Remark 3.16 According to the current standard cosmological model, dark energy is some mysterious substance which is responsible for the accelerated expansion of the universe and this is attributed to the cosmological constant Λ. Its physical dimension is m^{-2}, since the left-hand side of (3.52) has dimension s^{-2}. In spite of that, cosmologists describe it as the density of energy which has another physical dimension in the SI base units (International System of Units), namely kg $m^{-1}s^{-2}$. So can we talk about density of energy?

In the system $c = 1$, meters and seconds can be arbitrarily exchanged using a suitable multiplicative constant. However, the physical dimension of Λ again does not correspond to the density of energy, since kilograms also do not appear there.

Remark 3.17 Let us consider a special case of (3.52) when

$$T_{\mu\nu} = 0.$$

Similarly as in proof of Theorem 3.11 we find by (3.30) that

$$0 = g^{\mu\nu} R_{\mu\nu} + \left(\Lambda - \tfrac{1}{2}R\right) g^{\mu\nu} g_{\mu\nu} = R + 4\left(\Lambda - \tfrac{1}{2}R\right) = 4\Lambda - R,$$

i.e., in this case Einstein's equations (3.52) can be rewritten as

$$R_{\mu\nu} = \Lambda g_{\mu\nu}. \tag{3.53}$$

Remark 3.18 When Einstein's paper [94] appeared, then shortly afterwards several articles (cf. [153, p. 438], [374]) were published that looked for an exterior Schwarzschild-like solution of (3.53) using the standard spherical coordinates, see [167, Appendix C] for a detailed derivation,

$$g_{\mu\mu} = \operatorname{diag}\Big(-\Big(\frac{r-S}{r} - \frac{\Lambda r^2}{3}\Big), \Big(\frac{r-S}{r} - \frac{\Lambda r^2}{3}\Big)^{-1}, r^2 \sin^2\theta, r^2 \Big), \tag{3.54}$$

where S is given by (3.10), $r > S$, and $\theta \in [0, \pi]$. We observe that (3.54) reduces to the exterior Schwarzschild solution (3.6) for $\Lambda = 0$. If $\Lambda > 0$ and $S = 0$ (i.e., $M = 0$ by (3.10)), we get the de Sitter metric tensor [80] (or the anti-de Sitter metric tensor for $\Lambda < 0$). Moreover, if $\Lambda = 0$ and $S = 0$ then (3.54) changes into the Minkowski metric tensor (3.5).

Remark 3.19 Let $\Lambda > 0$ and let us define

$$L := \Lambda^{-1/2}. \tag{3.55}$$

Assume that $S > 0$ is such that $S < L/2$. Then

$$g_{00}(L) = -1 + \frac{S}{L} + \frac{1}{3} < -\frac{2}{3} + \frac{L}{2L} < 0,$$

$$g_{00}(2L) = -1 + \frac{S}{2L} + \frac{4\Lambda L^2}{3} = \frac{1}{3} + \frac{S}{2L} > 0. \tag{3.56}$$

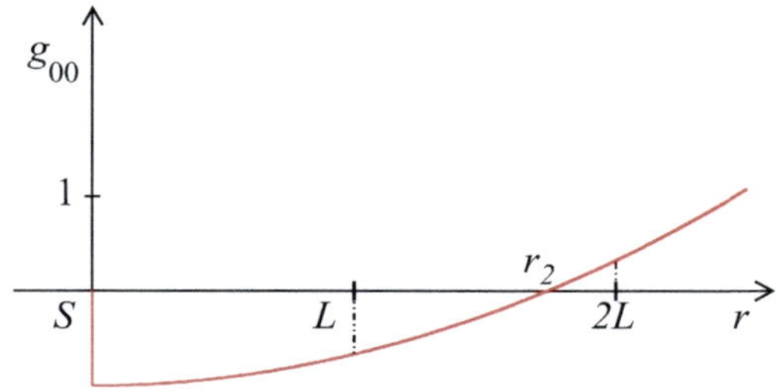

Fig. 3.6 Schematic illustration of the behavior of the metric coefficient $g_{00} = g_{00}(r)$ from (3.54) that changes its sign when $\Lambda > 0$

Hence, the continuous component $g_{00} = g_{00}(r)$ changes its sign on the interval $(L, 2L)$, i.e., there exists $\overline{r} \in (L, 2L)$ such that $g_{00}(\overline{r}) = 0$, see Fig. 3.6. Consequently, in the evaluation of the next component

$$g_{11}(r) = \left(\frac{r-S}{r} - \frac{\Lambda r^2}{3}\right)^{-1}$$

one divides by zero for $r = \overline{r}$ and the metric tensor breaks down. The tensor (3.54) is thus not well defined at cosmological distances even though the cosmological constant was invented just to treat extremely large distances in the universe. The component $g_{11}(r)$ attains arbitrarily large values when $r \to \overline{r}^-$ even though there is no physical barrier. Therefore, the quadratic form $g_{\mu\nu}(r)v^\mu v^\nu$ is also not well defined at $r = \overline{r}$ for any contravariant vector v^μ such that $v^1 \neq 0$. A different choice of coordinates does not help to avoid this singularity, since the form does not change its value for a given r.

Chapter 4
Numerical Analysis of Mercury's Perihelion Shift

4.1 A Brief Historical Overview

The perihelion shift of Mercury's orbit is thought to be one of the fundamental tests of the validity of the General Theory of Relativity [107, 249, 303, 312]. In the current astrophysical community, it is generally accepted that the additional relativistic perihelion shift (advance) of Mercury is the difference between its observed perihelion shift and the one predicted by the N-body problem, and that this difference equals $43''$ per century. However, as it results from the subtraction of two quite large and inexact numbers of almost equal magnitude which are not uniquely defined, this difference can be quite inexact. As such, the above accepted value $43''$ cy^{-1} (arc seconds per century) is highly uncertain and may not correspond to reality. In this chapter, we present a thorough mathematical and numerical analysis of diverse misconceptions about Mercury's perihelion problem (mostly taken from [90, 164, 172, 174, 356]).

The French astronomer Urbain Le Verrier (1811–1877) is best known for his prediction of the position of Neptune (at that time an unknown planet). He performed his calculations to explain certain irregularities in the orbit of Uranus. In 1859, Le Verrier also noticed some anomalies in the observed position of Mercury's perihelion compared to Newton's theory [220]. Again he started to look for another (still unknown) planet named Vulcan which should orbit the Sun more closely than Mercury. The existence of such a hypothetical planet deemed to be impossible at present, because we would have sometimes seen it when crossing the solar disk or during total solar eclipses.

Remark 4.1 According to [346, p. 36], Le Verrier needed to explain a shift of $38''$ per century in order to make the predicted position of Mercury's perihelion

M. Křížek, L. Somer, *Mathematical Aspects of Paradoxes in Cosmology*,
https://doi.org/10.1007/978-3-031-31768-2_4

using Newtonian mechanics to be in agreement with observations. In 1895, Simon Newcomb [260, Chapt. IX, p. 184] arrived at the comparably very small value of

$$E = 43.37'' \text{ per century.} \tag{4.1}$$

In 1915, Albert Einstein published in [90, p. 839] a formula for the relativistic perihelion shift, for one period, of

$$\varepsilon = 24\pi^3 \frac{a^2}{T^2 c^2 (1 - e^2)} = 5.012 \cdot 10^{-7} \text{ rad} \approx 0.1'', \tag{4.2}$$

where according to contemporary data

$T = 7.6005 \cdot 10^6$ s is the orbital period of Mercury,
$e = 0.2056$ is the (numerical) eccentricity of its elliptical orbit,
$a = 57.909 \cdot 10^9$ m is the length of its corresponding semimajor axis,
$c = 229{,}792{,}458$ m/s is the speed of light in a vacuum.

Substituting these values into (4.2), we obtain a value of

$$E = \varepsilon \frac{\tau}{T} \frac{180}{\pi} 3600'' = 43'' \quad \text{per century,} \tag{4.3}$$

that is in excellent agreement with the calculations of both Le Verrier and Newcomb (4.1). Here $\tau = 3{,}155{,}814{,}954$ s is the number of seconds in one century. In astrophysical community, the simple formula (4.2) is applied to test the validity of the General Theory of Relativity without any doubts how it was derived. Therefore, now we will point out several shortcomings which should be taken into account.

Remark 4.2 In 1898, Paul Gerber derived the following formula for the speed of light by means of retarded potentials (see [113])

$$c^2 = 24\pi^3 \frac{a^2}{T^2 (1 - e^2) \Phi},$$

where Φ is the perihelion shift of Mercury during one orbital period. We see that this formula is the same as (4.2) taking

$$\varepsilon = \Phi.$$

Therefore, the corresponding tests of General Theory of Relativity based on (3.4) or Einstein's system (3.1) + (3.3) yield after many approximations the same value (4.3) as tests of Gerber's theory of retarded potentials. So which theory do we test?

We should stress here that over the past 100 years Mercury's perihelion was shifted by a total amount of about 5600″ (see e.g. [66, p. 363], [231, p. 68]), of which approximately 5027″ is due to the precession of the Earth's axis. The vernal

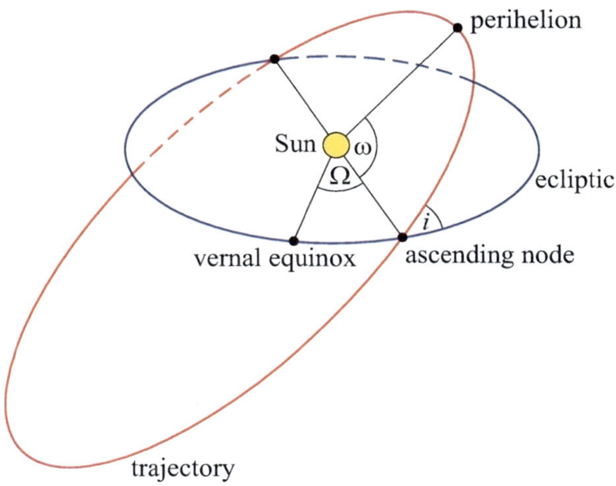

Fig. 4.1 Keplerian parameters of the elliptical orbit of a planet that determine its orientation in space are: the inclination i, the longitude of the ascending node Ω, and the argument of perihelion ω. They vary slightly in time due to the presence of other planets, precession and nutation of the Earth's axis. The ecliptic is the plane containing the Earth's orbit around the Sun

equinox (one of the two intersections of the ecliptic and the celestial equator, see Fig. 4.1), from which the equatorial right ascension is measured, moves along the ecliptic by, on average, $50.27''$ per year. This value is also continually distorted by the nutation of the Earth's axis of $\pm 9.21''$ with a period of 18.6 years which does not divide 100 years. The accurate determination of the position of the vernal equinox is therefore a very difficult task. Moreover, due to non-zero inclinations of planets (Jupiter 1.3°, Saturn 2.5°, etc.), the position of the ecliptic in space changes with time. Thus, the use of a time-dependent equatorial coordinate system with right ascension and declination is not appropriate.

Therefore, we will next consider a rectangular heliocentric system whose position is unchanged with respect to the fixed distant stars. In this system, the currently observed perihelion shift of Mercury due to the gravitational pull of the other planets is about $575''$ per century which is a more than ten times larger value than that in (4.3). On the other hand, according to the calculations of Le Verrier [220, p. 99], Mercury's perihelion shift is influenced by the other planets as presented in Table 4.1.

Remark 4.3 The total sum in the last line of Table 4.1 is to be understood only approximately, because the problem in question is nonlinear and such a sum is thus not justified. Additionally, although the other values in Table 4.1 are given in up to four significant digits, not all of them are accurate. For instance, at the time of Le Verrier, the masses of the planets were not precisely known: Mercury was believed to have double the mass assigned to it nowadays (using Kepler's third law and the accurately measured orbital period of the Messenger satellite which orbited Mercury

Table 4.1 The effect of particular planets on Mercury's perihelion shift per century by Le Verrier [220]

Planet	Its impact on Mercury's perihelion advance
Venus	280.6″
Earth	83.6″
Mars	2.6″
Jupiter	152.6″
Saturn	7.2″
Uranus	0.1″
Total	526.7″

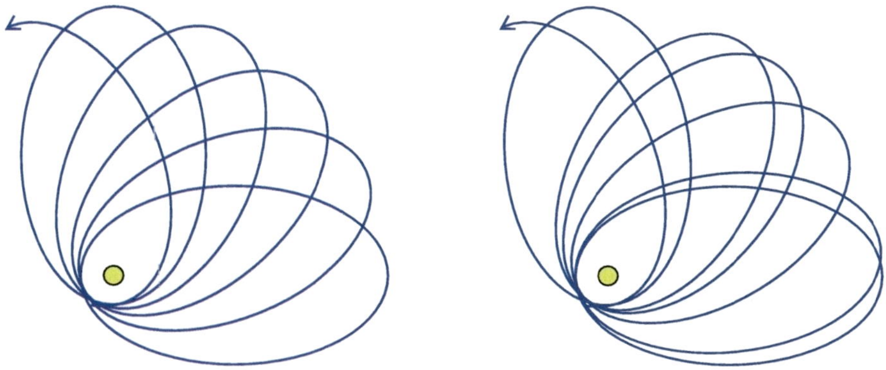

Fig. 4.2 Left: the idealized uniform perihelion shift of Mercury in the direction of circulation is shown. For clarity, a very high artificial eccentricity $e = 0.8$ is chosen. Right: an irregular perihelion shift caused by the gravitational tug of other planets is schematically depicted

between 2011 and 2015), see [328]. On the other hand, the mass of the Earth was underestimated as 0.937 of its currently accepted value [220, p. 19]. Therefore, Le Verrier was not able to derive reliable values for the gravitational forces between the planets. Neither did he solve the system of ordinary differential equations describing the N-body problem (see (1.14)), but made approximations using certain sums of finite series instead by means of his perturbation theory. Finally, he only had relevant data concerning the positions of the planets over the preceding 50 years.

Remark 4.4 The values in Table 4.1 are also affected by other sources of errors. For instance, the relative position of Jupiter in its orbit at the beginning of each century varies due to the simple fact that its orbital period of 11.861 years does not divide 100 years. As the same holds true for the other planets, they all influence Mercury in an irregular manner from varying positions (cf. the right part of Fig. 4.2). Thus, the observed perihelion shift can essentially differ century by century.

Our further aim will be to provide a more critical insight into the problem of Mercury's perihelion shift. In Sect. 4.2 we recall that the difference between two almost equally large numbers can be quite inaccurate. In Sects. 4.3 and 4.4 we illustrate why the observed and calculated values of Mercury's perihelion shift are highly imprecise. In Sect. 4.5 we show that the simple formula (4.2) due to Einstein was derived from the 10 nonlinear Einstein equations (3.1) by many approximations without any guaranteed error estimates. In Sect. 4.6, formula (4.2) is applied to other systems than the Sun-Mercury system. Finally, in Sect. 4.7 we repeat the main arguments concerning why the value (4.3) is questionable.

4.2 The Inaccuracy of the Difference Between Two Almost Equally Large Numbers

It is true that the manual calculations of Le Verrier and Newcomb (cf. (4.1)) led to values close to (4.3). However, they did not reach the accuracy of simple computer arithmetic in which each number is stored in six bytes, they originated from inexact data, and were obviously subject to various errors (see e.g. [137]). Actually, their modeling and discretization errors (see Fig. 1.9) were larger than the relativistic effect itself. Therefore, we restrict ourselves to the current status quo.

Denote by the letter O the observed value of Mercury's perihelion shift per century with respect to the Sun, and by C the calculated value per century in the heliocentric coordinates using Newtonian mechanics. At present, it is generally accepted by the current astrophysical community that the following equality is true

$$O - C = E, \tag{4.4}$$

where E is the value (4.3) predicted by Einstein. For instance, Narlikar and Rana [257, p. 657] claim that

$$O = 575''\text{per century and } C = 532''\text{per century} \tag{4.5}$$

which yields the value (4.3) by (4.4). However, they do not present any reference for the value O. Note that O cannot be by Remark 4.4 uniquely defined. The same is true for the calculated value C.

For simplicity, most of the perihelion shifts will be from now on rounded to integers given in arc seconds. We shall not present the corresponding error bars, since they are always less than $1''$ in recent literature. It is clear that if at least one term in relation (4.4) is not correctly established, then the proclaimed equality (4.4) cannot be properly used and the value E may differ from reality.

Numerical analysts know very well (see [45, 118]) that subtracting two almost equally large numbers (see (4.4)) in floating point computer arithmetic is burdened by a large resulting relative error. Although generally no rounding appears when subtracting such numbers, the mantissa of the difference contains only a few significant digits which inevitably leads to loss of precision. Let us present an illustrative example.

Example 4.1 Assume for simplicity that the mantissa has only 5 digits. Let us evaluate the difference $3.1416 \cdot 10^0 - 3.1415 \cdot 10^0$. Then the computer saves the number $1.0000 \cdot 10^{-4}$, where four zeros in the mantissa are not significant digits.

This means that the error of the difference moves from the originally last digit of the mantissa to the second digit.

The following three sections show that all three values in (4.4) are loaded with plenty of different errors, and thus the proposed equality (4.4) is not very likely to be numerically verifiable. From Remark 4.4 we know that the observed (and also calculated) perihelion shift must differ century by century, cf. the right part of Fig. 4.2 and Figs. 4.4 and 4.5 below. Furthermore, we see that the full angle has over a million arc seconds, namely,

$$u = 360 \cdot 3600'' = 1{,}296{,}000'',$$

while Eq. (4.2) gives less than one arc second per year. Mercury's perihelion separation from the Sun is $r_1 = a - ae = 46 \cdot 10^6$ km. According to (4.3), the additional perihelion shift is

$$\frac{2\pi r_1}{u} \cdot 0.43'' = 96 \, \text{km per year}, \tag{4.6}$$

while in [249, p. 1048] the additional perihelion shift (4.3) is presented quite inexactly as 120 km per year. For comparison, note that the orbital speed of Mercury is about 50 km/s. Hence, the time instant of the passage of Mercury through the perihelion should be established extremely precisely.

4.3 The Observed Perihelion Shift of Mercury

To determine the exact position of Mercury's actual perihelion is one of the most difficult tasks in contemporary positional astronomy and computational mathematics due to a multitude of reasons. In the case of a small eccentricity $0 \le e \ll 1$, we get an ill-conditioned problem, since for a circular path the perihelion occurs at each point. The eccentricity of Mercury's orbit is a relatively large number $e = 0.2056$ (see Table 1.1), but since the semiminor axis of Mercury has by (1.2) the length

$$b = a\sqrt{1 - e^2} = 0.98a,$$

the orbit is almost circular with the Sun at one of the foci.

Mercury can be seen only a few days per year in the projection on the shining celestial sphere. Hence, a 100 years ago, it was difficult to reliably determine from Earth its angular distances from neighboring stars on the celestial sphere. Since Mercury is mostly close to the horizon at sunrise or sunset, serious difficulties were also caused by the astronomical refraction of the atmosphere, see Table 4.2.

Moreover, the refraction depends on the air temperature, pressure, humidity, etc. Notice that the value in the last row of Table 4.2 is given in arc minutes, and not in arc seconds (cf. (4.3)). It is even larger than the angular diameter $30'$ of the Sun. Consequently, when the Sun touches the horizon at sunset and we would immediately remove the Earth's atmosphere, the Sun would already be below the

Table 4.2 Influence of the zenith distance on the mean atmospheric refraction

Zenith distance	Mean refraction
0°	0′
20°	0′22″
40°	0′50″
60°	1′43″
70°	2′43″
80°	5′30″
90°	35′

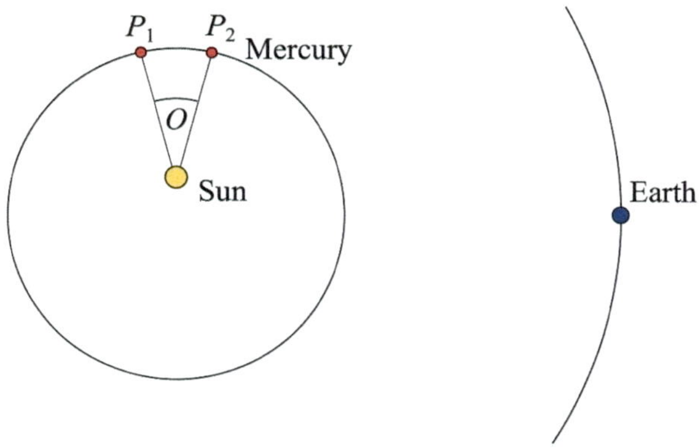

Fig. 4.3 Since the length of semiminor axis is 98 % of the semimajor axis, the orbit of Mercury is almost circular. The angle O corresponds to the observed Mercury's perihelion shift per century. For a better visualization, the angular distance O between two perihelia P_1 and P_2 in the figure is magnified 200 times. The observational data from the Earth must be recalculated to find the value of the angle O in the heliocentric coordinates

horizon and darkness would occur. Let us note that the refraction of the solar atmosphere is ignored in another test of GTR—called bending of light near the solar disk which is more than hundred times larger than the Earth.

Remark 4.5 To determine the exact present position of Mercury in heliocentric coordinates, we need to know precisely both the time-dependent right ascension and declination, but also the distance from the Earth, which currently can be obtained by using radar reflections [16, 143, 330]. However, these measurements were done in a time interval which is much shorter than one century, i.e., we do not have reliable data for one century. Moreover, we have to take into account that even the Earth moves during the measurements along a complicated nonelliptic orbit. The precise position of Mercury on the celestial sphere were also determined by satellite measurements (see Remark 4.3).

Since the vertex of O lies in the Sun, this angle seems always to be much smaller when it is observed from the Earth (see Fig. 4.3). In fact, the observed positions of Mercury in finitely many points have to be interpolated in a rather

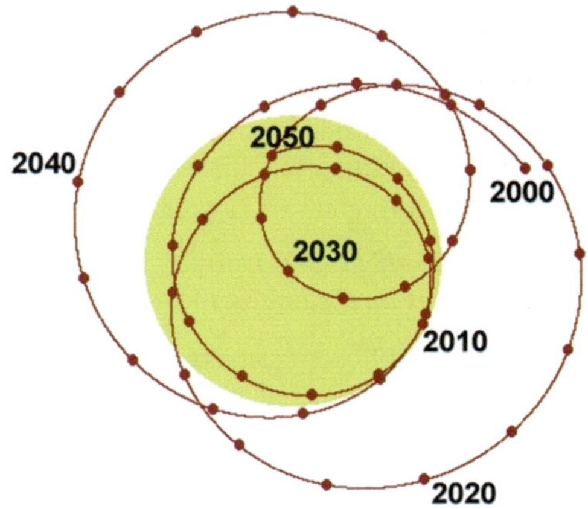

Fig. 4.4 Projection of the trajectory of the Newtonian barycenter of the Solar system onto the ecliptic for the period 2000–2050. The center of the Sun (whose diameter is almost 1.4 million km) is at the origin of the heliocentric system. The barycenter shifts each day by about 1000 km, while the additional relativistic perihelion shift of Mercury is on average only 96 km in a year due to (4.6)

complicated way and then recalculated to the heliocentric coordinate system which is fixed against distant quasars. Estimation of the acceleration of the Solar-system barycenter relative to a system of reference quasars is established in [347].

Nevertheless, here arises another serious problem, because the true orbit of Mercury is not elliptical. For illustration, consider first a simple two-body Sun-Jupiter system, where masses of the Sun and Jupiter are

$$M = 1.989 \cdot 10^{30}\text{kg}, \quad m = 1.899 \cdot 10^{27}\text{kg},$$

respectively. Assume that their trajectories are circular and that the barycenter of this system lies in the origin of Cartesian coordinates. Then the distances $R = 743,000$ km of the Sun to the origin and of Jupiter $r = 778 \cdot 10^6$ km to the origin satisfy the relation $MR = mr$ while the radius of the Sun is $R_\odot = 695{,}700$ km, see (3.23). Thus $R = mr/M > R_\odot$ and we observe that the barycenter of the Sun-Jupiter system lies outside the Sun, and the Sun and Jupiter orbit about it.

However, the Sun's trajectory is also deflected by other planets (especially those very distant and massive), i.e., the heliocentric system is not inertial. Figure 4.4 shows a two-dimensional projection of the trajectory of the Newtonian barycenter of the Solar system in the heliocentric system over the duration of half a century, starting from 2000. Thus, the actual trajectory of Mercury is not an ellipse, one of whose foci is placed at the Sun, but it is a very complicated nonplanar trajectory, which is influenced by the other planets and thus also by the barycenter of the Solar system. The barycenter shifts each day by about 1000 km which is a much larger value than the current shifts of the perihelion of Mercury's orbit given in (4.6).

Note that the largest elongations in Fig. 4.4 appear when Jupiter and Saturn are in conjunction with respect to the Sun every $P = 19.856$ years. Here $P = 1/(T_5^{-1} - T_6^{-1})$, and T_5 and T_6 are the orbital periods of Jupiter and Saturn, respectively, see Table 1.1.

It is therefore unclear how to define precisely e.g. the direction of the semimajor axis of Mercury (cf. Remark 4.17), which according to (4.2) should be shifted. It is also unclear how to determine an analogy of the Newtonian barycenter of the Solar system for a finite speed of propagation of gravity. In this case, the problem becomes non-potential, which is a little known fact.

Let us present other facts that need to be taken into consideration for determining the actual position of Mercury.

Remark 4.6 Recall that the average speed of Mercury is about 50 km/s. Mercury's diameter is nearly 5000 km and its angular diameter when crossing the solar disk is $12''$. Before information about its position reaches our Earth at the speed of light, Mercury will move more than three of its diameters, i.e. approx $40''$ and 17,000 km (cf. (4.3) and (4.6)). But astronomical tables indicate planets where they are observed and not where they really are.

To determine the actual position of Mercury it is necessary to take into account also the aberration of light, which for the Earth's orbit is $\approx 20''$. Terrestrial measurements are also slightly affected by the atmospheric dispersion of light. Furthermore, from precise observations we know that the measured secular acceleration of Mercury is proportional to its mean motion. Secular long-term deviations in the mean longitudes of this planet have reached $5''$ during the last 250 years [150, p. 884], i.e., after approximately 1000 periods. From this Igor N. Taganov [338, p. 72] derived that Mercury's orbit slightly expands at the rate comparable to the Hubble constant.

We should also take into consideration the calibration of instruments, which is essential when determining equatorial coordinates of Mercury and its distance from Earth. Determining the actual orbit of Mercury from astronomical observations is therefore an extremely difficult task which has the potential for different errors that have a significant impact on the value O from (4.4).

Remark 4.7 Interpolation of measured position data represents another important source of errors. The inclination of Mercury's trajectory with respect to the ecliptic is $i = 7°$, and therefore, we do not deal with a planar problem (see Fig. 4.1). From Sect. 4.1, we know that positions of planets are considered in rectangular heliocentric coordinates (X, Y, Z). They are interpolated by a finite series. For illustration we show how to express, for instance, the X-coordinate of the Earth's center in time t:

$$
\begin{aligned}
X =& 0.0056114 + 0.001234\,t + 0.9998293\cos(1.7534857 + 6283.075850\,t)\\
&+ 0.000011\,t\cos(2.02 + 6283.1\,t)\\
&+ \sum_{i=1}^{38} A_i \cos(B_i + C_i t) + \sum_{i=39}^{40} A_i t \cos(B_i + C_i t),
\end{aligned}
$$

where the 120 constants A_i, B_i, C_i are given by a table in [47, p. 313] and t is measured in thousands of years from the Julian date J2000.0. This is therefore an approximate analytical expression of X in the form of the sum of polynomials with trigonometric polynomials, which contributes to the total interpolation error

$|X| \to \infty$ for $t \to \infty$. We see that the remaining components Y and Z, and also the heliocentric coordinates of other planets are parametrized similarly. The constants A_i, B_i, C_i are corrected every few years so that they are consistent with the observed positions of the planets, cf. [46, 47]. However, in this way the true perihelion shift of Mercury is perturbed by interpolations.

4.4 Computed Perihelion Shift of Mercury

Figure 1.9 shows the general computational scheme for solving real-life (non-academic) problems of mathematical physics. We always produce three types of error: the modeling error e_0, the discretization error e_1, and computational error e_2. In our case, the physical reality is the Solar system. Its evolution is usually modeled by the problem of N bodies that interact with each other gravitationally. Bodies are replaced only by idealized mass points $m_i > 0$, $i = 1, \dots, N$, whose positions r_i satisfy the well-known system of second order differential equations (see (1.14))

$$\ddot{r}_i = G \sum_{j \neq i}^{N} \frac{m_j (r_j - r_i)}{|r_j - r_i|^3} \tag{4.7}$$

for $i = 1, \dots, N$ with given initial conditions on the positions $r_i(0)$ and velocities $\dot{r}_i(0)$ of all N bodies. Here $| \cdot |$ stands for the length of a vector and G is the gravitational constant.

Remark 4.8 Henri Poincaré proved [292] that for $N > 2$ the N-body problem generally does not have an analytic solution. Therefore, the solution of (4.7) is usually approximated by numerical methods. This leads to the so-called discrete finite dimensional model, which produces a nonzero error e_1. The N-body problem is not stable with respect to changes in initial conditions and continually acting perturbations. Thus each numerical integration method gives a considerable discretization error over long intervals, even if we use double (8 bytes) or extended (10 bytes) computer arithmetic.

Further, we note that the error e_2 depends among other things on the computer used, programming language, and also on the method of programming. For a catastrophic behavior of rounding errors in ill-conditioned problems, see e.g. [45, 162]. Numerical errors usually do not average out, but accumulate and grow exponentially in the course of calculation. Thus, each programmer (and also each observer) can get different results. For instance, in Russia completely different programs are used to calculate the ephemerides of planets [287, 288] than in France [47] or in the NASA Jet Propulsion Laboratory, Pasadena, USA, see e.g. [106, 329, 330].

The difference between the exact and approximate solution of a mathematical model is estimated in numerical mathematics by a priori or a posteriori error estimates. This step is very important. If we do not perform a reliable analysis of

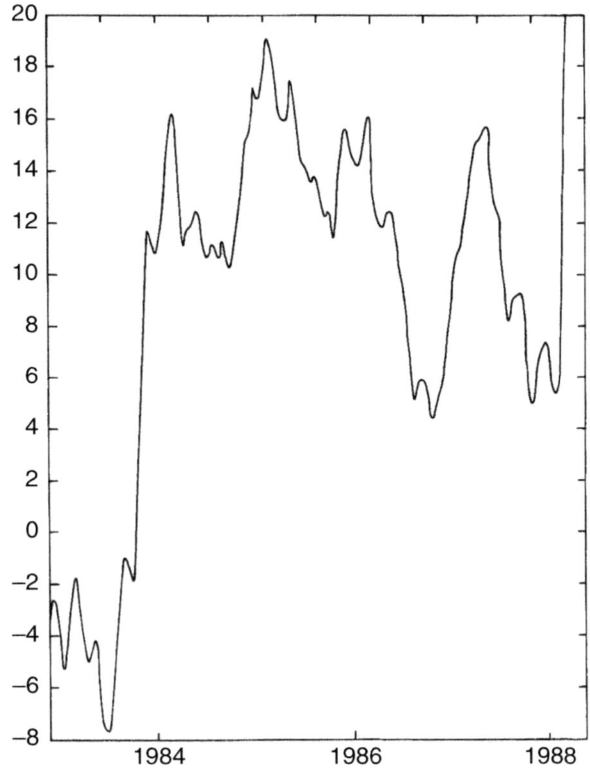

Fig. 4.5 The shift of the line of apsides of Mercury's orbit during 5 years only, calculated numerically from the N-body problem (adapted from [295]). The graph resembles deterministic chaos. The scale on the horizontal axis is given in years, and in the vertical axis the shift angle is in arc seconds

errors, we do not know what we have actually calculated, and how far the numerical solution is from the exact solution.

For elliptic orbits the line segment connecting the perihelion and aphelion is called the *line of apsides*. However, if the perihelion shifts, then we have two different lines of apsides, the first connects the perihelion with previous aphelion and the second connects the perihelion with next aphelion. As already mentioned, the orbital periods of the planets are not in a ratio of small integers, and therefore the "line of apsides" rotates quite irregularly. This follows from numerical simulations of the N-body problem. For instance, by Rana [295], the perihelion shift may increase about $24''$ or decrease about $11''$ within less than 1 year, which has a nonnegligible influence on the average shift per century. From the highly fluctuating and chaotic behavior of the shift of the line of apsides from Fig. 4.5 it is obvious that the value C from (4.4) cannot be determined with an accuracy better than several arc seconds per century. Therefore, in the literature several different values of C can be found, for instance,

$C = 532''$ per century (see [249, p. 1113]),
$C = 531''$ per century (see [312, p. 1]),

$C = 530''$ per century (see [356, p. 6]),
$C = 529''$ per century (see [295, p. 197]),
$C = 527''$ per century (see the last line of Table 4.1 taken from [220]).

So which value is correct? In spite of this the difference $E = O - C$ given by (4.5) has only one representation $E = 43''$ per century in the currently available literature (see e.g. Table in [286, p. 1175]) and the Einstein value (4.3) is presented as accurate.

Moreover, it is not clear what is the definition of the perihelion shift per century. If it refers to some kind of average value, then we have to define exactly what average. Over which time period do we perform averaging? We will illustrate this on Fig. 4.5, where the time interval is only 5 years long.

Let us point out that Fig. 4.5 corresponds only to a restricted N-body problem, where the Sun is fixed at the origin of heliocentric coordinates. In the real N-body problem (4.7), the Newtonian barycenter of the Solar system (see Fig. 4.4) is placed at the origin of Cartesian coordinates. These two problems have, of course, different modeling errors.

Elliptical orbits of test particles can only be obtained for the central force field, which is proportional to the gravitational potential $1/r$. Nevertheless, this does not hold in the Solar system. For instance, the total weight of interplanetary dust around the Sun, which causes zodiacal light, is estimated to 10^{16} and 10^{17} kg. The book [303] admits that its impact is almost insignificant on the Mercury's perihelion shift, together with the belt of asteroids, comets, etc. According to [312], the solar oblateness (quadrupole moment) contributes to the overall perihelion shift of Mercury only $0.0254''$ per century (see also [286]). On the other hand, a much larger value $3.4''$ per century is presented in Weinberg [369, p. 200]. He claims that the most probable reason for this effect is that the length of the rotational axis of the Sun oscillates. Also a differential rotation of the inner Sun may have a nonnegligible effect [81]. If the central core is an axially symmetric ellipsoid that rotates more rapidly than the surface of the Sun, then a part of the Mercury's perihelion shift could be explained in another way.

We should also take into account errors in the determination of physical constants (e.g. the gravitational constant G, the mass of the Sun and planets), the nonuniform influence of magnetic fields, tidal forces, gravitational aberration, etc. A large number of small errors can cause a nonnegligible total error $e_0+e_1+e_2$, see Fig. 1.9. Moreover, numerical results show that a finite speed of gravitational interaction also causes a shift of the line of apsides.

Example 4.2 Consider only the isolated system Sun-Mercury. From (4.3) we get that the total relativistic aphelion shift of Mercury after 750,000 years is equal to

$$750{,}000 \cdot 0.43'' \approx 90^\circ.$$

Using Table 1.1, we find that the distance between the Sun and the aphelion of Mercury's orbit is

$$a(1 + e) \approx 70 \cdot 10^6 \text{ km},$$

see the left part of Fig. 4.2. Here the aphelion of the Keplerian trajectory is the most right point. On the other hand, the most upper point corresponds to the relativistic aphelion shift after 750,000 years. The distance between these two points is $\sqrt{2} \cdot 70 \cdot 10^6$ km $\approx 100 \cdot 10^6$ km. This simple academic example shows that the difference between the modeling errors of the two-body problem and its relativistic analogue can be very large on short time intervals.

4.5 A Method of Albert Einstein

If the General Theory of Relativity well describes the planetary motion, then the perihelion shift of Mercury calculated from Einstein's equations for the whole Solar system should be close to the observed shift O from (4.5). However, Einstein's equations for the Solar system are unlikely to yield the observed perihelion shift of Mercury O from (4.4) exactly, since any equation of mathematical physics is always just an approximation of reality and thus possesses a nonzero modeling error e_0. Moreover, Einstein definitely did not solve his equations of General Theory of Relativity (3.1) with condition (3.3) for the Solar system analytically. These are represented by a very complicated nonlinear system of partial differential equations. Their exact solution is not even known for two mass bodies, since the left-hand side contains many thousand of terms (partial derivatives of scalar functions) as we shall see in Sect. 5.1. This is due to the fact that there are, in general, 10 components of the symmetric metric tensor and 40 Christoffel symbols (see Sect. 3.1).

Consequently, Einstein had to make a whole series of simplifications to get some value of the perihelion shift of Mercury (see (4.3)). In other words, formula (4.2) has not been derived as a consequence of Einstein's equations in terms of mathematical implications. On the other hand, it could give a good prediction, if the corresponding approximations were under control.

For instance, Einstein takes $\theta = 90°$ in spherical coordinates which corresponds to the equatorial plane and simplifies the calculation, cf. e.g. (3.6).

Remark 4.9 Mathematical implications are different from consequences used in physics that may involve approximations. Let us present a very trivial illustrative example: If $n = 8.999$, then n is not divisible by 3. Now, let us approximate n by $\tilde{n} = 9$. Then we cannot claim that $\tilde{n}$ is not divisible by 3.

Einstein assumed that the curvature of space around the Sun is time independent and spherically symmetric. That is, the considered curvature of space does not include the gravitational influence of Mercury, Jupiter, and other planets to get a pure relativistic effect (cf. Fig. 4.2). Einstein considered the zero right-hand side of his field equations outside the Sun. For simplicity, Mercury is substituted by a test particle with zero mass and its trajectory (geodesic) is described by

$$\frac{\mathrm{d}^2 x^\nu}{\mathrm{d}s^2} + \Gamma^\nu_{\sigma\tau}\frac{\mathrm{d}x^\sigma}{\mathrm{d}s}\frac{\mathrm{d}x^\tau}{\mathrm{d}s} = 0$$

for arbitrary curvilinear coordinates x^ν (see [90, p. 835], [107, p. 3]).

Example 4.3 Denote by T_0 the orbital period of Mercury with zero mass and by T the orbital period of Mercury with actual mass $m = 33.022 \cdot 10^{22}$ kg. Let us estimate the difference between their positions after one century. For $M = M_\odot$ by Kepler's third law $a^3/T^2 = \frac{1}{4}G(M+m)/\pi^2$ we get $T = 7{,}600{,}500$ s, cf. (4.2). Now from the relation $T^2(M+m) = T_0^2 M$ we find that

$$T_0 - T = 0.627\,\mathrm{s}.$$

During one century Mercury makes

$$n = \frac{\tau}{T} = 415.211 \tag{4.8}$$

revolutions, where τ and T have the same meaning as in (4.3). This corresponds to a total delay of $0.627n = 260.33$ s if Mercury is replaced by a point with zero mass. During this time interval the difference between those two positions will be more than 13,000 km for Mercury's speed of about 50 km/s.

Einstein described the gravitational field around the Sun by a solution of the so-called weak gravitational fields, which uses the Parametrized Post-Newtonian (PPN) formalism for velocities $v \ll c$, see [249, p. 1069]. The fact that the speed of Mercury is considerably less than the speed of light enabled Einstein to perform a series of further simplifications. For example, Einstein neglected higher order terms when calculating Christoffel symbols. Hence, formula (4.2) should not be applied to strong gravitational fields as it is sometimes done. After five pages [90, p. 833–837] of further approximations Einstein got (without any error estimates) an ordinary differential equation for the angle ϕ whose solution leads to the elliptic integral (see Fig. 4.6)

$$\phi = \big[1 + \alpha(\alpha_1 + \alpha_2)\big]\int_{\alpha_1}^{\alpha_2} \frac{\mathrm{d}x}{\sqrt{-(x-\alpha_1)(x-\alpha_2)(1-\alpha x)}}, \tag{4.9}$$

where $\phi > 180°$ is the angle between the radius-vector of perihelion and the radius-vector of aphelion of Mercury's orbit (see Fig. 4.7),

$$\alpha_1 = \frac{1}{a(1+e)} = 1.432 \cdot 10^{-11}\,\mathrm{m}^{-1}, \quad \alpha_2 = \frac{1}{a(1-e)} = 2.174 \cdot 10^{-11}\,\mathrm{m}^{-1} \tag{4.10}$$

Hierfür kann mit der von uns zu fordernden Genauigkeit gesetzt werden

$$\phi = [1 + \alpha(\alpha_1 + \alpha_2)] \cdot \int_{\alpha_1}^{\alpha_2} \frac{dx}{\sqrt{-(x-\alpha_1)(x-\alpha_2)(1-\alpha x)}},$$

oder nach Entwicklung von $(1 - \alpha x)^{-\frac{1}{2}}$

$$\phi = [1 + \alpha(\alpha_1 + \alpha_2)] \int_{\alpha_1}^{\alpha_2} \frac{\left(1 + \frac{\alpha}{2} x\right) dx}{\sqrt{-(x-\alpha_1)(x-\alpha_2)}}.$$

Die Integration liefert

$$\phi = \pi \left[1 + \frac{3}{4} \alpha(\alpha_1 + \alpha_2) \right],$$

oder, wenn man bedenkt, daß α_1 und α_2 die reziproken Werte der maximalen bzw. minimalen Sonnendistanz bedeuten,

$$\phi = \pi \left(1 + \frac{3}{2} \frac{\alpha}{a(1-e^2)} \right). \qquad (12)$$

Bei einem ganzen Umlauf rückt also das Perihel um

$$\varepsilon = 3\pi \frac{\alpha}{a(1-e^2)} \qquad (13)$$

im Sinne der Bahnbewegung vor, wenn mit a die große Halbachse, mit e die Exzentrizität bezeichnet wird. Führt man die Umlaufszeit T

Fig. 4.6 Page 838 from the original Einstein's paper [90] which contains an incorrect integration over the interval $[\alpha_1, \alpha_2]$

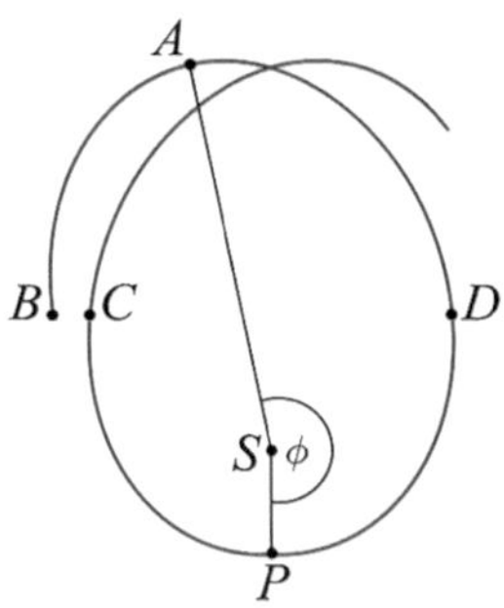

Fig. 4.7 The letters S, P, and A stand for the Sun, the perihelion, and the aphelion of Mercury's orbit, respectively. The angle $\angle PSA$ between the perihelion radius-vector and the aphelion radius-vector is denoted by ϕ

and by (3.27),

$$\alpha = \frac{2GM}{c^2} = 2953\,\mathrm{m} \tag{4.11}$$

is the Schwarzschild radius of the Sun.

Because the integral in (4.9) has no known analytical solution, Einstein used the linear part of the following Taylor expansion

$$\frac{1}{\sqrt{1-\alpha x}} = 1 + \frac{1}{2}\alpha x + \frac{3}{8}\alpha^2 x^2 + \dots$$

which is a fairly good approximation on the interval $[\alpha_1, \alpha_2]$, since $\alpha\alpha_i \ll 1$ for $i = 1, 2$. Hence, he expressed the angle ϕ from (4.9) as follows

$$\phi = [1 + \alpha(\alpha_1 + \alpha_2)] \int_{\alpha_1}^{\alpha_2} \frac{(1 + \frac{1}{2}\alpha x)\mathrm{d}x}{\sqrt{-(x-\alpha_1)(x-\alpha_2)}}$$

$$\stackrel{?}{=} \pi\Big[1 + \frac{3}{4}\alpha(\alpha_1 + \alpha_2)\Big] = \pi\Big[1 + \frac{3}{2}\frac{\alpha}{a(1-e^2)}\Big], \tag{4.12}$$

where the first equality should be replaced by $\approx$ as explained above and the last equality holds due to (4.10). However, Einstein does not present any details how the above integration denoted by $\stackrel{?}{=}$ was performed. We will postpone this important issue to Theorem 4.1. Note that the singularities near the end points α_1 and α_2 in (4.12) are integrable.

Remark 4.10 From (4.12), (4.11), and Kepler's third law $a^3/T^2 = GM/(4\pi^2)$ it follows that after one period (more precisely, between two successive perihelion passages) the perihelion shifts about the angle

$$\varepsilon = 2(\phi - \pi) = 3\pi \frac{\alpha}{a(1-e^2)} = 3\pi \frac{2GM}{ac^2(1-e^2)} = 24\pi^3 \frac{a^2}{T^2 c^2 (1-e^2)},$$

i.e. the relationship (4.2), which according to (4.3) yields the idealized value of the relativistic perihelion shift of Mercury $43''$ per century (cf. the left part of Fig. 4.2). Here one again subtracts two numbers of almost the same size, namely $\phi - \pi$, see Example 4.1.

All three of Kepler's laws hold only approximately, since there is always a modeling error. Nevertheless, a natural question arises: *Are we justified to use Kepler's third law in deriving relativistic formula (4.2)?* The radius vector of Mercury makes a larger angle than 2π between two passages of neighboring aphelia (or perihelia), see Fig. 4.7. Hence, T in (4.2) is not well defined.

Remark 4.11 It is curious that Einstein in [90, p. 839] writes that astronomers observe also an additional perihelion shift of Mars' trajectory of $9''$ per century, while his relationship (4.2) gives only $1''$ per century. This large discrepancy is not mentioned in the current literature. In other words, the questionable value for Mars is suppressed, whereas the value for Mercury that fits to (4.1) is emphasized.

Remark 4.12 A number of publications show (see e.g. [261, 312]) the value of (4.3) even to four significant digits $42.98''$ per century, when substituting more accurate values of a, e, and T to the simple algebraic Eq. (4.2). However, authors do not usually comment on the fact that this relationship was derived by many approximations of Einstein's equations, i.e., the errors e_0, e_1, and e_2 from Fig. 1.9 are ignored. Thus, the value of Mercury's perihelion shift (4.3) is not very reliable. Note that in [298, p. 1712] the value

$$E = 42.9773350296''\text{per century}$$

is presented, i.e., even to 12 significant digits!

In 2010, Anatoli Andrei Vankov pointed out that there is a numerical error in Einstein's evaluation of the integral appearing in (4.12), see [356, p. 21]. Further, we introduce a detailed calculation, so that everyone can check whether Vankov is right or not. The next statement is rather an example, but since it is important, we will call it a theorem. It yields a different value of ϕ than Einstein's value from (4.12), see also Fig. 4.6.

Theorem 4.1 *We have*

$$[1+\alpha(\alpha_1+\alpha_2)]\int_{\alpha_1}^{\alpha_2}\frac{(1+\frac{1}{2}\alpha x)\mathrm{d}x}{\sqrt{-(x-\alpha_1)(x-\alpha_2)}}=\pi\Big[1+\frac{5\alpha}{4}(\alpha_1+\alpha_2)+\frac{\alpha^2}{4}(\alpha_1+\alpha_2)^2\Big]. \tag{4.13}$$

Proof Define the linear transformation $\ell\colon [\alpha_1, \alpha_2] \to [-1, 1]$ by

$$\ell(x) = \frac{x-s}{h}, \tag{4.14}$$

where

$$s = \frac{\alpha_1+\alpha_2}{2} \tag{4.15}$$

is the midpoint of $[\alpha_1, \alpha_2]$ and $h = \alpha_2 - s$ half its width. Then $\ell(\alpha_1) = -1$ and $\ell(\alpha_2) = 1$, but also

$$\sqrt{-(x-\alpha_1)(x-\alpha_2)} = \sqrt{-(hy+h)(hy-h)} = h\sqrt{1-y^2},$$

where the substitution $y = \ell(x)$ with inverse

$$x = hy + s$$

was used. Hence, $\mathrm{d}x = h\mathrm{d}y$ and by (4.14) we get

$$\int_{\alpha_1}^{\alpha_2} \frac{(1+\frac{1}{2}\alpha x)\mathrm{d}x}{\sqrt{-(x-\alpha_1)(x-\alpha_2)}} = \Big(1+\frac{\alpha s}{2}\Big)\int_{-1}^{1} \frac{\mathrm{d}y}{\sqrt{1-y^2}} + \frac{\alpha h}{2}\int_{-1}^{1} \frac{y\,\mathrm{d}y}{\sqrt{1-y^2}}$$
$$= \Big(1+\frac{\alpha s}{2}\Big)\pi, \tag{4.16}$$

because the right-most integral exists over $[0, 1]$ and cancels the one over $[-1, 0]$, whereas the middle integral has $\arcsin y$ as antiderivative and thus evaluates to π.

Multiplying (4.16) by $[1 + \alpha(\alpha_1 + \alpha_2)]$ and applying (4.15), we find that

$$[1+\alpha(\alpha_1+\alpha_2)]\int_{\alpha_1}^{\alpha_2} \frac{(1+\frac{1}{2}\alpha x)\mathrm{d}x}{\sqrt{-(x-\alpha_1)(x-\alpha_2)}}$$
$$= [1+\alpha(\alpha_1+\alpha_2)]\pi\Big[1+\frac{\alpha}{4}(\alpha_1+\alpha_2)\Big]$$
$$= \pi\Big[1+\frac{5\alpha}{4}(\alpha_1+\alpha_2)+\frac{\alpha^2}{4}(\alpha_1+\alpha_2)^2\Big].$$

The proof is completed. □

Remark 4.13 The correct integration from (4.12) gives by (4.13) that

$$\phi = [1+\alpha(\alpha_1+\alpha_2)]\int_{\alpha_1}^{\alpha_2} \frac{(1+\frac{1}{2}\alpha x)\mathrm{d}x}{\sqrt{-(x-\alpha_1)(x-\alpha_2)}}$$
$$= \pi\Big[1+\frac{5\alpha}{4}(\alpha_1+\alpha_2)+\frac{\alpha^2}{4}(\alpha_1+\alpha_2)^2\Big]$$
$$\approx \pi\Big[1+\frac{5}{4}\alpha(\alpha_1+\alpha_2)\Big]$$

neglecting the quadratic term (of dimensionless order 10^{-15}). This differs from (4.12). Hence, by (4.10) and (4.11) we get

$$\varepsilon = 2(\phi-\pi) = \pi\frac{5\alpha}{2}(\alpha_1+\alpha_2) = 8.364\cdot 10^{-7}\ \mathrm{rad}$$

for one period which is a different value than that in (4.2). This is 5/3 times the value that Einstein computed, see [172].

However, Vankov in [90] proposes to replace the factor $1 + \alpha(\alpha_1 + \alpha_2)$ in (4.13) by $1 + \frac{1}{2}\alpha(\alpha_1 + \alpha_2)$ which then leads to the right-hand side of (4.12).

Remark 4.14 The angle ϕ given by (4.9) is an approximation of another integral

$$\phi \approx \int_{\alpha_1}^{\alpha_2} \frac{\mathrm{d}x}{\sqrt{\alpha x^3 - x^2 + Cx + D}}, \tag{4.17}$$

where C and D are some suitable constants. According to Einstein [90, p. 838], α_1 and α_2 are two roots of the cubic polynomial

$$\alpha x^3 - x^2 + Cx + D = \alpha(x - \alpha_1)(x - \alpha_2)(x - \beta),$$

where β is the third root. From this we get $-x^2 = -\alpha(\alpha_1 + \alpha_2 + \beta)x^2$ and thus

$$\beta = \frac{1}{\alpha} - \alpha_1 - \alpha_2.$$

Einstein then approximates this third root by $1/\alpha$ due to (4.10)–(4.11). Numerical integration based on the Chebyshev-Gauss quadrature formulae shows that

$$\phi = \left[1 + \frac{\alpha}{2}(\alpha_1 + \alpha_2)\right] \int_{\alpha_1}^{\alpha_2} \frac{\mathrm{d}x}{\sqrt{-(x - \alpha_1)(x - \alpha_2)(1 - \alpha x)}}$$

is a much better approximation of the integral in (4.17) than that of ϕ given by (4.9). This indicates that Vankov is right (see Remark 4.13 and also [139]). The equality sign $=$ in the above integral should be again considered as $\approx$.

Einstein was likely satisfied that his value $E = 43''$ per century agreed with the value obtained by Le Verrier (see Remark 4.1) and also Gerber (see Remark 4.2). Hundreds of publications claim that the value $E = 43''$ per century (see (4.3)) is very precise without any proof, that it nicely fits to astronomical observations, and thus it perfectly confirms the General Theory of Relativity, see e.g. [87, 107, 152, 154, 249, 257, 261, 268, 286, 295, 298, 303, 312]. Therefore, in the previous four sections we wanted to point out the main shortcomings which are not taken into account when investigating this ill-conditioned problem.

4.6 Further Applications and Critical Remarks

Einstein's formula (4.2) is often applied to other bodies than Mercury. For example, substituting the values $a = 9377\,\mathrm{km}$, $e = 0.0151$, and $T = 7.65\,\mathrm{h}$ corresponding to Phobos into (4.2), we can derive the quite large value of the pericenter shift of $23''$ per century. This Martian moon is perhaps another suitable candidate for testing the validity of General Theory of Relativity, since its trajectory is less disturbed by other bodies than for Mercury. A secular trend in Phobos' longitude was already observed in [209, p. 1082]. For the innermost satellites of some other planets the formula (4.2) produces much higher pericenter shifts, e.g., for Jupiter's moon Metis with $a = 127{,}969\,\text{km}$, $e \approx 0.04$, and $T = 7.075\,\text{h}$ we get $1.37°$ per century. This is $49''$ per year which is 115 times more than Mercury's perihelion shift. Similarly, the pericenter shift of Io is $15''$ per year. These large values are difficult to verify due to various resonances between the Galilean moons.

Einstein's formula (4.2) has also been applied to the star S2 orbiting the supermassive black hole Sgr A* at the center of our Galaxy with pericenter speed 12,000 km/s. Its orbit is seen in the projection to the celestial sphere and the trajectory during the first observed period of S2 was measured only very roughly since 1992. Therefore, it would be more precise to measure the apocenter shift instead of the pericenter shift. The orbital period T of S2 is approximately 16 years, $a \approx 970$ au, and $e \approx 0.885$. Substituting these values into (4.2), we get

$$\varepsilon = 11'. \tag{4.18}$$

The semimajor axis a was obtained by angular measurements. Nonetheless, since the distance 25.6 ly to Sgr A* is established only roughly, a more precise value of a is not known.

Remark 4.15 Notice that the input parameters a, T, and e to the formula (4.2) are squared. Therefore, this formula is very sensitive to their accurate measurements. For instance, we have

$$(a + \Delta a)^2 = a^2 + 2a\Delta a + (\Delta a)^2. \tag{4.19}$$

Hence, if the relative error in determining the value a is e.g. 1 %, then the error in a^2 is about 2 % (we usually have $|\Delta a| \ll a$). All errors depicted in Fig. 1.9 must also be considered. Furthermore, the rotation of the central black hole was not taken into account. There are, of course, many other errors coming from diverse sources like the pixel structure of CCD detectors, calibration errors, influence of neighboring stars, etc. Anyway, the authors of the paper [9] claim that the measured pericenter shift corresponds exactly to the value (4.18) and that this is another proof that General Relativity is valid (see also [392, 62 min]). It would be more precise to measure the apocenter shift. Note that (4.2) should be applied to a nonrotating central body while Sgr A* rotates.

Remark 4.16 It is curious that formula (4.2) always yields a positive perihelion shift $\varepsilon > 0$ even for the zero eccentricity $e = 0$. In this case, there are infinitely many perihelia, namely, the perihelion is at each point of a circular trajectory. In fact, by (4.2) we always have

$$\varepsilon \geq 24\pi^3 a^2 T^{-2} c^{-2} > 0.$$

However, this does not allow us to consider a retrograde perihelion shift with negative ε, i.e., when $\phi < \pi$ in Fig. 4.7. For instance, Andrea Ghez claimed that her team observed that the movement of the pericenter of the star S2 around the black hole Sgr A* is retrograde (see [391, 45 min]). Therefore, we suggest to wait several decades to find whether the formula (4.2) approximates reality well.

Remark 4.17 For Newtonian elliptic orbits, the eccentricity is defined as follows

$$e = \frac{\sqrt{a^2 - b^2}}{a},$$

where b is the length of the corresponding semiminor axis. However, what is the definition of e for non-elliptic orbits? (Cf. Fig. 4.4.) How do we define the semiminor axis b for the situation sketched in Fig. 4.7? Is $b = |BD|/2$ or $b = |CD|/2$? How to define B, C, and D at all? Should we define a as $a = |AP|/2$ or $a = (|AS| + |PS|)/2$ or by another manner? Formula (4.2) thus contains not well-defined quantities a and e. Moreover, the static space around the Sun is curved. So are α_1 and α_2 in (4.10) well defined?

Remark 4.18 Although formula (4.2) was derived for a massless test particle representing Mercury, it is often applied to close binaries whose components have positive masses M_1 and M_2 (see e.g. [18, 40, 121, 131, 334, 336, 372]). Note that PSR J0737-3039 is the only known double pulsar. By Lyne [228] it has a relativistic periastron advance. The mass $M = M_\odot$ in Remark 4.10 is usually replaced by the sum $M_1 + M_2$ using the PPN formalism, see e.g. [377, p. 47]. However, the exact solution of Einstein's nonlinear equations for two bodies with positive masses is not known. Hence, we cannot verify if such a simple approximation for large orbital velocities and strong gravitational fields is correct.

Remark 4.19 Einstein's approach does not take into account the influence of tidal forces, since Mercury is replaced by a point and the gravitational field around the "nonrotating" Sun is supposed to be spherically symmetric (it is approximately described by the exterior Schwarzschild metric tensor (3.6)). Therefore, no tidal bulges are possible.

Tidal forces are not taken into account also in the classical N-body problem which yields the calculated value C. The actual trajectory of Mercury can be slightly deformed due to the locking resonance $2 : 3$ between the orbital and rotational period. This resonance may have a nonnegligible influence on the actual perihelion shift as well.

4.7 Conclusions

In 1915, Albert Einstein derived the value of the additional relativistic perihelion shift of Mercury $43''$ per century. In his article [90, p. 831] he claims that Le Verrier needed approximately $45''$ per century to explain the difference between the observed and calculated perihelion position. This caused a sensation and Einstein's General Theory of Relativity suddenly became famous. The value of relativistic shift (4.3), however, applies only to the very idealized case, which may be quite different from reality. It was obtained by many approximations, assumptions and also several errors, see [90] for their list.

According to [139], in 1913 Einstein wrote a manuscript on the motion of the perihelion of Mercury together with his friend Michele Besso. Although Besso corrected many errors in their common calculations, Einstein does not mention him in the final version [90].

Hence, Einstein worked on the problem of Mercury's perihelion shift for at least 2 years. Nonetheless, when [90] has appeared, he also claimed that: *Ich habe die Gerbersche Arbeit ursprünglich schon deshalb nicht erwähnt, weil ich sie nicht kannte, als ich meine Arbeit über die Perihelbewegung des Merkur schrieb*, which can be translated as follows: *I have not mentioned the work by Gerber originally, because I did not know it when I wrote my work on the perihelion motion of Mercury.* It is, however, interesting that Einstein used a very similar notation as Gerber in [113], namely, the angle ϕ from Fig. 4.7 satisfies $\phi = \Phi/2$, where Φ appears in Gerber's formula, see Remark 4.2.

It is very difficult to separate relativistic effects from effects of Newtonian mechanics of similar size, and many other approximations. In formula (4.4), two almost equally large numbers are subtracted that are, in addition, perturbed by various errors. In other words, we do not evaluate the difference $O - C$, but $\tilde{O} - \tilde{C}$ without any guaranteed error estimates, where $\tilde{O} \approx O$ and $\tilde{C} \approx C$ which are also not uniquely defined. The estimated difference of Mercury's perihelion shift obtained from astronomical observations and numerical solution of the problem of N bodies is therefore ill-conditioned, see [356]. According to [312], formula (4.2) represents a weak experimental and theoretical confirmation of General Relativity.

The main objections against the proposed value (4.3) can be summarized as follows:

1. The input parameters a, e, T from (4.2) are not well defined, see Remarks 4.10 and 4.17.
2. The formula (4.2) is quite sensitive to the accuracy of all input data due to the relation (4.19), see Remark 4.15.
3. Since Mercury's orbit with $b = 0.98a$ is almost circular, to determine accurately the exact position of perihelion is an ill-conditioned problem, see Sect. 4.3.
4. Since the proposed relativistic perihelion shift $0.43''\ \mathrm{yr}^{-1}$ is a quite small number, all error estimates in deriving this value should be done very carefully. However, this is not done in the literature.
5. The modeling error (see Fig. 1.9) of Einstein's equations is not known.
6. The observed perihelion shift O from (4.4) is not uniquely defined, see Sect. 4.3.

7. The calculated perihelion shift C from (4.4) is also not uniquely defined, see Sect. 4.4. Moreover, the subtraction $O - C$ of two numbers of a similar size is burdened by a large relative error, see Sect. 4.2.
8. The formula (4.2) does not allow a nonpositive shift, see Remark 4.16.
9. The tidal forces of Mercury and the Sun are ignored, see Remark 4.19.
10. The barycenter of the Solar system shifts every day by about 1000 km, whereas the additional relativistic perihelion shift of Mercury is on average only 96 km in a year, see Fig. 4.4.

The Einstein value E from (4.3) was derived under many simplifications and approximations without any error estimates, see Sect. 4.5. The list of these simplifications would be too long, e.g., a special form of the metric tensor is prescribed, the real rotational axis of the Sun is not perpendicular to Mercury's orbit, Lense-Thirring precession effect is ignored, the spacetime curved by Mercury, Jupiter etc. is replaced by Euclidean space, the PPN formalism is used for velocities $v \ll c$, higher order terms are neglected. Retardation effects of gravitational interaction on the perihelion shift of Mercury are studied in [382].

In this chapter, we primarily wanted to point out that we should not indiscriminately take the value (4.3) and spread it further without checking how it was derived.

Chapter 5
Computational Problems of Einstein's Equations

5.1 On the Explicit Form of the First Einstein Equation

In this section, we want to point out the extreme complexity of Einstein's equations which does not allow us to find their analytical solution of the N-body problem for $N > 1$. Applicability of Einstein's equations in the relativistic community is therefore overrated.

In (3.29), the dependence of the Ricci scalar R and the Ricci tensor $R_{\mu\nu}$ on the metric tensor $g_{\mu\nu}$ is not indicated. Consequently, Einstein's equations (3.29) seem to be quite simple. In [249, p. 42], one even can read:

No equation of physics can be written more simply.

To avoid this deceptive appearance, we will show now how to derive an explicit form of the first Einstein equation of system (3.29). The form (3.29) is evidently not suitable for computer implementation.

First we shall consider only the case when $T_{\mu\nu} = 0$ with zero cosmological constant which is due to Theorem 3.11 given by the relation

$$R_{\mu\nu} = 0. \tag{5.1}$$

Nevertheless, this apparently very simple equation takes an extremely complicated form when the unknown metric tensor $g_{\mu\nu}$ is expressed explicitly as we shall see below. The reason is that mathematicians mark all unknowns in the equations they solve, whereas in (5.1) no unknowns and also no coordinates are indicated.

So now we will rewrite (5.1) such that the metric tensor explicitly appears there. Using (3.4), we can express the first Einstein equation of (5.1) as follows

$$\begin{aligned} 0 = R_{00} &= \Gamma^{\varkappa}_{00,\varkappa} - \Gamma^{\varkappa}_{0\varkappa,0} + \Gamma^{\lambda}_{00}\Gamma^{\varkappa}_{\lambda\varkappa} - \Gamma^{\lambda}_{0\varkappa}\Gamma^{\varkappa}_{0\lambda} \\ &= \underline{\Gamma^{0}_{00,0}} + \Gamma^{1}_{00,1} + \Gamma^{2}_{00,2} + \Gamma^{3}_{00,3} - \underline{\Gamma^{0}_{00,0}} - \Gamma^{1}_{01,0} - \Gamma^{2}_{02,0} - \Gamma^{3}_{03,0} \end{aligned}$$

M. Křížek, L. Somer, *Mathematical Aspects of Paradoxes in Cosmology*,
https://doi.org/10.1007/978-3-031-31768-2_5

$$
\begin{aligned}
&+ \underline{\Gamma^0_{00}(\Gamma^0_{00}} + \Gamma^1_{01} + \Gamma^2_{02} + \Gamma^3_{03}) + \underline{\Gamma^1_{00}(\Gamma^0_{10}} + \Gamma^1_{11} + \Gamma^2_{12} + \Gamma^3_{13})\\
&+ \underline{\Gamma^2_{00}(\Gamma^0_{20}} + \Gamma^1_{21} + \Gamma^2_{22} + \Gamma^3_{23}) + \underline{\Gamma^3_{00}(\Gamma^0_{30}} + \Gamma^1_{31} + \Gamma^2_{32} + \Gamma^3_{33})\\
&- \underline{\Gamma^0_{00}\Gamma^0_{00}} - \Gamma^0_{01}\Gamma^1_{00} - \Gamma^0_{02}\Gamma^2_{00} - \Gamma^0_{03}\Gamma^3_{00}\\
&- \underline{\Gamma^1_{00}\Gamma^0_{01}} - \Gamma^1_{01}\Gamma^1_{01} - \Gamma^1_{02}\Gamma^2_{01} - \Gamma^1_{03}\Gamma^3_{01}\\
&- \underline{\Gamma^2_{00}\Gamma^0_{02}} - \Gamma^2_{01}\Gamma^1_{02} - \Gamma^2_{02}\Gamma^2_{02} - \Gamma^2_{03}\Gamma^3_{02}\\
&- \underline{\Gamma^3_{00}\Gamma^0_{03}} - \Gamma^3_{01}\Gamma^1_{03} - \Gamma^3_{02}\Gamma^2_{03} - \Gamma^3_{03}\Gamma^3_{03},
\end{aligned}
$$

where the underlined terms cancel. Hence, the first Einstein equation can be rewritten by means of the Christoffel symbols of the second kind in the following way:

$$
\begin{aligned}
0 &= R_{00}\\
&= \Gamma^1_{00,1} + \Gamma^2_{00,2} + \Gamma^3_{00,3} - \Gamma^1_{01,0} - \Gamma^2_{02,0} - \Gamma^3_{03,0}\\
&\quad + \Gamma^0_{00}(\Gamma^1_{01} + \Gamma^2_{02} + \Gamma^3_{03}) + \Gamma^1_{00}(-\Gamma^0_{10} + \Gamma^1_{11} + \Gamma^2_{12} + \Gamma^3_{13})\\
&\quad + \Gamma^2_{00}(-\Gamma^0_{20} + \Gamma^1_{21} + \Gamma^2_{22} + \Gamma^3_{23}) + \Gamma^3_{00}(-\Gamma^0_{30} + \Gamma^1_{31} + \Gamma^2_{32} + \Gamma^3_{33})\\
&\quad - 2\Gamma^1_{02}\Gamma^2_{01} - 2\Gamma^1_{03}\Gamma^3_{01} - 2\Gamma^2_{03}\Gamma^3_{02} - (\Gamma^1_{01})^2 - (\Gamma^2_{02})^2 - (\Gamma^3_{03})^2.
\end{aligned}
\tag{5.2}
$$

Using (3.2) and the symmetry of $g_{\mu\nu}$, we obtain

$$
\begin{aligned}
2\Gamma^{\varkappa}_{\mu\nu} &= g^{\varkappa 0}(g_{\mu 0,\nu} + g_{\nu 0,\mu} - g_{\mu\nu,0}) + g^{\varkappa 1}(g_{\mu 1,\nu} + g_{\nu 1,\mu} - g_{\mu\nu,1})\\
&\quad + g^{\varkappa 2}(g_{\mu 2,\nu} + g_{\nu 2,\mu} - g_{\mu\nu,2}) + g^{\varkappa 3}(g_{\mu 3,\nu} + g_{\nu 3,\mu} - g_{\mu\nu,3})
\end{aligned}
$$

and thus by (5.1) we can express the first Einstein equation $R_{00} = 0$ by means of the metric coefficients and their first and second order derivatives as follows:

$$
\begin{aligned}
0 &= 4R_{00}\\
&= 2\big[g^{10}_{,1} g_{00,0} + g^{11}_{,1}(2g_{01,0} - g_{00,1}) + g^{12}_{,1}(2g_{02,0} - g_{00,2})\\
&\quad + g^{13}_{,1}(2g_{03,0} - g_{00,3})\\
&\quad + g^{10} g_{00,01} + g^{11}(2g_{01,01} - g_{00,11}) + g^{12}(2g_{02,01} - g_{00,21})\\
&\quad + g^{13}(2g_{03,01} - g_{00,31})
\end{aligned}
$$

$$
\begin{aligned}
&+ g^{20}_{,2} g_{00,0} + g^{21}_{,2}(2g_{01,0} - g_{00,1}) + g^{22}_{,2}(2g_{02,0} - g_{00,2}) + g^{23}_{,2}(2g_{03,0} - g_{00,3})\\
&+ g^{20} g_{00,02} + g^{21}(2g_{01,02} - g_{00,12}) + g^{22}(2g_{02,02} - g_{00,22})\\
&+ g^{23}(2g_{03,02} - g_{00,32})\\
&+ g^{30}_{,3} g_{00,0} + g^{31}_{,3}(2g_{01,0} - g_{00,1}) + g^{32}_{,3}(2g_{02,0} - g_{00,2}) + g^{33}_{,3}(2g_{03,0} - g_{00,3})\\
&+ g^{30} g_{00,03} + g^{31}(2g_{01,03} - g_{00,13}) + g^{32}(2g_{02,03} - g_{00,23})\\
&+ g^{33}(2g_{03,03} - g_{00,33})\\
&- g^{10}_{,0} g_{00,1} - g^{11}_{,0} g_{11,0} - g^{12}_{,0}(g_{02,1} + g_{12,0} - g_{01,2})\\
&- g^{13}_{,0}(g_{03,1} + g_{13,0} - g_{01,3})\\
&- g^{10} g_{00,10} - g^{11} g_{11,00} - g^{12}(g_{02,10} + g_{12,00} - g_{01,20})\\
&- g^{13}(g_{03,10} + g_{13,00} - g_{01,30})\\
&- g^{20}_{,0} g_{00,2} - g^{21}_{,0}(g_{01,2} + g_{21,0} - g_{02,1}) - g^{22}_{,0} g_{22,0}\\
&- g^{23}_{,0}(g_{03,2} + g_{23,0} - g_{02,3})\\
&- g^{20} g_{00,20} - g^{21}(g_{01,20} + g_{21,00} - g_{02,10}) - g^{22} g_{22,00}\\
&- g^{23}(g_{03,20} + g_{23,00} - g_{02,30})\\
&- g^{30}_{,0} g_{00,3} - g^{31}_{,0}(g_{01,3} + g_{31,0} - g_{03,1}) - g^{32}_{,0}(g_{02,3} + g_{32,0} - g_{03,2})\\
&- g^{33}_{,0} g_{33,0}\\
&- g^{30} g_{00,30} - g^{31}(g_{01,30} + g_{31,00} - g_{03,10}) - g^{32}(g_{02,30} + g_{32,00} - g_{03,20})\\
&- g^{33} g_{33,00}\Big]\\
&+ (g^{00} g_{00,0} - g^{01} g_{00,1} - g^{02} g_{00,2} - g^{03} g_{00,3})\\
&\times \Big[g^{10}(2g_{10,1} - g_{11,0}) + g^{11} g_{11,1} + g^{12}(2g_{12,1} - g_{11,2}) + g^{13}(2g_{13,1} - g_{11,3})\\
&+ g^{20}(g_{10,2} + g_{20,1} - g_{12,0}) + g^{21} g_{11,2} + g^{22} g_{22,1}\\
&+ g^{23}(g_{13,2} + g_{23,1} - g_{12,3})\\
&+ g^{30}(g_{10,3} + g_{30,1} - g_{13,0}) + g^{31} g_{11,3} + g^{32}(g_{12,3} + g_{32,1} - g_{13,2})\\
&+ g^{33} g_{33,1}\Big]
\end{aligned}
$$

$$
\begin{aligned}
&+ (g^{10}g_{00,0} + g^{11}g_{11,1} - g^{12}g_{11,2} - g^{13}g_{11,3})\\
&\times \big[- g^{00}g_{00,1} - g^{01}g_{11,0} - g^{02}(g_{12,0} + g_{02,1} - g_{10,2})\\
&- g^{03}(g_{13,0} + g_{03,1} - g_{10,3})\\
&+ g^{10}(2g_{10,1} - g_{11,0}) + g^{11}g_{11,1} + g^{12}(2g_{12,1} - g_{11,2}) + g^{13}(2g_{13,1} - g_{11,3})\\
&+ g^{20}(g_{10,2} + g_{20,1} - g_{12,0}) + g^{21}g_{11,2} + g^{22}g_{22,1}\\
&+ g^{23}(g_{13,2} + g_{23,1} - g_{12,3})\\
&+ g^{30}(g_{10,3} + g_{30,1} - g_{13,0}) + g^{31}g_{11,3} + g^{32}(g_{12,3} + g_{32,1} - g_{13,2})\\
&+ g^{33}g_{33,1}\big]
\end{aligned}
$$

$$
\begin{aligned}
&+ \big[g^{20}g_{00,0} + g^{21}(2g_{01,0} - g_{00,1}) + g^{22}(2g_{02,0} - g_{00,2}) + g^{23}(2g_{03,0} - g_{00,3})\big]\\
&\times \big[- g^{00}g_{00,2} - g^{01}(g_{21,0} + g_{01,2} - g_{20,1}) - g^{02}g_{22,0}\\
&- g^{03}(g_{23,0} + g_{03,2} - g_{20,3})\\
&+ g^{10}(g_{20,1} + g_{10,2} - g_{21,0}) + g^{11}g_{11,2} + g^{12}g_{22,1} + g^{13}(g_{23,1} + g_{13,2} - g_{21,3})\\
&+ g^{20}(2g_{20,2} - g_{22,0}) + g^{21}(2g_{21,2} - g_{22,1}) + g^{22}g_{22,2} + g^{23}(2g_{23,2} - g_{22,3})\\
&+ g^{30}(g_{20,3} + g_{30,2} - g_{23,0}) + g^{31}(g_{21,3} + g_{31,2} - g_{23,1}) + g^{32}g_{22,3} + g^{33}g_{33,2}\big]
\end{aligned}
$$

$$
\begin{aligned}
&+ \big[g^{30}g_{00,0} + g^{31}(2g_{01,0} - g_{00,1}) + g^{32}(2g_{02,0} - g_{00,2})\\
&+ g^{33}(2g_{03,0} - g_{00,3})\big]\\
&\times \big[- g^{00}g_{00,3} - g^{01}g_{01,3} - g^{02}(g_{32,0} + g_{02,3} - g_{30,2}) - g^{03}g_{33,0}\\
&+ g^{10}(g_{30,1} + g_{10,3} - g_{31,0}) + g^{11}g_{11,3} + g^{12}(g_{32,1} + g_{12,3} - g_{31,2}) + g^{13}g_{33,1}\\
&+ g^{20}(g_{30,2} + g_{20,3} - g_{32,0}) + g^{21}(g_{31,2} + g_{21,3} - g_{32,1}) + g^{22}g_{22,3} + g^{23}g_{33,2}\\
&+ g^{30}(2g_{30,3} - g_{33,0}) + g^{31}(2g_{31,3} - g_{33,1}) + g^{32}(2g_{32,3} - g_{33,2}) + g^{33}g_{33,3}\big]
\end{aligned}
$$

$$- 2\big[g^{10}g_{00,2} + g^{11}(g_{01,2} + g_{21,0} - g_{02,1}) + g^{12}g_{22,0}$$
$$+ g^{13}(g_{03,2} + g_{23,0} - g_{02,3})\big]$$
$$\times \big[g^{20}g_{00,1} + g^{21}g_{11,0} + g^{22}(g_{02,1} + g_{12,0} - g_{01,2}) + g^{23}(g_{03,1} + g_{13,0} - g_{01,3})\big]$$

$$- 2\big[g^{10}g_{00,3}+g^{11}(g_{01,3} + g_{31,0} - g_{03,1})+g^{12}(g_{02,3} + g_{32,0} - g_{03,2})+g^{13}g_{33,0}\big]$$
$$\times \big[g^{30}g_{00,1} + g^{31}g_{11,0} + g^{32}(g_{02,1} + g_{12,0} - g_{01,2}) + g^{33}(g_{03,1} + g_{13,0} - g_{01,3})\big]$$

$$- 2\big[g^{20}g_{00,3}+g^{21}(g_{01,3} + g_{31,0} - g_{03,1})+g^{22}(g_{02,3} + g_{32,0} - g_{03,2})+g^{23}g_{33,0}\big]$$
$$\times \big[g^{30}g_{00,2} + g^{31}(g_{01,2} + g_{21,0} - g_{02,1}) + g^{32}g_{22,0} + g^{33}(g_{03,2} + g_{23,0} - g_{02,3})\big]$$

$$- \big[g^{10}g_{00,1}+g^{11}g_{11,0}+g^{12}(g_{02,1} + g_{12,0} - g_{01,2})+g^{13}(g_{03,1} + g_{13,0} - g_{01,3})\big]^2$$
$$- \big[g^{20}g_{00,2}+g^{21}(g_{01,2} + g_{21,0} - g_{02,1})+g^{22}g_{22,0}+g^{23}(g_{03,2} + g_{23,0} - g_{02,3})\big]^2$$
$$- \big[g^{30}g_{00,3}+g^{31}(g_{01,3} + g_{31,0} - g_{03,1})+g^{32}(g_{02,3} + g_{32,0} - g_{03,2})+g^{33}g_{33,0}\big]^2. \tag{5.3}$$

Now we should substitute all entries of (5.3) with double upper indices to (3.32). For instance, the entry g^{11} in the fourth line of (5.3) could be rewritten by means of the Sarrus rule for 3×3 symmetric algebraic adjoints g^*_{11} by

$$g^{11} = \frac{g^*_{11}}{\det(g_{\mu\nu})}$$
$$= \frac{g_{00}g_{22}g_{33} + 2g_{02}g_{03}g_{23} - g_{00}(g_{23})^2 - g_{22}(g_{03})^2 - g_{33}(g_{02})^2}{\sum_{\pi\in S_4}(-1)^{\operatorname{sgn}\pi}\, g_{0\nu_0}g_{1\nu_1}g_{2\nu_2}g_{3\nu_3}}, \tag{5.4}$$

where the sum in the denominator contains $4! = 24$ terms. Note that the optimal expression for the minimum number of arithmetic operations to calculate the inverse of a 4×4 matrix is not known, yet. The other nine entries g^{00}, g^{01}, g^{02}, g^{03}, g^{12}, g^{13}, g^{22}, g^{23}, and g^{33} can be expressed similarly.

However, we have to evaluate also the first derivatives of $g^{\mu\nu}$. Consider for instance the entry $g^{11}_{,1}$ in the second line of (5.3). Then by (5.4) we get

$$
\begin{aligned}
g^{11}_{,1} &= \frac{\partial}{\partial x_1}\Big(\frac{g^*_{11}}{\det(g_{\mu\nu})}\Big) \\
&= \Bigg(\frac{1}{\sum_{\pi\in S_4}(-1)^{\operatorname{sgn}\pi} g_{0\nu_0}g_{1\nu_1}g_{2\nu_2}g_{3\nu_3}}\big(g_{00}g_{22}g_{33} + 2g_{02}g_{03}g_{23} \\
&\quad - g_{00}(g_{23})^2 - g_{22}(g_{03})^2 - g_{33}(g_{02})^2\big)\Bigg)_{,1} \\
&= \Big[\big(g_{00,1}g_{22}g_{33} + 2g_{02,1}g_{03}g_{23} - g_{00,1}(g_{23})^2 - g_{22,1}(g_{03})^2 - g_{33,1}(g_{02})^2 \\
&\quad + g_{00}g_{22,1}g_{33} + 2g_{02}g_{03,1}g_{23} + g_{00}g_{22}g_{33,1} + 2g_{02}g_{03}g_{23,1} - 2g_{00}g_{23,1} \\
&\quad - 2g_{22}g_{03,1} - 2g_{33}g_{02,1}\big)\Big(\sum_{\pi\in S_4}(-1)^{\operatorname{sgn}\pi} g_{0\nu_0}g_{1\nu_1}g_{2\nu_2}g_{3\nu_3}\Big) \\
&\quad - \big(g_{00}g_{22}g_{33} + 2g_{02}g_{03}g_{23} - g_{00}(g_{23})^2 - g_{22}(g_{03})^2 - g_{33}(g_{02})^2\big) \\
&\quad \times \sum_{\pi\in S_4}(-1)^{\operatorname{sgn}\pi}\big(g_{0\nu_0,1}g_{1\nu_1}g_{2\nu_2}g_{3\nu_3} + g_{0\nu_0}g_{1\nu_1,1}g_{2\nu_2}g_{3\nu_3} \\
&\quad + g_{0\nu_0}g_{1\nu_1}g_{2\nu_2,1}g_{3\nu_3} + g_{0\nu_0}g_{1\nu_1}g_{2\nu_2}g_{3\nu_3,1}\big)\Big] \\
&\quad \times \Big(\sum_{\pi\in S_4}(-1)^{\operatorname{sgn}\pi} g_{0\nu_0}g_{1\nu_1}g_{2\nu_2}g_{3\nu_3}\Big)^{-2}. \qquad (5.5)
\end{aligned}
$$

Remark 5.1 Substituting all $g^{\mu\nu}$ and also their first derivatives into (5.3), we get the explicit form of the first Einstein equation $R_{00} = 0$ of the second order for the 10 unknowns $g_{00}, g_{01}, g_{02}, \ldots, g_{33}$. It is evident that such an equation is extremely complicated. Relation (5.2) takes only 5 lines, relation (5.3) takes more than 50 lines and after substitution of all entries with determinants given by (5.4), (5.5), etc., into (5.3), the Eq. (5.1) for the component R_{00} of the Ricci tensor will occupy more than 10 pages. The other nine equations $R_{\mu\nu} = 0$ can be expressed similarly.

Remark 5.2 The explicit expression of the left-hand side of (3.29) for a given covariant divergence-free tensor $T_{\mu\nu} \neq 0$ in terms of the unknown components of $g_{\mu\nu}$ is even more complicated. Using (3.30) and (5.3)–(5.5), we still have to express the term $-\frac{1}{2}Rg_{\mu\nu}$ similarly. As of 2022, nobody has calculated how many terms Einstein's equations really contain, in general.

Einstein's equations are nonlinear in the first derivatives of the metric tensor, since the Ricci tensor (3.4) contains a number of multiplications of two Christoffel symbols. From (3.4) it seems that these equations are linear in the second derivatives of the metric tensor. Now we prove that this is not true.

Theorem 5.1 *Einstein's equations (3.52) are not linear in the second derivatives of the metric tensor.*

Proof Obviously, it is enough to deal only with the Ricci tensor defined by (3.4). From the third line of (5.3) we observe that the second derivative $g_{01,01}$ is multiplied by the entry g^{11} of the inverse metric tensor. From (5.4) we find that this entry depends highly nonlinearly on the metric tensor. Moreover, the second derivative $g_{01,01}$ appears in (5.3) only once, so it cannot be canceled by another term.

Similar considerations can be made if Einstein's equations are written in mixed or contravariant from. Hence, we again cannot avoid to include the entries of the inverse metric tensor, since

$$R^{\mu}_{\nu} = g^{\mu\varkappa} R_{\varkappa\nu}, \qquad R^{\mu\nu} = g^{\mu\varkappa} g^{\lambda\nu} R_{\varkappa\lambda}.$$

The proof is completed. □

5.2 Difficulties with Initial and Boundary Conditions

Einstein's equations (3.29) without initial and boundary conditions may have more different solutions (see e.g. Theorem 3.10). Since (3.29) represents a system of hyperbolic equations with second order time derivatives, two initial conditions should be prescribed. In particular, one should prescribe initial conditions for all 10 components $g_{\mu\nu}$ and all their 40 first derivatives. However, this is almost impossible in general case.

There are similar large problems with boundary conditions and also with the divergence-free right-hand side (3.31) of Einstein's equations (3.29). Moreover, Einstein's equations are highly nonlinear. Therefore, the existence and uniqueness of a weak solution of Einstein's equations with some initial and boundary conditions is a serious problem. Note, for example, that the one-dimensional steady-state nonlinear heat conduction equations with any prescribed type of boundary conditions does not have a unique weak solution even when the heat conduction coefficient depends continuously on the temperature, see [132, p. 182].

It is usually difficult, if not impossible, to prescribe explicitly any appropriate initial and boundary conditions for non-spherically symmetric regions for $g_{\mu\nu}$ satisfying (3.29). The reason is that *matter tells spacetime how to curve and spacetime tells matter how to move* [249]. This is the so-called back reaction. Hence, the initial space manifold is a priori not known except for some special cases when an analytical solution of Einstein's equations is known. More precisely, the left-hand side of (3.29) corresponds to geometry and the Einstein tensor $G_{\mu\nu}$ tells spacetime how to curve. The right-hand side of (3.29) corresponds to physics and the tensor $T_{\mu\nu}$ tells matter how to move.

Remark 5.3 Another non-negligible problem lies in the nonuniqueness of topology. The reason is that the knowledge of the metric tensor $g_{\mu\nu}$ does not determine uniquely the topology and shape of the associated space-time manifold. For instance, Euclidean space $\mathbb{E}^3$ has obviously the same metric tensor $g_{\mu\nu} = \delta_{\mu\nu}$, $\mu, \nu = 1, 2, 3$, as the infinite cylinder $\mathbb{S}^1 \times \mathbb{E}^2$, but a different topology and shape for the time-independent case with $T_{\mu\nu} = 0$ in (3.29). Here $\mathbb{S}^1$ stands for the unit circle. Another manifold $\{(x, y) \in \mathbb{E}^2 \mid y = \sin x\} \times \mathbb{E}^2$ has a Euclidean geometry, too. Hence, having the metric tensor, which satisfies Einstein's equations, does not mean that we know what the corresponding spacetime manifold looks like. Its geometry is fully described by the Riemann curvature tensor defined in Sect. 6.5 that has 20 independent components. Several other examples can be found in [249, p. 725].

5.3 Extreme Computational Complexity of Einstein's Equations

According to (5.3)–(5.5), we observe that Einstein's equations are highly nonlinear. From Remark 5.1 we find that the explicit form of all 10 equations (3.29) will occupy more than one hundred pages. For comparison note that the Laplace equation $\Delta u = 0$ has only three terms $\partial^2 u/\partial x_i^2$ on its left-hand side, $i = 1, 2, 3$, and the famous Navier-Stokes equations have 24 terms.

Let N denote the number of mass bodies. If $N = 0$, then the simplest solution of Einstein's equations is the Minkowski metric tensor (3.5). If $N = 1$, then there are several other simple solutions of (3.29) that use spherical or axial symmetry of one body, e.g. the Schwarzschild solutions (3.6) and (3.16), or the Kerr solution (see Remark 3.11 and [249, p. 878]). However, these solutions are local and not global solutions (cf. Theorem 3.9). Moreover, for two or more mass bodies no analytical solution of (3.29) is known, cf. Fig. 5.1. Thus we have a serious problem to verify whether Einstein's equations describe well the N-body problem with $N > 1$ (e.g. for the Solar system). For instance, Jupiter cannot be replaced by a test particle with zero mass as Einstein did for Mercury, compare with Example 4.3. Therefore, we are unable to reliably calculate any part of Jupiter's trajectory from Einstein's equations.

There are many numerical methods for solving partial differential equations such as the finite difference method, the finite volume method, the boundary element method, the finite element method (see e.g. [44, 191]), etc. For numerical solution of Einstein's equations we have to include back all arguments of the functions $g_{\mu\nu} = g_{\mu\nu}(x^0, x^1, x^2, x^3)$, $g_{\mu\nu,\varkappa} = g_{\mu\nu,\varkappa}(x^0, x^1, x^2, x^3)$, and $g_{\mu\nu,\varkappa\lambda} = g_{\mu\nu,\varkappa\lambda}(x^0, x^1, x^2, x^3)$ appearing in (5.3)–(5.5) for all $\mu, \nu, \varkappa, \lambda = 0, 1, 2, 3$. For simplicity, we will not take into account that Einstein's equations are mutually dependent, see Remark 3.9.

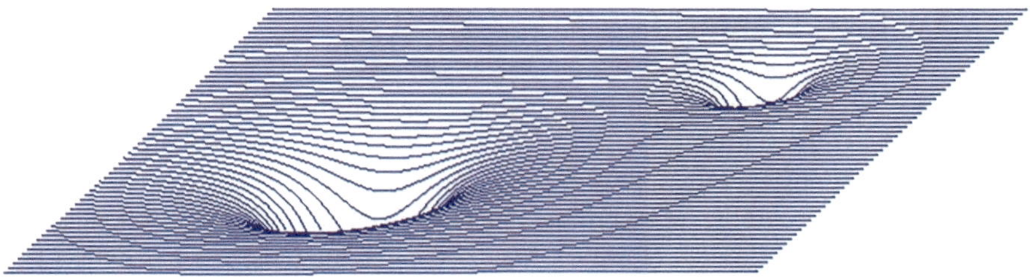

Fig. 5.1 Schematic illustration of space deformations by two interacting bodies

Consider, for instance, the following simplest setting of the finite difference method. One has to establish a four-dimensional regular space-time mesh, e.g., with n^4 mesh points

$$(x_i^0, x_j^1, x_k^2, x_l^3) \quad \text{for } i, j, k, l = 1, 2, \dots, n.$$

Then the 10 values $g_{\mu\nu}(x^0, x^1, x^2, x^3)$, their $40 = 10\times 4$ first derivatives and $100 = 10\times(1+2+3+4)$ second derivatives (of the Hessian) appearing in (5.3)–(5.5) have to be replaced by finite differences at all mesh points. For instance, the following second derivative appearing in the second line of (5.3) can be approximated by the standard central difference as

$$g_{00,11}(x_i^0, x_j^1, x_k^2, x_l^3)$$
$$\approx \frac{g_{00}(x_i^0, x_j^1 + h, x_k^2, x_l^3) - 2g_{00}(x_i^0, x_j^1, x_k^2, x_l^3) + g_{00}(x_i^0, x_j^1 - h, x_k^2, x_l^3)}{h^2},$$

where $h = n^{-1}$ is the discretization parameter. The huge system of nonlinear partial differential equations described in Sect. 5.1 would be then replaced by a much larger system of nonlinear algebraic equations for approximate values of the metric tensor at all mesh points at a given time level. In particular, at each mesh point the corresponding discrete Einstein equations will be much longer than Einstein's equations themselves written explicitly. Hence, if e.g. $n \approx 100$, then the discrete system will occupy millions of pages of extremely complicated and highly nonlinear algebraic equations.

It is well known that explicit numerical methods for solving evolution problems are unstable. Therefore, one should apply implicit methods. Nevertheless, up to now, we do not know any convergent and stable method that would yield a realistic numerical solution of the above system with guaranteed error bounds of discretization, iteration, and rounding errors.

5.4 Further Arguments

Remark 5.4 Classical relativistic tests are based on verification of very simple algebraic formulae (see e.g. (2.22), (3.7), and (4.2)) derived by various simplifications and approximations of the Schwarzschild solution (3.6) of Einstein's equations (3.29). This very special solution corresponds only to the exterior of one spherically symmetric nonrotating body. However, we cannot test good approximation properties of Einstein's equations (3.29) by means of one particular exterior Schwarzschild solution. Such an approach could be used only to disprove their good modeling properties of reality. Analogously, good modeling properties of the Laplace equation $\Delta u = 0$ cannot be verified by testing some of its trivial linear solution, since there exist infinitely many other nontrivial solutions and there are infinitely many other equations whose solutions satisfy $\Delta u = 0$. Similarly, there exist infinitely many equations having the Schwarzschild exterior solution (3.6), cf. [167, p. 269]. So which equations do we test?

From Sect. 1.5 we know that no equation of mathematical physics describes reality absolutely exactly on any scale. Therefore, each mathematical model has only a limited scope of its application. In particular, Einstein's equations with cosmological constant $\Lambda \neq 0$ (see (3.52)) also should not be applied to the entire universe as is often done, since they are nonlinear and thus not scale invariant (cf. Sect. 7.2). Note that the observable universe is at least 15 orders of magnitude larger than one astronomical unit.

Remark 5.5 In case of a curved spacetime it is not possible, in general, to define the global coordinates (x^0, x^1, x^2, x^3).

We should always state precisely in which admissible set we shall look for a solution of a particular problem. Let us illustrate it in the following very simple example.

Example 5.1 The equation $x^4 = 1$ has no solution on the set of even numbers, it has exactly one solution $x = 1$ on the set of positive integers, it has two solutions $x = \pm 1$ on the set of integers, and it has four solutions $\{\pm 1, \pm \mathrm{i}\}$ on the set of complex numbers.

Therefore, for solving problems of mathematical physics various function spaces were developed. For instance, the solution of the Poisson equation (see Example 3.8) with some boundary conditions is searched for in the Sobolev space H^1 of square integrable functions whose first generalized derivatives are square integrable, too (see e.g. [191]). Similarly, the solution of the Navier-Stokes equations is searched for in the space of divergence-free vector functions whose components are in H^1. On the other hand, in most works on the GTR it is not specified in which admissible set one should look for a solution, cf. e.g. [107, 249, 270, 272, 332, 337, 369, 371].

Remark 5.6 The left-hand side of Einstein's equations (3.52) contains the second derivatives of the metric tensor. However, will such an expression produce reliable results on cosmological distances? If the metric tensor of the physical spacetime can be more faithfully described by higher derivatives or another nonlinearity, we get a discrepancy of equations (3.52) with reality on cosmological scales.

Finally let us emphasize that Einstein's equations are fully deterministic whereas the universe does not operate solely gravitationally due to quantum phenomena. Their effects can be observed not only on microscopic scales. For instance, we may decide whenever we wish to change the trajectory of an asteroid in an arbitrary direction by the famous kinetic impactor method. The evolution of the physical world is thus quite unstable with respect to initial conditions, since small quantum fluctuations in our minds can produce large consequences. Such processes are not described by Einstein's equations. This reminds us of the classical philosophical problem of free will versus determinism.

Remark 5.7 Einstein's equations do not describe quantum phenomena governed by the *Pauli exclusion principle* which states that two or more identical particles with half-integer spins (i.e. fermions) cannot occupy the same quantum state within a quantum system simultaneously. The neutron is a spin 1/2 particle, that is, it is a fermion.

The main objections to the practical use of Einstein's equations for more than one axially symmetric body can be summarized as follows:

1. Their extreme complexity, see Sect. 5.1.
2. The impossibility to prescribe reliable initial conditions, see Sect. 5.2.
3. The impossibility to prescribe reliable boundary conditions, see Sect. 5.2.
4. The modeling error (see Fig. 1.9) of Einstein's equations is not known. Is it small or large?
5. The difficulty with the left-hand side of Einstein's equations which is divergence-free, see Remark 5.2.
6. The metric tensor does not uniquely define the topology of the corresponding spacetime manifold, see Remark 5.3.
7. The extreme numerical computational complexity, see Sect. 5.3.
8. No significant digit of the cosmological constant is known, see Remark 3.15. So what will we substitute for Λ when solving (3.52)?
9. Einstein's equations are fully deterministic whereas the physical world is stochastic, see Remark 5.6.
10. Since the classical relativistic tests are based only on the exterior Schwarzschild solution, we cannot guarantee that Einstein's equations describe the physical universe well, especially on very large or very small scales. These tests are based on verification of very simple algebraic formulae which are not mathematical consequences of Einstein's equations, see Remark 5.4.

Chapter 6
Friedmann Equation

6.1 Nonuniqueness of the Notion Universe

In this chapter we first derive in detail the Friedmann equation from Einstein's equations applied to the entire universe. Then we point out various drawbacks of this approach (see also the next chapter).

In 1584, Giordano Bruno wrote the treatise [50], where among other things he conjectured that the universe is infinite. From that time, opinions of the shape of the universe have often changed. Isaac Newton and many others understood the universe as Euclidean space $\mathbb{E}^n$ for $n = 3$.

In 1900, however, the German mathematician and physicist Karl Schwarzschild proposed (see [320, p. 66]) that the universe at any fixed time might be non-Euclidean and even finite, i.e. having a finite volume. He envisioned it as a three-dimensional manifold.

Note that non-Euclidean geometry arose in the first half of the nineteenth century during many attempts to understand the axiomatic construction of Euclidean geometry—especially in proving the independence of Euclid's fifth postulate of parallels. We include among its founders Carl Friedrich Gauss, Nikolai I. Lobachevskii, János Bolyai, Bernhard Riemann, Sophus Lie, Felix Klein, and many others. The history of the development of non-Euclidean geometries is described in detail in the review article [54].

By the *Copernican principle* humans on the Earth are not privileged observers of the universe. Thus, we are not at some umbilic point [192, p. 530] as in the tip of an egg which satisfies the required isotropy only at this special point and its antipodal point. According to *Einstein's cosmological principle*, our universe on each isochrone is homogeneous and isotropic on large scales. Roughly speaking, its curvature is constant at any point and in any direction.

More precisely, *homogeneity* is the assumed property of the universe that at every fixed time instant and on large spatial scales the universe appears the same

M. Křížek, L. Somer, *Mathematical Aspects of Paradoxes in Cosmology*,
https://doi.org/10.1007/978-3-031-31768-2_6

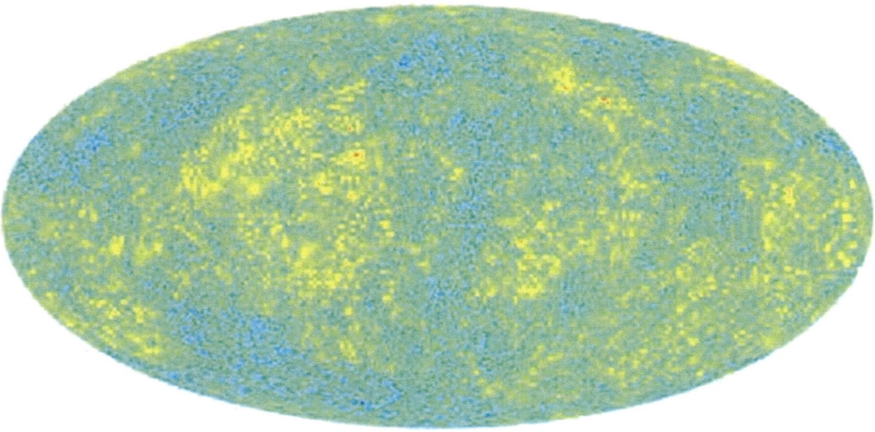

Fig. 6.1 Fluctuations in temperature ≈ 2.726 K of the cosmic microwave background (CMB) radiation that corresponds to the redshift $z = 1089$. This radiation arose when the space was 1090 times smaller than at present and its mean temperature was almost 3000 K (photo from the Planck satellite)

to all observers, wherever they are. In other words, at any fixed time the translation symmetry of the universe is required. This is, in fact, the Copernican principle.

Similarly, *isotropy* is the assumed property of the universe in which the universe at large spatial scales would seem to an observer at any point to be the same in all directions, i.e., for each fixed time the rotational symmetry of the universe is required.

Note that the term "Einstein's cosmological principle" was first introduced by E. A. Milne in 1935, see [247]. Already in 1922, C. Charlier writes that the cosmological principle was invented by Einstein, although he did not call it that.

Remark 6.1 The isotropy is continually verified by astronomers. For instance, it is confirmed by the cosmic microwave background radiation (CMB) which possesses tiny fluctuations of order 10^{-4} K from its mean temperature 2.7260 K on the whole celestial sphere, see Fig. 6.1. Also the well-known γ-ray bursts show almost uniform distribution on the celestial sphere. Pictures of the Hubble Deep Field, the Hubble Deep Field-South, the Hubble Ultra-Deep Field, the Hubble eXtreme Deep Field, Webb's First Deep Field, etc., illustrate this isotropy on very large cosmological scales as well.

Remark 6.2 In the Hubble test of the homogeneity of the universe, one has to measure the apparent magnitude (energy flux f from a given galaxy). By [56] the number of observed galaxies in the sky brighter than f should vary as $f^{-3/2}$, see also [25, 272], [369, Chapt. 14]. For a modification of this test to γ-ray bursts see [221]. There is no homogeneity in time due to the existence of the CMB and since quasars exist only at cosmological distances and galaxies evolve.

Note that isotropy at all points implies homogeneity at all points on the space manifold, see e.g. [370]. The converse statement is not true. For instance, a crystal of $CaCO_3$ is homogeneous but anisotropic. For the Gödel homogeneous and anisotropic universe we refer to [332, p. 384].

Note also that the universe on scales of about 100 million light-years consists of large-scale structures as giant sheets and long fibers (filaments) of galaxies. For example, a fiber consisting of many thousand galaxies with a total length of 1.37 billion light years was discovered, the so-called *Sloan Great Wall*. There is also the KBC void (roughly spherical) with diameter 2 billion light years and low mass density. This indicates that the assumption of homogeneity is not entirely correct.

Einstein's equations were not developed for a dynamical evolution of the universe. This was done later in 1922 by Alexander Friedmann [108] who derived from Einstein equations a nonlinear ordinary differential equation for the *expansion function* $a = a(t)$. He assumed that the universe can be described by an expanding three-dimensional sphere

$$\mathbb{S}_a^3 = \{(x, y, z, w) \in \mathbb{E}^4 \mid x^2 + y^2 + z^2 + w^2 = a^2\}$$

which enabled him to avoid boundary conditions. The symbol t stands for a *cosmic time* which is a measure of time by a physical clock with zero peculiar velocity in the absence of matter over-/under-densities (to prevent time dilation due to relativistic effects or confusions caused by expansion of the universe).

The current standard cosmological model of the evolution of our universe is based just on the Friedmann equation. However, Friedmann's description was very brief. Therefore, the main objective of this chapter is a detailed derivation of the Friedmann equation for an unknown expansion function $a = a(t)$ representing the radius of the universe. Note that the expansion function is sometimes also called the *scaling parameter* or *expansion factor*. A natural question arises whether we can apply Einstein's equations that are tested on scales of the Solar system to the entire universe, which is a 15 orders of magnitude larger object than the astronomical unit au.

Remark 6.3 The expansion of our universe is usually described by the hot Big Bang model. The term *Big Bang* was first used by Fred Hoyle in 1949. Nevertheless, the idea that the universe could have "zero radius" in the very distant past was formulated by Friedmann already in 1922. In the English translation of his article [108], footnote 11 states: *The time since the creation of the world is the time that has flowed from that instant when the space was one point* ($R = 0$) *until the present state* ($R = R_0$); *this time may also be infinite.* See the origin of coordinates in Fig. 6.2. Georges Lemaître came to the same conclusion 5 years later, see [218].

The term "universe" is used in cosmology with various meanings: true spacetime, true space (i.e. a part of spacetime at any fixed time instant), and the observable universe, which is seen as a projection on the celestial sphere. These are three different objects. Their mathematical models are also three completely different

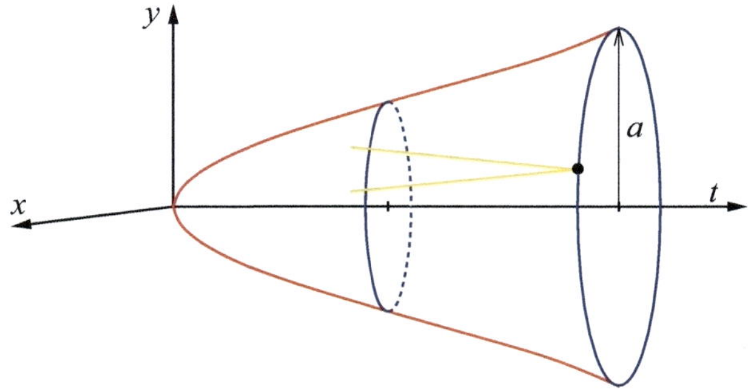

Fig. 6.2 Schematic illustration of three different manifolds that are used in the Big Bang model our universe with positive curvature index. For simplicity, the space dimensions are reduced by two. Hence, the sphere $\mathbb{S}_a^3$ with radius $a = a(t) > 0$ at any fixed time instant t is replaced only by its great (blue) circle $\mathbb{S}_a^1$ for $z = w = 0$. This is the model of the space (universe) with positive constant curvature $1/a$. The model of spacetime can be obtained by rotation of the (red) graph of the expansion function about the time axis t. The observable universe is marked by the (yellow) past light cone whose vertex corresponds to an observer. This light cone is deformed near the origin, but we do not know how. Each of these three models has a different center, see Remark 6.6

manifolds (see Fig. 6.2). Thus altogether we have $6 = 3 + 3$ meanings of the problematic notion "universe" for which the terminology is not fixed yet. The first three contain real matter, whereas the other three are only abstract mathematical idealizations of reality.

In accordance with the Einstein cosmological principle, we shall understand by the *universe* a cross-section of spacetime at any fixed time instant, i.e., the universe will be an isochrone in spacetime for constant t. For instance, if the curvature is positive (for a negative curvature see Theorem 6.5), then the corresponding model of the universe is the sphere $\mathbb{S}_a^3$ for some fixed radius $a = a(t) > 0$, which is a three-dimensional manifold in the four-dimensional Euclidean space $\mathbb{E}^4$ (cf. Fig. 6.3 below). The associated model of spacetime in $\mathbb{E}^5$ has dimension four and the model of the observable universe has dimension three (cf. Fig. 6.2).

All six above-mentioned objects have to be carefully distinguished; otherwise we may come to various confusions. For instance, the observable universe is not homogeneous, since for different cosmological redshifts z, it has a different mass density. Thus, it is an entirely different object than the universe as a space. From a similar reason the spacetime is also not homogeneous. Therefore, the expansion of the universe is a completely different notion than the expansion of the observable universe.

Albert Einstein is a part of the actual spacetime which is modeled by the red manifold in Fig. 6.2. However, he is not a part of the current physical universe not even of the observable universe modeled by the blue and yellow manifolds, respectively. We shall see that the Friedmann equation (6.24) describes the time behavior of the blue manifold, whereas all data were taken from the observable universe modeled by the yellow manifold.

Note that some galaxies are seen duplicate via gravitational lenses that corresponds to two different time instants. This nonuniqueness should be taken into account locally.

In the observable universe we cannot measure angles α, β, γ in some large triangle to ascertain the spherical, Euclidean, or hyperbolic geometry of the universe by means of their sum $\alpha+\beta+\gamma$. Such a triangle has to be considered in the universe (space), in which we see only our very close neighborhood (strictly speaking, only the one point at which we are situated). This limits our ability to perform such measurements.

Remark 6.4 Let us emphasize that the observable universe cannot be modeled by the three-dimensional Euclidean space, since it is a cone whose interior does not belong to it. The whole situation is sketched in Fig. 6.2, where two spatial dimensions are removed. We see that the observable universe can be modeled by a light cone which is deformed towards the origin of the time coordinate. Instead of spheres $\mathbb{S}_a^3$ for $a = a(t)$ we may consider only their great circles $\mathbb{S}_a^1$ having centers on the time axis.

Remark 6.5 It would be a mistake to believe that the known map of the cosmic microwave background (CMB) radiation from Fig. 6.1 shows the entire universe, how it looked like 380,000 years after the Big Bang. This map shows only a two-dimensional slice of a three-dimensional manifold corresponding to the universe for $z \approx 1089$ when its radius was $z + 1 = 1090$ times smaller than at present. This radius is the same as the radius of any great circle (e.g. equator) at the time when relict photons were emitted. Moreover, we observe everything only on the projection on the celestial sphere. For example, the relict radiation produced at that time in our neighborhood is not on the map of the cosmic microwave background radiation. We also do not see any relict radiation from the places where are to date all 10^{12} galaxies in the observable universe. At each of these galaxies we would observe at present completely different maps of the cosmic microwave background fluctuations, see [198, p. 277]. So on the Earth, an observer may have an idea about how only a tiny part of the early universe looked like.

Remark 6.6 We often hear that the universe has no center. This is similar to the statement that a circle has no center. The circle $x^2 + y^1 = 1$, of course, has its center even though it does not belong to it. In Fig. 6.2, the centers of the blue circles lie on the axis t. Therefore, also the model of the universe $\mathbb{S}_a^3$ has its center at the origin $(0, 0, 0, 0)$ of coordinates of the space $\mathbb{E}^4$ although $(0, 0, 0, 0) \notin \mathbb{S}_a^3$. On the other hand, the Earth is at the center (see the yellow vertex in Fig. 6.2) of the observable universe which is finite for any $k \in \{-1, 0, 1\}$. Its horizon can be modeled by a two-dimensional sphere. The center of the red model of the expanding universe corresponds to the Big Bang at an initial time (see also Fig. 9.10). The Big Bang thus happened everywhere.

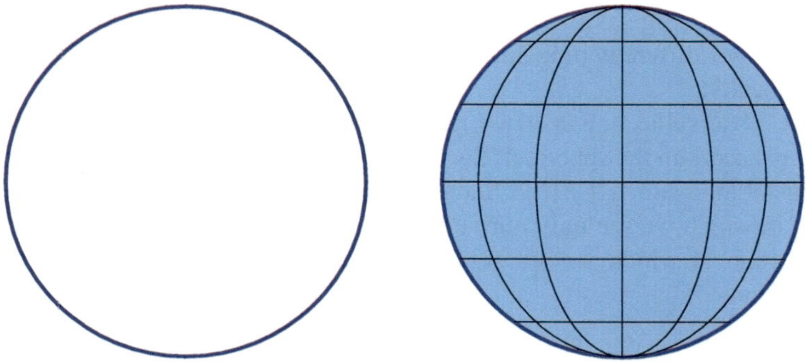

Fig. 6.3 The unit circle on the left illustrates the sphere $\mathbb{S}^1 = \{(x, y) \in \mathbb{E}^2 \mid x^2 + y^2 = 1\}$. The surface of the unit ball on the right is the sphere $\mathbb{S}^2 = \{(x, y, z) \in \mathbb{E}^3 \mid x^2 + y^2 + z^2 = 1\}$. It can be expressed using polar coordinates as follows: $x = r \cos \phi$, $y = r \sin \phi$, and the remaining third coordinate will be, by the Pythagorean Theorem, equal to $z = \pm\sqrt{1 - r^2}$, where $r \in [0, 1]$ and $\phi \in [0, 2\pi)$

For every $n = 1, 2, \ldots$ define an *n-dimensional sphere (hypersphere)* with radius $r > 0$ by

$$\mathbb{S}_r^n = \{(x_1, \ldots, x_{n+1}) \in \mathbb{E}^{n+1} \mid x_1^2 + \cdots + x_{n+1}^2 = r^2\}.$$

Note that an isometry is a continuous mapping, whose inverse exists and is also continuous, and which preserves distances on a given manifold.

Theorem 6.1 *For $r \leq R$ the sphere $\mathbb{S}_r^n$ can be isometrically imbedded into $\mathbb{S}_R^{n+1}$.*

The proof is constructive, but we will not present it, even though it is easy. For instance, on the right part of Fig. 6.3 there are two circles corresponding to parallels at $\pm 30°$ isometrically imbedded into the two-dimensional sphere $\mathbb{S}^2$. Thus, each sphere can be isometrically imbedded into a larger sphere if the dimension is increased by one. In other words, the sections of the sphere $\mathbb{S}_r^n$ by the hyperplanes $x_{n+1} = C$, where C is constant and $|C| < r$, are spheres of dimension $n - 1$.

6.2 How to Imagine the Sphere $\mathbb{S}^3$?

In accordance with the original article [108] published by Friedmann in 1922, we assume that at any fixed time t, the universe can be modeled by a three-dimensional sphere. From Sect. 6.1 we know that Karl Schwarzschild already in 1900 conjectured [320] that the universe can be described as a huge three-dimensional sphere. Albert Einstein also assumed that the gravitational interaction of mass causes a positive curvature and that the universe can be modeled by a three-dimensional sphere $\mathbb{S}_a^3$ with unchanging radius a, see [94, p. 152]. This manifold is maximally symmetric and the corresponding group of symmetries is the orthogonal group $O(4)$. This expresses the fact that our universe is homogeneous and isotropic on large scales. If $a = 1$ we will, for simplicity, omit the subscript a.

Contrary to e.g. a cylindrical surface, a two-dimensional sphere (see the right part of Fig. 6.3)

$$\mathbb{S}^2 = \{(x, y, z) \in \mathbb{E}^3 \,|\, x^2 + y^2 + z^2 = 1\}$$

cannot be isometrically unrolled into the Euclidean plane $\mathbb{E}^2$. Therefore, all maps of the Earth's surface have to be more or less distorted. Similarly, the curved sphere $\mathbb{S}^3$ cannot be mapped into the flat space $\mathbb{E}^3$ without any distortion of distances.

Further, we will present 10 independent ways to imagine the unit sphere $\mathbb{S}^3$, see [188]:

1. The sphere $\mathbb{S}^3$ is actually a three-dimensional surface of a four-dimensional ball, see Fig. 6.3 for its lower-dimensional analogues.
2. The second way uses parallel cuts. In the Euclidean space $\mathbb{E}^4$ consider a point with coordinates

$$\begin{aligned} x &= r \sin\theta \cos\phi, \\ y &= r \sin\theta \sin\phi, \\ z &= r \cos\theta, \\ w &= \pm\sqrt{1 - r^2}, \end{aligned} \tag{6.1}$$

   where $r \in [0, 1]$, $\theta \in [0, \pi]$, and $\phi \in [0, 2\pi)$. Then we easily find that

$$x^2 + y^2 + z^2 + w^2 = 1,$$

   and thus $(x, y, z, w) \in \mathbb{S}^3$. For clarity, we now cut[1] the sphere $\mathbb{S}^3$ by parallel planes $w = $const. for $|w| \leq 1$. This yields two-dimensional spheres with radii $r = \sqrt{1 - w^2}$ and centers $(0, 0, 0, w)$. Their union is $\mathbb{S}^3$. The points $(x, y, z) \in \mathbb{S}^2_r$ are then expressed in standard spherical coordinates defined by the first three equations of (6.1). The angle $\theta = 0$ corresponds to the North Pole and to the degenerated radius $r = 0$ (cf. Figs. 6.3 and 6.6).
3. The third way uses the following *hyperspherical coordinates*

$$\begin{aligned} x &= \sin\chi \sin\theta \cos\phi, \\ y &= \sin\chi \sin\theta \sin\phi, \end{aligned}$$

[1] If we similarly cut the unit sphere $\mathbb{S}^2$ by parallel planes $z = $const. for $|z| \leq 1$, we get circles (two of them are degenerated) with radii $r = \sqrt{1 - z^2}$ and centers $(0, 0, z)$, see the parallels in the right part of Fig. 6.3. The union of these parallels (circles) is the sphere $\mathbb{S}^2$.

$$z = \sin\chi\cos\theta,$$

$$w = \cos\chi,$$

where $\chi, \theta \in [0, \pi]$ and $\phi \in [0, 2\pi)$. It is easy to see that the equality $x^2 + y^2 + z^2 + w^2 = 1$ holds again. We observe that the hyperspherical coordinates are a natural generalization of the standard spherical coordinates for the unit sphere $\mathbb{S}^2$ taking $\chi = \pi/2$ (cf. the first three lines of (6.1) for $r = 1$).

Note that the manifold $\mathbb{S}^3$ with radius 1 can be divided into infinitely many spherical shells with radii $\sin\chi$ for $\chi \in [0, \pi]$. The corresponding surface areas are $4\pi\sin^2\chi$. Using the above hyperspherical coordinates, we find that

$$\mathrm{vol}(\mathbb{S}^3) = 4\pi\int_0^\pi \sin^2\chi\,\mathrm{d}\chi = 2\pi\Big[\chi - \frac{1}{2}\sin(2\chi)\Big]_0^\pi = 2\pi^2,$$

which leads to

$$\mathrm{vol}(\mathbb{S}_a^3) = 2\pi^2 a^3 \tag{6.2}$$

for an arbitrary radius $a > 0$ (cf. also [94, p. 152]).

4. The fourth way relies on local orthogonal projections. For simplicity, we first investigate the sphere $\mathbb{S}^2$. On $\mathbb{S}^2$ consider a small curved "square neighborhood" whose center is at the point $(0, 1, 0)$ and whose sides are parts of great circles passing through the poles $(0, 0, \pm 1)$ and the points $(\pm 1, 0, 0)$ on the equator. The left part of Fig. 6.4 shows the orthogonal projection of this neighborhood to the tangent plane $y = 1$, e.g. meridians are projected to converging vertical arcs.

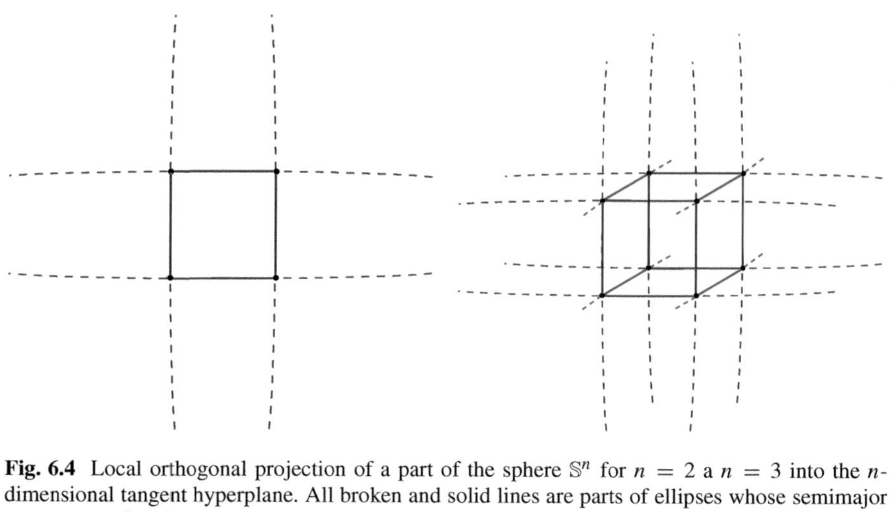

Fig. 6.4 Local orthogonal projection of a part of the sphere $\mathbb{S}^n$ for $n = 2$ a $n = 3$ into the n-dimensional tangent hyperplane. All broken and solid lines are parts of ellipses whose semimajor axes are equal to 1

Similarly, on the sphere $\mathbb{S}^3$ we may consider a small curved cube, whose edges are parts of great circles. The right part of Fig. 6.4 shows the orthogonal projection of this cube to the three-dimensional hyperplane tangent to $\mathbb{S}^3$ at the center of the curved cube.

5. The fifth way is the well-known stereographic projection. First, we again consider only the sphere $\mathbb{S}^2$. We will project from the North Pole $N = (0, 0, 1)$ all the remaining points of $\mathbb{S}^2$ onto the plane tangential to the South Pole $S = (0, 0, -1)$. Then the meridians of the sphere (see Fig. 6.3 right) are mapped to lines passing through S and the parallels are mapped on circles. Note that such a stereographic projection was used in the construction of the astronomical dial representing the celestial sphere at the Prague Astronomical Clock (Horologe), see [187, 194, 203]. This projection is a one-to-one mapping from $\mathbb{S}^2 \setminus N$ onto $\mathbb{E}^2$.

Similarly, for the sphere $\mathbb{S}^3$ we select two opposite poles $N = (0, 0, 0, 1)$ and $S = (0, 0, 0, -1)$. Then all points from $\mathbb{S}^3 \setminus N$ are projected from the North Pole N into a three-dimensional hyperplane tangent to $\mathbb{S}^3$ at the South Pole S. Great circles passing through N and S will be again mapped on lines passing through S, etc.

6. The sixth way uses the Gaussian plane of complex numbers $\mathbb{C}$. Setting

$$\boldsymbol{x} = x + \mathrm{i}y, \;\; \boldsymbol{z} = z + \mathrm{i}w \in \mathbb{C},$$

the sphere

$$\mathbb{S}^3 = \{(x, y, z, w) \in \mathbb{E}^4 \mid |x|^2 + |y|^2 + |z|^2 + |w|^2 = 1\}$$

can be expressed as a manifold resembling the unit circle in complex variables

$$\mathbb{S}^3 \cong \{(\boldsymbol{x}, \boldsymbol{z}) \in \mathbb{C}^2 \mid |\boldsymbol{x}|^2 + |\boldsymbol{z}|^2 = 1\}.$$

Moreover, we have $|\boldsymbol{x}|^2 = |\boldsymbol{x}^2|$ and $|\boldsymbol{z}|^2 = |\boldsymbol{z}^2|$.

7. The seventh way employs the quaternions which is a noncommutative algebraic field. Similarly as for the complex plane we can find that the sphere $\mathbb{S}^3$ is isomorphic to the quaternions with norm 1 (see [333]).

8. The eighth way uses special regular tessellations of $\mathbb{S}^3$ to imagine this manifold. We can tile it with congruent bounded regular "polyhedral" cells. These cells are curved, since the manifold is not Euclidean. Edges of these tiles are geodesics and their faces consist of families of geodesics. Regular tessellations then represent a certain mesh that shows how the manifold is bent.

Let $\mathbb{S}^2$ be the circumscribed sphere of the regular tetrahedron, octahedron, and icosahedron. Then the radial (i.e. central orthographic) projection from the center of $\mathbb{S}^2$ maps all edges onto $\mathbb{S}^2$. In this way we obtain regular triangular tessellations of $\mathbb{S}^2$, see Fig. 6.5. The sphere $\mathbb{S}^3$ can be tessellated similarly by five curved regular tetrahedra, or 16 curved regular tetrahedra, or 600 curved

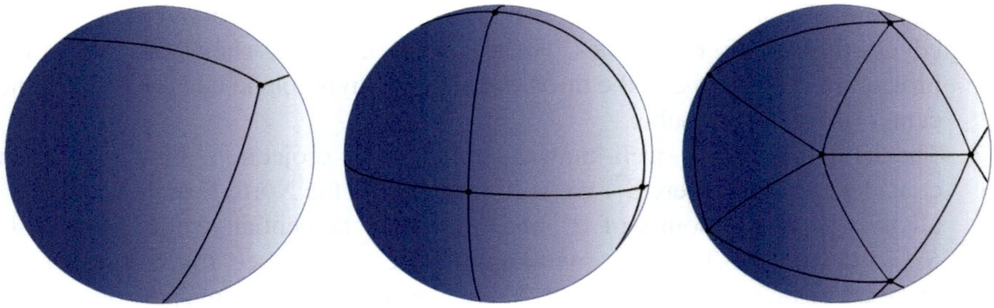

Fig. 6.5 Regular simplicial tessellations of the sphere $\mathbb{S}^2$: the projected tetrahedron, the projected octahedron, and the projected icosahedron. The corresponding vertex angles are 120°, 90°, and 72°, respectively

regular tetrahedra. Their dihedral angles at edges are again 120°, 90°, and 72°, respectively.

9. The sphere $\mathbb{S}^3$ can also be very roughly approximated by the three-dimensional surfaces of the 6 four-dimensional regular polytopes in $\mathbb{E}^4$. They are the 4-simplex, the 4-cube, the 4-orthoplex, the 24-cell, the 120-cell, and the 600-cell, see [44, p. 281].

10. The regular 120-cell whose three-dimensional surface consists of 120 regular dodecahedra quite well approximates $\mathbb{S}^3$ and it can be nicely modeled by soap bubbles, see [333].

Remark 6.7 Assume again that the universe can be modeled by the sphere $\mathbb{S}_a^3$ with fixed $a > 0$. Then the farthest point from the Earth in space (the so-called *antipodal point* [44]) is located at the intersection of all the great circles of $\mathbb{S}_a^3$ that pass through the Earth. Its distance from us is $\pi a \approx 3a$, i.e., it is approximately $3\times$ further than the present radius of the space.

Note that Kurt Gödel worked on a theory of a rotating universe [116]. For any odd dimension $n = 1, 3, 5, \dots$ the sphere $\mathbb{S}^n$ may rotate around its center such that all points have the same speed $|v| > 0$ (because $\mathbb{S}^n$ can be "combed" if n is odd). It is not a rotation around an axis as e.g. for $n = 2$. For $n = 3$ the corresponding centrifugal force could then contribute to the expansion of the universe (as in case of a rotating circle for $n = 1$).

Remark 6.8 The present total mean density of our universe is about $\rho \approx 10^{-26}$ kg/m^3, which can be achieved by having only a few protons per cubic meter. If the universe can be modeled by $\mathbb{S}^3$, its present radius is usually very roughly estimated to be $a \approx ct_0 \approx 10^{26}$ m, where $t_0 \approx 13.82$ Gyr is the present age of the universe according to [290]. Then by (6.2) its total mass would be

$$M = 2\pi^2 a^3 \rho \approx 2 \cdot 10^{53} \text{ kg}$$

and during its expansion its radius increases with time, but its topology cannot change. So during the Big Bang the topology of the cosmic (cosmological) egg should also be $\mathbb{S}_{a_0}^3$. If there exists the maximum possible mass density e.g. $\approx 10^{18}$ kg/m^3 (cf. (12.2) below), then the corresponding initial radius would be $a_0 \approx 2 \cdot 10^{11}$ m ≈ 1.3 au at the beginning. The expansion function $a = a(t)$ can also be oscillating with minimum a_0. Such a model does not contradict the Principle of causality as the Big Bang model does.

Recall that a *white hole* is a hypothetical region of spacetime that cannot be entered from outside (although energy-matter, light, and information can escape from it). It is the reverse of a black hole which can be entered only from outside. Some cosmologists compare the Big Bang to a white hole, since it cannot be entered from outside because we cannot travel against time. However, such a white hole would have the same topology as the entire space, i.e., a different topology than an ordinary black hole.

6.3 Metric Tensor for the Sphere $\mathbb{S}_a^3$

The metric tensor of spacetime has in general 10 independent entries. Now we show that due to the high symmetry of the sphere $\mathbb{S}_a^3$ all the non-diagonal entries of this tensor vanish.

From Sect. 6.2 we know that the unit sphere $\mathbb{S}^3$ is actually a three-dimensional hypersurface in four-dimensional Euclidean space $\mathbb{E}^4$. Let $P \in \mathbb{S}^3$ and let $P = P(r, \theta, \phi)$ whose coordinates satisfy (6.1). Similarly to [296, p. 253], the components of local coordinate vectors for $r < 1$ are

$$p_1(r, \theta, \phi) := \frac{\partial}{\partial r} P(r, \theta, \phi) = (\sin\theta\cos\phi, \sin\theta\sin\phi, \cos\theta, -r(1-r^2)^{-1/2}),$$

$$p_2(r, \theta, \phi) := \frac{\partial}{\partial \theta} P(r, \theta, \phi) = (r\cos\theta\cos\phi, r\cos\theta\sin\phi, -r\sin\theta, 0),$$

$$p_3(r, \theta, \phi) := \frac{\partial}{\partial \phi} P(r, \theta, \phi) = (-r\sin\theta\sin\phi, r\sin\theta\cos\phi, 0, 0),$$

where the vector p with components p_1, p_2, p_3 lies in the hyperplane tangent at the point $(x, y, z, w) \in \mathbb{S}^3$ (cf. Fig. 6.6 for a two-dimensional sphere).

The covariant coordinates of the symmetric metric tensor corresponding to the unit sphere $\mathbb{S}^3$ are given by the relations

$$\tilde{g}_{\alpha\beta} = p_\alpha \cdot p_\beta \quad \text{for} \quad \alpha, \beta = 1, 2, 3,$$

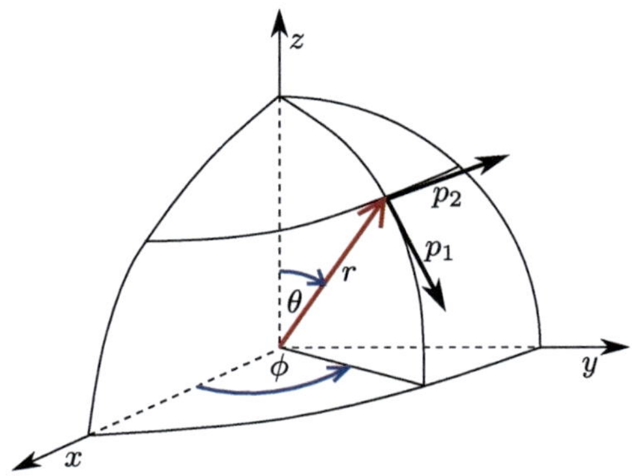

Fig. 6.6 Local coordinate orthogonal vectors p_1 and p_2 in the plane tangent at the point $(x, y, z) \in \mathbb{S}^2$ given in the standard spherical coordinates (r, ϕ, θ), cf. (6.1)

where $\cdot$ denotes the scalar product. Hence, it follows that

$$\tilde{g}_{11} = \sin^2\theta\cos^2\phi + \sin^2\theta\sin^2\phi + \cos^2\theta + \frac{r^2}{1-r^2} = 1 + \frac{r^2}{1-r^2} = \frac{1}{1-r^2},$$
$$\tilde{g}_{22} = r^2\cos^2\theta\cos^2\phi + r^2\cos^2\theta\sin^2\phi + r^2\sin^2\theta = r^2,$$
$$\tilde{g}_{33} = r^2\sin^2\theta\sin^2\phi + r^2\sin^2\theta\cos^2\phi = r^2\sin^2\theta,$$

and the other entries vanish, because $p_1 \cdot p_2 = p_1 \cdot p_3 = p_2 \cdot p_3 = 0$ due to the symmetry of $\mathbb{S}^3$. The metric tensor corresponding to the unit sphere $\mathbb{S}^3$ is therefore given by the diagonal matrix

$$\tilde{g}_{ij} = \begin{pmatrix} \frac{1}{1-r^2} & 0 & 0 \\ 0 & r^2 & 0 \\ 0 & 0 & r^2\sin^2\theta \end{pmatrix}, \quad i, j = 1, 2, 3, \tag{6.3}$$

where $r \in [0, 1)$, $\theta \in [0, \pi]$, and $\phi \in [0, 2\pi)$. Note that the Schwarzschild exterior solution (3.6) is spherically symmetric in the spatial coordinates with respect to one point only. On the other hand, (6.3) is spherically symmetric at all points and this describes the isotropy of the model $\mathbb{S}^3$. Note that the metric tensor of $\mathbb{S}^3$ in the hyperspherical coordinates is given in [178, p. 115].

The metric tensor associated to the entire spacetime manifold is then also given by the diagonal matrix

$$g_{\alpha\beta} = \begin{pmatrix} 1 & 0 & 0 & 0 \\ 0 & \frac{-a^2}{1-r^2} & 0 & 0 \\ 0 & 0 & -a^2r^2 & 0 \\ 0 & 0 & 0 & -a^2r^2\sin^2\theta \end{pmatrix}, \quad \alpha, \beta = 0, 1, 2, 3, \tag{6.4}$$

with the opposite signature than in (6.3) (cf. also (3.5)). The zero index corresponds to the time coordinate and the remaining three indices correspond to the spatial coordinates.

In the sequel, the Latin indices will run through the set $\{1, 2, 3\}$, while the Greek indices through the set $\{0, 1, 2, 3\}$.

Remark 6.9 In the theory of relativity, the metric associated to (6.4) is usually written using infinitesimally small quantities as follows:

$$ds^2 = c^2 dt^2 - a^2(t)\Big[\frac{dr^2}{1-r^2} + r^2(d\theta^2 + \sin^2\theta \, d\phi^2)\Big], \tag{6.5}$$

where the time coordinate fulfills

$$x_0 = ct$$

and $r \in [0, 1)$ is dimensionless. The relation (6.5) of the space-time manifold modeling the evolution of the universe for a special expansion function $a = a(t)$ first appeared in the article by Howard Percy Robertson [302, p. 826] in 1929. The same formula was also published by Arthur Geoffrey Walker [367, p. 121] in 1936. Therefore, the relation (6.5) is called the *Robertson-Walker metric*. Sometimes it is also called the Friedmann-Lemaître-Robertson-Walker metric or just *FLRW-metric* due to the pioneering papers [108] and [218]. Note that Walker does not cite Robertson's article [302], but only some of Robertson's later works from Astrophys. J. 82 (1935), 284–301; 83 (1936), 187–201, 257–271. Consequently, the priority belongs to Robertson.

For the entries of the covariant metric tensor (6.4) we thus have

$$g_{00} = 1, \quad g_{0i} = 0, \quad g_{ij} = -a^2(t)\tilde{g}_{ij} \quad \text{for } i, j = 1, 2, 3, \tag{6.6}$$

where $\tilde{g}_{ij}$ is the metric tensor (6.3) of the unit sphere $\mathbb{S}^3$, i.e., the tilde actually corresponds only to the expression in brackets in relation (6.5).

Since the inverse matrix to (6.4) is also diagonal, for the entries of the contravariant symmetric metric tensor we have

$$g^{00} = 1, \quad g^{0i} = 0, \quad g^{ij} = -\frac{1}{a^2}\tilde{g}^{ij}, \tag{6.7}$$

where

$$\tilde{g}^{11} = 1 - r^2, \quad \tilde{g}^{22} = \frac{1}{r^2}, \quad \tilde{g}^{33} = \frac{1}{r^2 \sin^2\theta}, \quad \tilde{g}^{ij} = 0 \quad \text{for} \quad i \neq j,$$

$r \neq 0$, and $\theta \neq 0$.

6.4 Calculation of the Christoffel Symbols

Entries of the Ricci tensor which appears in Einstein's equations (3.29) are defined by means of the Christoffel symbols (cf. (3.4)). In the covariant form (i.e. only with lower indices), the Christoffel symbols of the first kind are by (3.8) defined as follows

$$\Gamma_{\alpha\beta\gamma} = \frac{1}{2}(g_{\alpha\beta,\gamma} + g_{\gamma\alpha,\beta} - g_{\beta\gamma,\alpha}), \tag{6.8}$$

where the index after the comma indicates, for simplicity, the position number of a variable according to which we differentiate. By (3.9) the Christoffel symbols are symmetric in the second and third index,

$$\Gamma_{\alpha\beta\gamma} = \Gamma_{\alpha\gamma\beta} \quad \text{for } \alpha, \beta, \gamma = 0, 1, 2, 3. \tag{6.9}$$

Therefore, there are generally only $4 \times (1 + 2 + 3 + 4) = 40$ Christoffel symbols. Now let us calculate all of them for the sphere $\mathbb{S}_a^3$.

From (6.8) and (6.4) we see that

$$\Gamma_{000} = \frac{1}{2} g_{00,0} = 0.$$

Since the off-diagonal entries in (6.4) vanish and the derivative of a constant function is zero, we obtain by (6.8) for $i = 1, 2, 3$,

$$\Gamma_{0i0} = \Gamma_{00i} = \frac{1}{2}(g_{0i,0} + g_{00,i} - g_{i0,0}) = 0.$$

Similarly, we get

$$\Gamma_{i00} = \frac{1}{2}(g_{i0,0} + g_{0i,0} - g_{00,i}) = 0. \tag{6.10}$$

All of the 7 Christoffel symbols, in which at least two indices are zeros, are thus equal to 0.

Next, assume that just one index is zero. Then, according to (6.6) for $i, j = 1, 2, 3$ we get only 6 Christoffel symbols due to the symmetry (6.9),

$$\Gamma_{0ij} = \frac{1}{2}(g_{0i,j} + g_{j0,i} - g_{ij,0}) = -\frac{1}{2} g_{ij,0} = a a_{,0} \tilde{g}_{ij},$$

where

$$a_{,0} = \frac{\partial a}{\partial x_0} = \frac{1}{c}\frac{\partial a}{\partial t} \tag{6.11}$$

is a dimensionless quantity. For the additional $3 \times 3 = 9$ Christoffel symbols we have by (6.6)

$$\Gamma_{i0j} = \Gamma_{ij0} = \frac{1}{2}(g_{ij,0} + g_{0i,j} - g_{j0,i}) = \frac{1}{2}g_{ij,0} = -aa_{,0}\tilde{g}_{ij}, \tag{6.12}$$

since we again differentiate with respect to the time variable.

Finally, assume that all indices are nonzero, i.e., we will differentiate only with respect to spatial coordinates. Then by (6.8) and (6.6), we have

$$\Gamma_{ijk} = \frac{1}{2}(g_{ij,k} + g_{ki,j} - g_{jk,i}) = -\frac{a^2}{2}(\tilde{g}_{ij,k} + \tilde{g}_{ki,j} - \tilde{g}_{jk,i}) =: -a^2\tilde{\Gamma}_{ijk}$$

and the number of these remaining Christoffel symbols is $3 \times 6 = 18$. Hence, we have established in total $7 + (6 + 9) + 18 = 40$ values of the Christoffel symbols (which are generally different).

Next we shall use Einstein's summation convention for summation over repeating upper and lower index. To get once contravariant and twice covariant Christoffel symbol (i.e. with one upper and two lower indices), we use the following definition (cf. [249, p. 340])

$$\Gamma^{\alpha}{}_{\beta\gamma} = g^{\alpha\lambda}\Gamma_{\lambda\beta\gamma}$$

with summation over $\lambda = 0, 1, 2, 3$. In particular, we find that

$$\Gamma^{0}{}_{\beta\gamma} = g^{0\lambda}\Gamma_{\lambda\beta\gamma} = g^{00}\Gamma_{0\beta\kappa} = \Gamma_{0\beta\gamma},$$

since by (6.7) the off-diagonal entries of the contravariant metric tensor vanish and $g^{00} = 1$. Similarly for $i, j = 1, 2, 3$ we have by (6.6)

$$\Gamma^{i}{}_{\beta\gamma} = g^{i\alpha}\Gamma_{\alpha\beta\gamma} = g^{ij}\Gamma_{j\beta\gamma} = -\frac{1}{a^2}\tilde{g}^{ij}\Gamma_{j\beta\gamma},$$

as $g^{i0} = 0$. Furthermore, from the equality (6.10) we have

$$\Gamma^{i}{}_{00} = 0$$

and by (6.12) we obtain

$$\Gamma^{i}{}_{0j} = -\frac{1}{a^2}\tilde{g}^{ik}\Gamma_{k0j} = -\frac{1}{a^2}\tilde{g}^{ik}(-aa_{,0})\tilde{g}_{jk} = \delta^{i}{}_{j}\frac{a_{,0}}{a}, \tag{6.13}$$

where $\delta^{i}{}_{k}$ is the Kronecker delta, i.e. $\delta^{1}{}_{1} = \delta^{2}{}_{2} = \delta^{3}{}_{3} = 1$ and $\delta^{i}{}_{j} = 0$ for $i \neq j$. From relations (6.7) it follows that

$$\Gamma^{0}{}_{00} = 0, \quad \Gamma^{0}{}_{0i} = 0, \quad \Gamma^{i}{}_{00} = g^{i\alpha}\Gamma_{\alpha 00} = 0, \quad \Gamma^{0}{}_{ij} = aa_{,0}\tilde{g}_{ij} \tag{6.14}$$

and

$$\Gamma^i{}_{jk} = g^{i\beta}\Gamma_{\beta jk} = g^{i\ell}\Gamma_{\ell jk} = -\frac{1}{a^2}\tilde{g}^{i\ell}(-a^2\tilde{\Gamma}_{\ell jk}) = \tilde{g}^{i\ell}\tilde{\Gamma}_{\ell jk} = \tilde{\Gamma}^i{}_{jk}. \tag{6.15}$$

We see that in calculating the Christoffel symbols (6.15), the expansion function a is canceled out.

6.5 Calculation of the Ricci Tensor

First note that

$$\tilde{g}^{ij}\tilde{g}_{jk} = \delta^i{}_k \quad \text{for} \quad i, j, k = 1, 2, 3,$$

where $\delta^i{}_k$ is the Kronecker delta. Using Einstein's summation convention, we thus have

$$\delta^i{}_i = 3 \qquad \text{and} \qquad \delta^i{}_j\delta^j{}_i = 3, \tag{6.16}$$

which can be seen by a step-by-step substitution for $i, j = 1, 2, 3$.

Further recall the definition of the Ricci tensor (3.4)

$$R_{\alpha\beta} = R_{\beta\alpha} = R^\gamma{}_{\alpha\gamma\beta} = \Gamma^\gamma{}_{\alpha\beta,\gamma} - \Gamma^\gamma{}_{\alpha\gamma,\beta} + \Gamma^\lambda{}_{\alpha\beta}\Gamma^\gamma{}_{\lambda\gamma} - \Gamma^\lambda{}_{\alpha\gamma}\Gamma^\gamma{}_{\beta\lambda}, \tag{6.17}$$

where $R^\gamma{}_{\alpha\lambda\beta}$ is the *Riemann curvature tensor* defined by

$$R^\gamma{}_{\alpha\mu\beta} = \Gamma^\gamma{}_{\alpha\beta,\mu} - \Gamma^\gamma{}_{\alpha\mu,\beta} + \Gamma^\lambda{}_{\alpha\beta}\Gamma^\gamma{}_{\lambda\mu} - \Gamma^\lambda{}_{\alpha\mu}\Gamma^\gamma{}_{\beta\lambda}.$$

From this and (6.13)–(6.16) we come to

$$\begin{aligned} R_{00} &= \Gamma^\gamma{}_{00,\gamma} - \Gamma^\gamma{}_{0\gamma,0} + \Gamma^\lambda{}_{00}\Gamma^\gamma{}_{\lambda\gamma} - \Gamma^\lambda{}_{0\gamma}\Gamma^\gamma{}_{0\lambda} = -\Gamma^i{}_{0i,0} - \Gamma^i{}_{0j}\Gamma^j{}_{0i} \\ &= -\Big(\frac{a_{,0}}{a}\delta^i{}_i\Big)_{,0} - \Big(\frac{a_{,0}}{a}\delta^i{}_j\Big)\Big(\frac{a_{,0}}{a}\delta^j{}_i\Big) \\ &= -3(a_{,0}/a)_{,0} - 3(a_{,0}/a)^2 = -3\frac{a_{,00}}{a}, \end{aligned} \tag{6.18}$$

where $a_{,00}$ denotes the second derivative with respect to time variable x_0, and

$$\begin{aligned} R_{0i} &= \Gamma^\gamma{}_{0i,\gamma} - \Gamma^\gamma{}_{0\gamma,i} + \Gamma^\lambda{}_{0i}\Gamma^\gamma{}_{\lambda\gamma} - \Gamma^\lambda{}_{0\gamma}\Gamma^\gamma{}_{i\lambda} \\ &= \Gamma^j{}_{0i,j} - \Gamma^j{}_{0j,i} + \Gamma^j{}_{0i}\Gamma^k{}_{jk} - \Gamma^j{}_{0k}\Gamma^k{}_{ij} \\ &= \Gamma^j{}_{0i}\Gamma^k{}_{jk} - \Gamma^j{}_{0k}\Gamma^k{}_{ij} = \frac{a_{,0}}{a}\delta^j{}_k\tilde{\Gamma}^k{}_{jk} - \frac{a_{,0}}{a}\delta^j{}_k\tilde{\Gamma}^k{}_{ij} \\ &= \frac{a_{,0}}{a}(\tilde{\Gamma}^k{}_{ik} - \tilde{\Gamma}^j{}_{ij}) = 0, \end{aligned}$$

where we used the fact that ${\Gamma^j}_{0i,j} = {\Gamma^j}_{0j,i} = 0$, because derivatives with respect to spatial coordinates vanish due to (6.13). Note that we will not employ the equality of $R_{0i} = 0$ to derive the Friedmann equation.

Finally, according to (6.13)–(6.17) for $\alpha, \beta = 1, 2, 3$ we have

$$\begin{aligned} R_{ij} &= {\Gamma^\gamma}_{ij,\gamma} - {\Gamma^\gamma}_{i\gamma,j} + {\Gamma^\lambda}_{ij}{\Gamma^\gamma}_{\lambda\gamma} - {\Gamma^\lambda}_{i\gamma}{\Gamma^\gamma}_{j\lambda} \\ &= {\Gamma^0}_{ij,0} + \underline{{\tilde\Gamma^k}_{ij,k}} - \underline{{\tilde\Gamma^k}_{ik,j}} + {\Gamma^0}_{ij}{\Gamma^k}_{0k} \\ &\quad + \underline{{\tilde\Gamma^k}_{ij}{\tilde\Gamma^\ell}_{k\ell}} - {\Gamma^0}_{ik}{\Gamma^k}_{j0} - {\Gamma^k}_{i0}{\Gamma^0}_{jk} - \underline{{\tilde\Gamma^k}_{i\ell}{\tilde\Gamma^\ell}_{jk}} \\ &= \tilde R_{ij} + (a_{,00}a + a_{,0}^2)\tilde g_{0j} + 3a_{,0}^2\tilde g_{ij} - a_{,0}^2\tilde g_{ij} - a_{,0}^2\tilde g_{ij} \\ &= \tilde R_{ij} + (a_{,00}a + 2a_{,0}^2)\tilde g_{ij} = (a_{,00}a + 2a_{,0}^2 + 2)\tilde g_{ij}, \end{aligned} \tag{6.19}$$

where the sum of the underlined terms defines the Ricci tensor $\tilde R_{ij} = 2\tilde g_{ij}$ (see [369, p. 383]), which has nonzero entries only on the diagonal due to the maximal symmetry of the manifold $\mathbb{S}^3$, see (6.3).

Remark 6.10 We see that the Ricci tensor contains no spatial derivatives of the metric tensor, although according to (6.17) it is defined by the first derivatives of the Christoffel symbols and these are by (6.8) defined by the first derivatives of the metric tensor. The reason is the very large symmetry of the sphere $\mathbb{S}^3$ (see [369]).

6.6 The Friedmann Equation for the Sphere $\mathbb{S}_a^3$

Now we derive the Friedmann equation from Einstein's equations (3.52). For simplicity, suppose first that the cosmological constant is $\Lambda = 0$. Then Einstein's equations have the form

$$R_{\mu\nu} - \frac{1}{2}Rg_{\mu\nu} = \frac{8\pi G}{c^4}T_{\mu\nu}, \tag{6.20}$$

where the left-hand side is the Einstein tensor, $R = g^{\mu\nu}R_{\mu\nu}$ is the Ricci scalar, the entries of the metric tensor $g_{\mu\nu}$ are unknown *gravitational potentials*, and $T_{\mu\nu}$ is the tensor of density of energy and momentum. Since $\mathbb{S}_a^3$ is maximally symmetric, all the tensors in (6.20) are diagonal and (see [107, p. 176])

$$T_{00} = T_0^0 = T^{00} = \rho c^2, \quad T_1^1 = T_2^2 = T_3^3 = -p,$$

where the mass density $\rho = \rho(t)$ and the isotropic pressure $p = p(t)$ are supposed to be continuous and independent of the space coordinates and ρc^2 is the density of energy, see e.g. [222, p. 108], [249, p. 728]. This corresponds to a perfect cosmic fluid (with zero viscosity). From this and (3.33) we have

$$R_{\mu\nu} = \frac{8\pi G}{c^4}\left(T_{\mu\nu} - \frac{1}{2}T g_{\mu\nu}\right), \tag{6.21}$$

where

$$T = T_0^0 + T_i^i = \rho c^2 - 3p$$

is the trace and $R = -8\pi GT/c^4$. The 00-entry of these specially modified Einstein's equations (6.21) takes by (6.18) and (6.6) the form

$$-3\frac{a_{,00}}{a} = \frac{8\pi G}{c^4}\Big(\rho c^2 - \frac{1}{2}g_{00}(\rho c^2 - 3p)\Big) = \frac{4\pi G}{c^2}\Big(\rho + \frac{3p}{c^2}\Big).$$

From this and (6.20), we get the linear differential equation of the 2nd order

$$\frac{\ddot{a}}{a} = -\frac{4\pi G}{3}\Big(\rho + \frac{3p}{c^2}\Big), \tag{6.22}$$

where the dot denotes the time derivative. This equation is sometimes called the *linear Friedmann equation*, even though it does not appear in the original paper [108] by Friedmann.

Einstein's equations (6.20) are nonlinear, in general, due to (6.17). Notice how the nonlinearity elegantly disappears in (6.18) for the sphere $\mathbb{S}_a^3$.

Note that the other three equations $R_{0i} = 8\pi Gc^{-4}\big(T_{0i} - \frac{1}{2}Tg_{0i}\big) = 0$ for $i = 1, 2, 3$ do not give us any useful relationship for the unknown functions a, ρ, p.

Finally, according to (6.19), (6.7), and (6.21), we have

$$\begin{aligned}(a_{,00}a + 2a_{,0}^2 + 2)\tilde{g}_{ij} = R_{ij} &= \frac{8\pi G}{c^4}\Big(- p g_{ij} - \frac{1}{2}g_{ij}(\rho c^2 - 3p)\Big)\\ &= \frac{4\pi G}{c^4}g_{ij}(p - \rho c^2) = \frac{4\pi G}{c^4}(-a^2\tilde{g}_{ij})(p - \rho c^2).\end{aligned}$$

Thus from (6.11) we obtain

$$(a\ddot{a} + 2\dot{a}^2 + 2c^2)\tilde{g}_{ij} = \frac{4\pi Ga^2}{c^2}(\rho c^2 - p)\tilde{g}_{ij}.$$

This is altogether six equations, but the three corresponding to the diagonal entries are the same and the remaining three corresponding to non-diagonal entries vanish and lead to the unusable identity $0 = 0$. So we obtain only one equation

$$a\ddot{a} + 2\dot{a}^2 + 2c^2 = 4\pi G a^2\Big(\rho - \frac{p}{c^2}\Big). \tag{6.23}$$

Hence, together with (6.22), we get two equations for three unknowns a, ρ, p. As the third equation we may consider, e.g. the state equation $p = p(\rho)$.

Now multiply Eq. (6.22) by the function a^2 and subtract it from (6.23). Then we find that the second derivative $\ddot{a}$ disappears and

$$2\dot{a}^2 + 2c^2 = 4\pi G a^2\Big(\rho - \frac{p}{c^2}\Big) + \frac{4\pi G a^2}{3}\Big(\rho + \frac{3p}{c^2}\Big),$$

which after a simple adjustment leads to the sought nonlinear *Friedmann equation* of the first order

$$\frac{\dot{a}^2}{a^2} + \frac{c^2}{a^2} = \frac{8\pi G\rho}{3} \tag{6.24}$$

which does not contain the pressure p.

Note that (6.22)–(6.24) are three equations, but only two of them are independent. Let us further point out that Lemaître in [218, p. 52] derived the same equation (6.24) without citing Friedmann.

Remark 6.11 The case of a nonzero cosmological constant $\Lambda \neq 0$ can be formally rewritten by appropriate transformations (6.25) to the case $\Lambda = 0$. For $\Lambda \neq 0$ let us set

$$\overline{p} := p - \frac{\Lambda c^4}{8\pi G} \quad \text{and} \quad \overline{\rho} := \rho + \frac{\Lambda c^2}{8\pi G}, \tag{6.25}$$

i.e., the pressure $\overline{p}$ can also attain negative values. We see that $\overline{p} + \overline{\rho}c^2 = p + \rho c^2$ and that the equation

$$R_{\mu\nu} - \frac{1}{2}Rg_{\mu\nu} - \Lambda g_{\mu\nu} = \frac{8\pi G}{c^4}T_{\mu\nu} = \frac{8\pi G}{c^4}((p + \rho c^2)U_\mu U_\nu - pg_{\mu\nu}),$$

where $U_\mu = (1, 0, 0, 0)$, changes into the form (6.20) without the cosmological constant

$$R_{\mu\nu} - \frac{1}{2}Rg_{\mu\nu} = \frac{8\pi G}{c^4}((\overline{p} + \overline{\rho}c^2)U_\mu U_\nu - \overline{p}g_{\mu\nu}).$$

Then we obtain like in (6.22)

$$\frac{\ddot{a}}{a} = -\frac{4\pi G}{3}\Big(\overline{\rho} + \frac{3\overline{p}}{c^2}\Big) = -\frac{4\pi G}{3}\Big(\rho + \frac{3p}{c^2} - \frac{\Lambda c^2}{4\pi G}\Big)$$

and instead of the Friedmann equation (6.24) we get

$$\frac{\dot{a}^2}{a^2} + \frac{c^2}{a^2} = \frac{8\pi G\overline{\rho}}{3} = \frac{8\pi G\rho}{3} + \frac{\Lambda c^2}{3}. \tag{6.26}$$

The corresponding graph of the expansion function $a = a(t)$ is given in [197, p. 70].

6.7 Division by Zero in Cosmological Parameters

For a given smooth positive expansion function $a = a(t)$ define the *Hubble parameter* by

$$H(t) := \frac{\dot{a}(t)}{a(t)}. \tag{6.27}$$

It can be thus expressed as the time derivative of the natural logarithm in the following way

$$H(t) = \frac{\mathrm{d}\left(\ln\frac{a(t)}{a_0}\right)}{\mathrm{d}t} = \frac{\mathrm{d}(\ln a(t))}{\mathrm{d}t},$$

where $a_0 > 0$ is an arbitrary length constant and thus the argument $a(t)/a_0$ is dimensionless.

Since the Friedmann equation (6.26) contains too small terms, we divide it by the square $H^2 \geq 0$ as is usually done in the literature on cosmology, i.e., without any preliminary warning that we may possibly divide by zero. This may lead to various paradoxes (see e.g. Remarks 6.12 and 6.13 below, and [173, 196, 198]). Then we get the so-called *normalized Friedmann equation*

$$1 = \Omega_{\mathrm{M}}(t) + \Omega_\Lambda(t) + \Omega_{\mathrm{K}}(t) \tag{6.28}$$

for three normalized cosmological parameters, which are defined as follows

$$\Omega_{\mathrm{M}}(t) := \frac{8\pi G\rho(t)}{3H^2(t)} > 0, \quad \Omega_\Lambda(t) := \frac{\Lambda c^2}{3H^2(t)}, \quad \Omega_{\mathrm{K}}(t) := -\frac{kc^2}{\dot{a}^2(t)}, \tag{6.29}$$

where $k \in \{-1, 0, 1\}$ is the *curvature index* ($k = 1$ in the case of sphere). Saul Perlmutter calls the parameter Ω_{M} the *mass density* and Ω_Λ the *vacuum energy density* (see [279]). The parameter Ω_{K} is called the *curvature parameter* in [289] (or the curvature density). The parameter Ω_{M} is also called the density of dark and baryonic matter and Ω_Λ the density of dark energy. Note that (6.28) is really

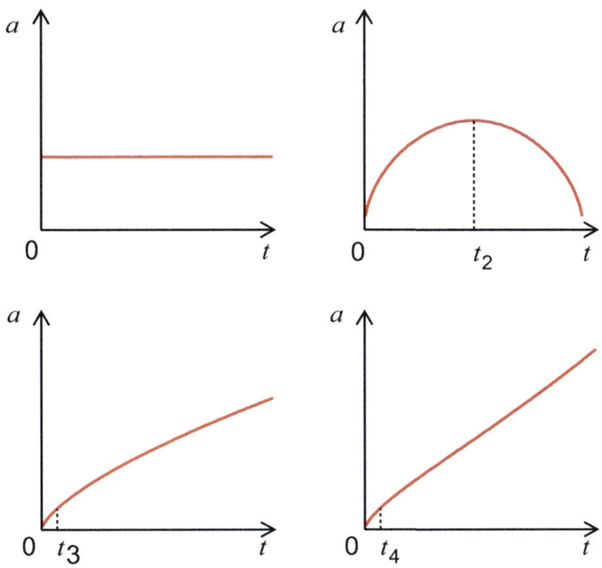

Fig. 6.7 The expansion function for Einstein's static universe, the cyclic universe (the expansion stops at some time $t_2 > 0$ and then starts to shrink), the universe with zero cosmological constant, and for the currently proposed expansion of the universe with a positive cosmological constant. Here $t_4 = \tau$ denotes the time instant of the origin of the cosmic microwave background radiation

a differential equation, since the derivative $\dot{a}$ is hidden in the Hubble parameter appearing in (6.29).

Remark 6.12 The radius a of Einstein's static universe follows directly from the Friedmann equation (6.26). For $\Lambda > 0$ the Einstein static solution (see Fig. 6.7)

$$a(t) \equiv \Lambda^{-1/2}$$

solves the Friedmann equation for $\rho = \Lambda c^2/(4\pi G)$. Since $\dot{a}(t) \equiv 0$ the mass density parameter Ω_M attains by (6.27) and (6.29) an infinite value, even though nothing unusual happens. We should properly say that this parameter is not well defined as its denominator is zero. We divide by zero also in the case of the other two parameters from (6.29).

Remark 6.13 By (6.29) we see that

$$\dot{a}(t) \to 0 \quad \Longrightarrow \quad \Omega_M(t) \to \infty \quad \text{and} \quad \Omega_\Lambda(t) \to \pm\infty \quad \text{for} \quad \Lambda \neq 0,$$

corresponding e.g. to an oscillating universe, which is also called a *cyclic universe* (see Fig. 6.7), where the expansion function repeatedly increases and then decreases [348]. The reason is that the derivative $\dot{a}$ vanishes at the point of maximum. Hence,

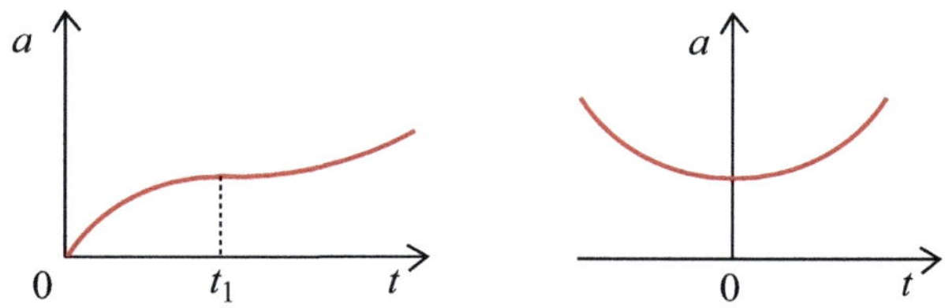

Fig. 6.8 Graph of the expansion function $a = a(t)$ for (1) a loitering universe for which there exists $t_1 > 0$ such that $\dot{a}(t_1) = 0$, and (2) a bouncing universe, where $\dot{a}(0) = 0$

when the density of dark energy reaches infinity, the universe starts to shrink which is absurd. The names of cosmological parameters are thus confusing. Is there really an infinite density of dark and baryonic matter and vacuum dark energy, when $a = a(t)$ reaches its extremal values?

Let us emphasize that the measured density of baryonic matter is always finite. This shows that the Friedmann model is strange, since the ratio between dark and baryonic matter may attain arbitrarily large values. Cosmological parameters have a strange behavior also in other cases [196].

Similarly one divides by zero also for the loitering universe (see Fig. 6.8) or in the de Sitter solution $a(t) = \alpha \cosh(ct/\alpha)$ for $\rho = 0$ and a suitable constant $\alpha > 0$, see e.g. [59, 80, 94, 211]. Note that in the latter case the expansion rate of $a(t)$ is superluminal for large t.

Remark 6.14 At present there is a large number of papers on the so-called *precise cosmology* that categorically state (see Fig. 6.9):

> The universe is composed of 26.8% of dark matter, 4.9% of baryonic matter, and 68.3% of dark energy.

Rightly they should claim:

> According to the standard cosmological model based on the Friedmann equation, the universe could consist of about 27% of dark matter, 5% of baryonic matter, and 68% of dark energy.

It is very important to realize the fundamental difference between these two statements, since the Friedmann equation (6.28) was derived by excessive extrapolations, see e.g. [196, 198, 360, 361]. Therefore, we will now look in more detail as to how the above percentages were obtained.

Remark 6.15 The current values of cosmological parameters are determined, for example, by very distant explosions of type Ia supernovae, which are treated as so-called standard candles (for details see [279, 299]). Type Ia supernovae have about 10–15% less luminosity than if the expansion of the universe were to decelerate only by gravity. From this it can be concluded that the supernova light spreads into a larger volume, and therefore the expansion of the universe is *accelerating*, i.e.,

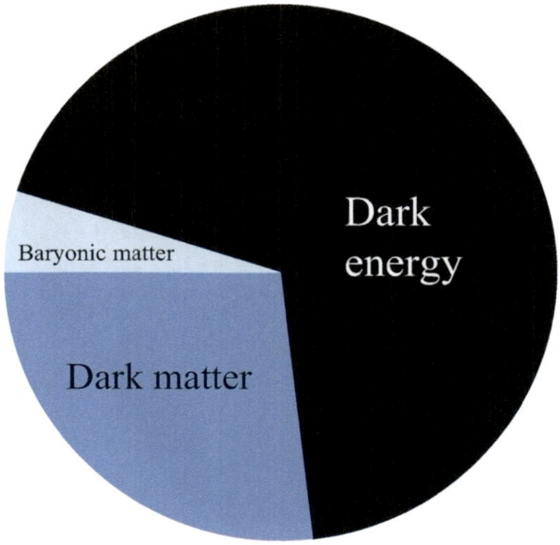

Fig. 6.9 Results of the Planck satellite [289] are interpreted in such a way that our universe consists of about 26.8% of dark matter, 4.9% of baryonic matter, and 68.3% of dark energy. However, they are based on the normalized Friedmann equation (6.28), which was derived using excessive cosmological extrapolations by many orders of magnitude, leading to the relation $\Omega_M + \Omega_\Lambda \approx 1$. Cosmological parameters are sought so that the solution of (6.28) is as close as possible to the measured data. The amount of baryonic matter is estimated by the amount of luminous matter [371, p. 74]

$a = a(t)$ is a strictly convex function. Another explanation is given in [358, 359] by Václav Vavryčuk.

Remark 6.16 By [22, 305] we should consider the extinction of light by host galaxies. The measured luminosity depends considerably on whether the supernova lies inside or on the edge of the galaxy, the direction of its rotation axis, etc. Thus, type Ia supernovae are standard candles only very roughly.

To determine the distances of supernovae (see [146], [299, p. 1021]) the following formula from [59, p. 511] derived from the Friedmann equation for the so-called *luminosity distance* is used (see (7.16) and Remark 7.17 below for more details)

$$d_L = \frac{c(1+z)}{H_0\sqrt{|\Omega_K|}} \sin\left(\sqrt{|\Omega_K|}\int_0^z \frac{d\overline{z}}{\sqrt{(1+\overline{z})^2(1+\Omega_M\overline{z}) - \overline{z}(2+\overline{z})\Omega_\Lambda}}\right). \tag{6.30}$$

Here

$$H_0 = H(t_0) \approx 70 \text{ km s}^{-1}\text{Mpc}^{-1} = \frac{70}{3.086 \cdot 10^{19}} \text{ s}^{-1} = 2.27 \cdot 10^{-18} \text{ s}^{-1} \tag{6.31}$$

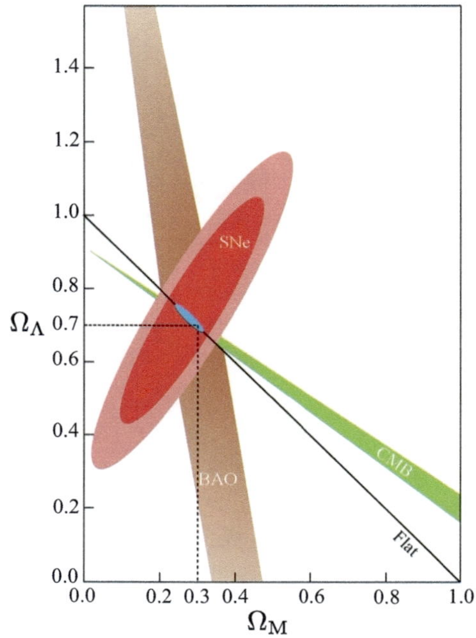

Fig. 6.10 Admissible values of the cosmological parameters Ω_M and Ω_Λ, obtained by three different methods: BAO (Baryonic Acoustic Oscillations), CMB (Cosmic Microwave Background), and SNe (Supernova explosions). However, these methods are not independent, because they are all based on the Friedmann equation (6.26) derived by questionable cosmological extrapolations

is the current value of the Hubble parameter [339] at time $t = t_0$ which corresponds to the present. We will call it the *Hubble-Lemaître constant*, since by the IAU 2018 Resolution B4 in Vienna, the expansion of the universe should be referred to as the *Hubble-Lemaître law.* Moreover, $\Omega_K = 1 - \Omega_M - \Omega_\Lambda$ by (6.28), where for brevity the actual values of the cosmological parameters are denoted by

$$\Omega_M = \Omega_M(t_0), \quad \Omega_\Lambda = \Omega_\Lambda(t_0), \quad \text{and} \quad \Omega_K = \Omega_K(t_0).$$

From this, the measured values of the redshift z of absorption spectral lines of silicon, and the luminosity of type Ia supernovae we can determine particular values of cosmological parameters (6.29) by the χ^2 test [299, p. 1021]. The resulting set of admissible values of cosmological parameters is marked by SNe in Fig. 6.10. In fact, the questionable Friedmann equation is solved by three different methods [279].

Remark 6.17 In 1927, Lemaître [218, p. 56] obtained the value $H_0 = 625$ km s^{-1} Mpc^{-1} which is much larger than that of (6.31), since he had no exact data. In 1929 Hubble [134] got $H_0 = 500$ km s^{-1} Mpc^{-1}.

Remark 6.18 By Remark 7.4 below the second denominator in (6.30) is positive for $\Omega_M \geq 0$ and $\Omega_\Lambda \leq 1$. Furthermore, note that Weinberg [369, pp. 421, 461] presents different relations of the luminosity distance independent of the Friedmann equation.

Remark 6.19 The sum of the measured values Ω_M and Ω_Λ is approximately equal to 1 (see (6.28) and Fig. 6.9). However, this does not allow us to claim that the true space is flat (i.e. infinite Euclidean) as it is often stated at present. Even if the sum were to be

$$\Omega_M + \Omega_\Lambda = 1.0000000000001,$$

we would still have a bounded hyperspherical (i.e. elliptic Riemannian) universe that can be described by the sphere $\mathbb{S}^3_a$ with an incredibly large radius r. Such a space is locally almost Euclidean, but finite. There is a big difference between a bounded and unbounded space. Moreover, the sphere $\mathbb{S}^3_a$ has an entirely different topology than the flat space $\mathbb{E}^3$ which is promulgated by cosmologists at present. Testing the equality $\Omega_M + \Omega_\Lambda = 1$ resembles an inconvenient and incorrect computer test $X = Y$ in single-precision arithmetic (4 bytes), where X and Y are declared to be real (not integer).

Remark 6.20 The fact that $H = H(t)$ is a decreasing function does not contradict the statement that the expansion of the universe could accelerate, i.e., that the second derivative $\ddot{a} > 0$ on some interval, where a is a strictly convex function. To see this, take e.g. $a(t) = t^2$ on $I = (0, \infty)$. Then $\dot{a}(t) = 2t$, $\ddot{a}(t) = 2 > 0$, and $H(t) = \dot{a}(t)/a(t) = 2/t$ is a decreasing function on I.

Remark 6.21 There are several serious drawbacks of the cosmological parameters defined by (6.29), in particular:

1. For the Einstein stationary universe we have $\dot{a} \equiv 0$, see Fig. 6.7. Hence, $H(t) = 0$ for any time instant and the density of dark and baryonic matter $\Omega_M = \infty$ which is absurd.
2. We encounter division by zero as stated in Remark 6.13.
3. For a parabolic space the first derivative of the expansion function $\dot{a} = \dot{a}(t)$ is decreasing, but then the function $t \mapsto |\Omega_K(t)|$ is increasing in time. Is not it strange?
4. If the expansion function was first strictly concave and then strictly convex (since at present the universe accelerates by the standard cosmological model), then $\dot{a} = \dot{a}(t)$ is not a monotone function. However, then the curvature parameter $t \mapsto \Omega_K(t)$ is also not monotone. This is also strange, since the universe is expanding all the time.
5. Another paradox concerns the Big Bang itself. Since $\dot{a}(t) \to \infty$ for $t \to 0^+$, we find that $\Omega_K(t) \to 0$. However, $\Omega_K(t_0) = 0$ is proposed by cosmologists also at the present time t_0. Moreover, note that the Big Bang event contradicts the principle of causality.

Remark 6.22 There are also essential problems with predicted and measured values of the σ_8 cosmological parameter. This parameter describes the effective value of the mass density fluctuations on a sphere of radius $8h^{-1}$ Mpc for $h \in (0.5, 0.75)$, where h depends on the currently accepted value of the Hubble-Lemaître constant H_0. For a substantial discordance in estimating σ_8 see [130].

Remark 6.23 In 2016, the Planck Collaboration stated from CMB data that

$$H_0 = 66.93 \pm 0.62 \text{ km s}^{-1}\text{Mpc}^{-1}.$$

This present value is based on questionable long-term extrapolations applied to the standard cosmological model (i.e. Einstein's equations), while the Gaia and Hubble Space Telescope measurements of Cepheids and RR Lyrae yield a 10% larger value

$$H_0 = 73.52 \pm 1.62 \text{ km s}^{-1}\text{Mpc}^{-1},$$

see [301]. We again get a disagreement between observations and the theory based on the Friedmann equation.

6.8 Maximally Symmetric Manifolds

By Einstein's cosmological principle our universe at any fixed time seems to be isotropic and homogeneous on very large scales. Since its global curvature is not definitely established yet, it is usually modeled by one of the following three three-dimensional maximally symmetric manifolds: the sphere $\mathbb{S}^3$, the Euclidean space $\mathbb{E}^3$, or the pseudosphere $\mathbb{H}^3$.

Theorem 6.2 *For any dimension $n > 1$ there exist exactly three maximally symmetric manifolds up to scaling, namely $\mathbb{S}^n$, $\mathbb{E}^n$, and $\mathbb{H}^n$.*

For the proof see e.g. [369, Chapt. 13] and [277, p. 721]. The curvature index of the above three maximally symmetric manifolds is $k = 1$, $k = 0$, and $k = -1$, respectively.

Remark 6.24 Already in 1900, Karl Schwarzschild [320] speculated that our universe could be possibly hyperbolic. In 1924, Alexander Friedmann (1888–1925) published another famous paper [109], where a negative curvature index $k = -1$ is considered, cf. (6.29). He suggests that only negative mass density is possible for a static universe (see [109, p. 2006]) and thus it is not clear how to satisfy such a paradoxical requirement. Let us also point out that Friedmann did not consider the case $k = 0$ corresponding to the three-dimensional Euclidean space $\mathbb{E}^3$.

From Einstein's equation we can derive (see Remark 6.11) the following ordinary differential equation for the expansion function $a = a(t)$

$$\dot{a}^2 = \frac{8\pi G\rho a^2}{3} + \frac{\Lambda c^2 a^2}{3} - kc^2. \tag{6.32}$$

This is the *Friedmann equation* for any curvature index $k \in \{-1, 0, 1\}$, Λ is the cosmological constant, and $\rho = \rho(t) > 0$ is the mean mass density. At present it is assumed that (6.32) should be considered only for

$$t > \tau,$$

where

$$\boxed{\tau \approx 380\,000 \text{ yr}} \tag{6.33}$$

is the time of decoupling of the cosmic microwave background radiation (CMB) when matter started to dominate over radiation, see e.g. [97, 290]. It is estimated that this decoupling lasted approximately 10,000 years.

Remark 6.25 Let us emphasize that the Friedmann equation (6.32) was derived exactly from Einstein's equations (3.52) without any approximations for a homogeneous and isotropic universe which is described by maximally symmetric manifolds. Therefore, Alan Guth [122, p. 348] calls the Friedmann equations (6.22) and (6.24) the *Einstein equations.*

Remark 6.26 For $n = 1$ the space $\mathbb{E}^1$ can be isometrically imbedded into $\mathbb{H}^1$ and vice versa. Nontrivial hyperbolic geometries thus exist only in dimensions $n > 1$.

Now we present further relations between Euclidean space and the pseudosphere (see [42, p. 3]).

Theorem 6.3 *The space $\mathbb{E}^n$ can be isometrically imbedded into $\mathbb{H}^{n+1}$.*

Converse imbeddings are not so simple. The hyperbolic plane is an unbounded two-dimensional manifold which curves away from itself at every point. It has a somewhat counter-intuitive geometry which is difficult to imagine. In 1901, David Hilbert proved that the hyperbolic plane $\mathbb{H}^2$ cannot be isometrically imbedded into $\mathbb{E}^3$ (see [128]). His statement is usually formulated as follows:

Theorem 6.4 (Hilbert) *There is no complete regular two-dimensional manifold of negative constant Gaussian curvature imbedded into $\mathbb{E}^3$.*

In 1955, Danilo Blanuša proved that the hyperbolic plane $\mathbb{H}^2$ can be isometrically imbedded into the space $\mathbb{E}^6$ (see [34]). Therefore, if we would transform ourselves into six-dimensional beings in $\mathbb{E}^6$ (with Euclidean distance), we could admire the beauty of the symmetry of the pseudosphere $\mathbb{H}^2$, similarly to the way in which as three-dimensional beings we enjoy the beautiful symmetry of the sphere $\mathbb{S}^2$. For the time being it is not known whether the dimension six can be reduced. Blanuša's assertion was generalized by David Brander [42] as follows:

Theorem 6.5 *For $n > 1$ the pseudosphere $\mathbb{H}^n$ can be isometrically imbedded into $\mathbb{E}^{6n-6}$.*

It is also not known whether the exponent $6n - 6$ can be reduced. Brander [43] however proves that there exists a local isometric imbedding from $\mathbb{H}^n$ to $\mathbb{E}^{2n-1}$. The manifold $\mathbb{S}^3$, resp. $\mathbb{H}^3$, which possibly model our universe at a fixed time, can by previous statements be isometrically imbedded to the Euclidean space $\mathbb{E}^4$, resp. $\mathbb{E}^{12}$, i.e.

$$\boxed{\mathbb{S}^3 \hookrightarrow \mathbb{E}^4, \quad \mathbb{H}^3 \hookrightarrow \mathbb{E}^{12}.} \tag{6.34}$$

Here the symbol $\hookrightarrow$ denotes the isometric imbedding. In 12-dimensional Euclidean space the distances in $\mathbb{H}^3$ will be undeformed. The pseudosphere $\mathbb{H}^3$ is a rather exotic object and as 12-dimensional beings we could convince ourselves about its maximal symmetry. Visualization and construction of the pseudosphere $\mathbb{H}^3$ without any deformations of distances is therefore extremely difficult. On the other hand, it is easier to imagine the manifolds $\mathbb{E}^3$ and $\mathbb{S}^3$ which are further candidates for the model of the universe and which do not contain any privileged points and directions.

In [192], we prove the theorem that any n-dimensional manifold in $\mathbb{E}^{n+1}$ which has the same curvature at any point and any principal direction is necessarily the sphere $\mathbb{S}^n_r$ or the hyperplane in $\mathbb{E}^{n+1}$. Since the pseudosphere $\mathbb{H}^n$ cannot be imbedded into $\mathbb{E}^{n+1}$ for $n > 1$, the assumptions of this theorem are not satisfied for the manifold $\mathbb{H}^n$.

Remark 6.27 The manifolds $\mathbb{E}^3$ and $\mathbb{H}^3$ are used to model an infinite isotropic and homogeneous universe in the standard cosmological model. However, the case that this unbounded universe would have at every point and every time instant almost the same temperature, density, pressure, curvature, etc., on large scales would be a miracle. This would require an infinite speed of information transfer. Therefore, the popular theory of inflation [122] cannot explain such a homogeneity and isotropy of unbounded infinite universes. On the other hand, the manifold $\mathbb{S}^3_a$ is always bounded (finite) and no inflation is necessary to assume, e.g. if $a(t) \approx \alpha \exp(ct/\alpha)$ for some constant $\alpha > 0$.

Remark 6.28 In [182, p. 9], we present another argument against Euclidean and hyperbolic geometry of the physical universe, namely, for the interior of a homogeneous mass ball we find by (3.17)–(3.19) that

$$\mathrm{d}l^2 = \frac{1}{1 - r^2 s^2}\,\mathrm{d}r^2 + r^2 \mathrm{d}\theta^2 + r^2 \sin^2\theta\,\mathrm{d}\phi^2, \tag{6.35}$$

where $rs \in [0, 1)$, $\theta \in [0, \pi]$, and $\phi \in [0, 2\pi)$. This relation is very similar to

$$\mathrm{d}l^2 = \frac{1}{1 - kr^2}\,\mathrm{d}r^2 + r^2 \mathrm{d}\theta^2 + r^2 \sin^2\theta\,\mathrm{d}\phi^2$$

for the unit sphere $\mathbb{S}^3$ with the curvature parameter $k = 1$ (see [369, p. 403] and also (6.3)). Here $r \in [0, 1)$ is a dimensionless parameter, $\theta \in [0, \pi]$, and $\varphi \in [0, 2\pi)$. We observe that the cases $k = 0$ and $k = -1$ do not match (6.35) when $s > 0$. Thus, a homogeneous mass distribution causes a positive curvature globally and the most probable model of our universe at any fixed time is $\mathbb{S}^3_a$.

6.9 Regular Tessellations of Hyperbolic Manifolds

Due to Theorem 6.5, a natural question arises: How can we imagine hyperbolic manifolds $\mathbb{H}^n$? One way is to tile them with congruent bounded regular polyhedral cells. These cells are curved, of course. For simplicity, we shall often omit the term "regular" in what follows. First consider $n = 2$ corresponding to the hyperbolic plane.

Remark 6.29 The following often presented model of a hyperbolic plane

$$\tilde{H}^2 = \{(x_1, x_2, x_3) \in \mathbb{E}^3 \mid x_1^2 + x_2^2 - x_3^2 = -1\}$$

is confusing and thus not suitable for our purpose, since the variables x_1, x_2, x_3 are not equivalent and the used Minkowski-like pseudometric is not natural. Another often sketched representation of the hyperbolic plane as the graph of a saddle-point is also not suitable, since this graph has different curvatures at different points and different directions, in general.

According to Hilbert's Theorem 6.4, the hyperbolic plane should be deformed somehow to get some idea of how its regular tessellations look like. There are at least six basic ways to perform such a deformation (see [54]). One way is to use the well-known *Poincaré disc* in which all angles between geodesics in $\mathbb{H}^2$ are preserved, but distances are not preserved. This model comes from a stereographic projection of the hyperboloid in $\mathbb{E}^{n+1}$ onto the hyperplane $x_{n+1} = 0$. The hyperbolic plane is represented by the interior of the unit circle k in $\mathbb{E}^2$. Geodesics are then circular arcs that are orthogonal to k, see Figs. 6.11 and 6.12. The corresponding metric is defined in [184, p. 273].

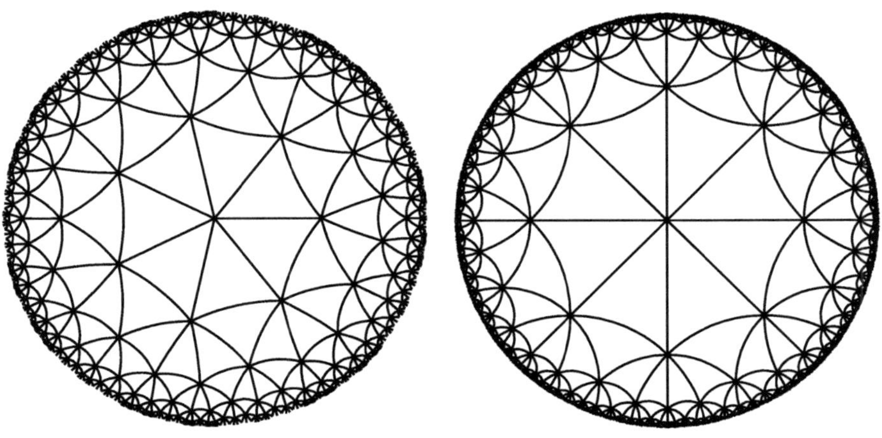

Fig. 6.11 The regular tessellations of the hyperbolic plane by equilateral hyperbolic triangles with vertex degree $d = 7$ and $d = 8$. All sides on the left (resp. right) picture have the same length in the corresponding metric (Picture courtesy of Petr Kůrka)

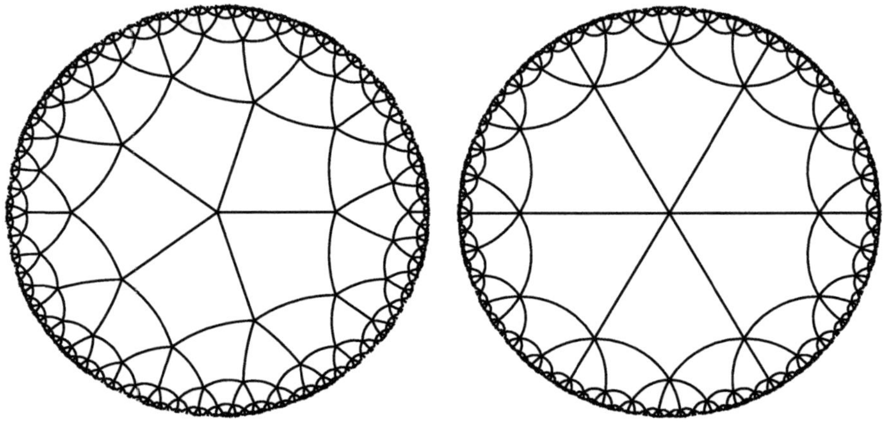

Fig. 6.12 The regular tessellations of the hyperbolic plane by hyperbolic squares with vertex degree $d = 5$ and $d = 6$. Every vertex angle is equal to 72° and 60°, respectively, and all sides have the same length in the corresponding metric. (Picture courtesy of Petr Kůrka)

Theorem 6.6 *There exist infinitely many different tessellations by equilateral hyperbolic triangles of the hyperbolic plane* $\mathbb{H}^2$. *In particular, for every* $d \geq 7$ *there exists one equilateral hyperbolic triangle with angles* $2\pi/d$ *that tessellates the hyperbolic plane.*

The cases $d = 7$ and $d = 8$ are illustrated in Fig. 6.11.

There are also infinitely many different regular tessellations of the hyperbolic plane $\mathbb{H}^2$ by other curved hyperbolic regular polygons, see e.g. Fig. 6.12. In particular, in addition to Theorem 6.6 we can tessellate $\mathbb{H}^2$ by:

- by hyperbolic squares with $d \geq 5$ (see Fig. 6.12),
- by hyperbolic pentagons with $d \geq 4$,
- by hyperbolic hexagons with $d \geq 4$, and
- by hyperbolic heptagons, octagons, etc., with $d \geq 3$.

In all these cases, d is also the number of hyperbolic polygons meeting at each vertex.

Concerning the hyperbolic space $\mathbb{H}^3$, it can only be tessellated by several types of curved congruent regular polyhedra up to scaling, translation, mirroring, and rotation. We call them *regular tessellations*. They were found by Harold Scott MacDonald Coxeter.

Theorem 6.7 (Coxeter) *There are only four regular tessellations of* $\mathbb{H}^3$, *namely, by a hyperbolic cube, by one of the two types of hyperbolic dodecahedra, and by a hyperbolic icosahedron.*

For the proof see [72]. For the corresponding space-fillers see Figs. 6.13 and 6.14. The next theorem is proved in [44, p. 167].

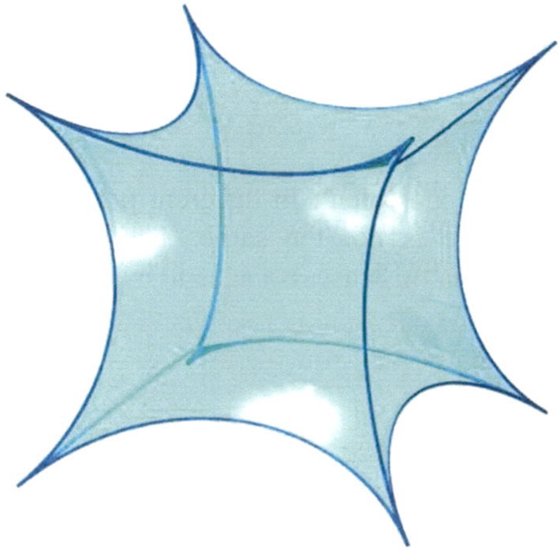

Fig. 6.13 A schematic illustration of the cubic space-filler cell of $\mathbb{H}^3$ whose dihedral angles are 72° (From Wolfram MathWorld)

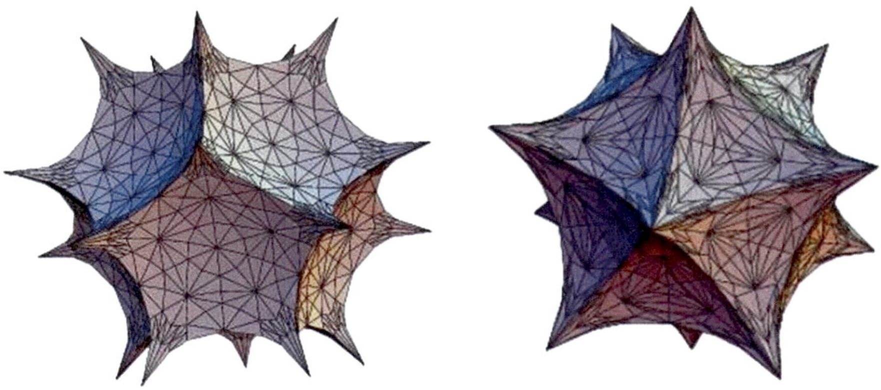

Fig. 6.14 Schematic illustrations of hyperbolic dodecahedral and icosahedral space-filler cells of $\mathbb{H}^3$ (From Wolfram MathWorld)

Theorem 6.8 *There are no regular simplicial tessellations of* $\mathbb{E}^n$ *and* $\mathbb{H}^n$ *for* $n \geq 3$.

Remark 6.30 A modified version of the hyperspherical coordinates from Sect. 6.2 is sometimes used to express the unit hyperbolic pseudosphere $\mathbb{H}^3$, in particular,

$$\begin{aligned} x &= \sinh\chi\,\sin\theta\cos\phi, \\ y &= \sinh\chi\,\sin\theta\sin\phi, \\ z &= \sinh\chi\,\cos\theta, \\ w &= \cosh\chi, \end{aligned}$$

where $\chi \in [0, \infty)$, $\theta \in [0, \pi]$, and $\phi \in [0, 2\pi)$. However, this expression is again not a very intuitive model of our isotropic universe at a fixed time, since

$$x^2 + y^2 + z^2 - w^2 = -1$$

and these variables are not equivalent. In different points and different directions the corresponding curvature is not the same. The use of the Minkowski-like pseudometric is not natural in this case, because w is not a time variable.

Chapter 7
Excessive Extrapolations From the Friedmann Equation

7.1 Typical Examples

The contemporary cosmological model which is based on the Friedmann equation (6.28) does not give acceptable results. For instance, it admits a division by zero, see Remark 6.13 and Figs. 6.7 and 6.8. It also leads to a wide range of paradoxes and has other serious problems, in particular, the problem of the existence of an invisible exotic dark matter and dark energy (see Sect. 1.1). The main reason is that physical reality is identified with a very simple mathematical model (see (7.21) below) ignoring the modeling error which can be quite large.

Now let us look at these facts in detail. In this section, we present several illustrative examples that show why we should be very careful in making diverse unsubstantiated and unjustified extrapolations in cosmology.

Example 7.1 Applying Newtonian theory to the entire homogeneous and isotropic universe leads to the well-known Jeans paradox. The corresponding Newtonian potential $\phi = \text{const.}$ (see [141]) should satisfy the Poisson equation

$$\Delta\phi = 4\pi G\rho,$$

where $\Delta = \partial^2/\partial x^2 + \partial^2/\partial y^2 + \partial^2/\partial z^2$ is the standard Laplace operator and $\rho > 0$ is a constant mass density. However, we immediately see that the left-hand side is zero while the right-hand side is positive. Einstein knew about this paradox. That is why he was inspired by the equation $\Delta\phi - \Lambda\phi = 4\pi G\rho$ which has a non-constant solution for $\Lambda > 0$. In this way he arrived at equation (3.52).

Example 7.2 Consider a unit cube with edge $e = 1$ m (see Fig. 7.1) and solve a steady-state heat conduction problem

$$-\Delta u = f$$

M. Křížek, L. Somer, *Mathematical Aspects of Paradoxes in Cosmology*,
https://doi.org/10.1007/978-3-031-31768-2_7

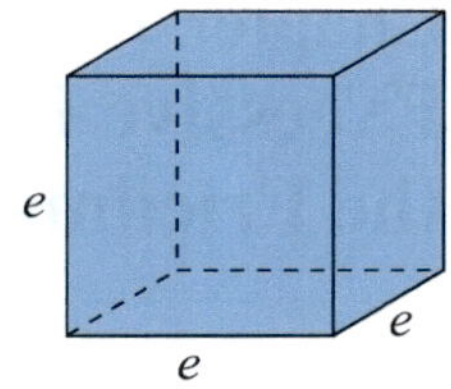

Fig. 7.1 A homogeneous isotropic cube with edge of length e

with some boundary conditions, where u is the temperature and f is proportional to the density of heat forces. This elliptic problem approximates very well the true temperature in homogeneous isotropic solids which can be verified by direct measurements [191]. However, in applying the heat equation on the atomic level in the cube with edge $e = 10^{-10}$ m, we get nonsensical numbers, since it is not clear how to define the temperature on such a small scale.

Example 7.3 We also obtain meaningless numbers for $e = 10^{10}$ m. Such a large cube from Fig. 7.1 with density $\rho \approx 1000$ kg/m^3 would immediately collapse into a black hole, since the diameter $\sqrt{3}e$ of this cube is about ten times larger than the diameter $1.4 \cdot 10^9$ m of the Sun. We can, of course, solve the above steady-state heat conduction problem on an arbitrarily large cube. However, the question is for which e do we still get acceptable results, and when the resulting temperatures have nothing to do with reality.

Other equations of mathematical physics (such as supraconductivity equations, Navier-Stokes equations for fluids, Korteweg-de Vries equations, linear elasticity equations, Maxwell's equations, semiconductor equations, magneto-hydro-dynamic equations, and so on) are subjected to analogous restrictions.

Remark 7.1 Time scales cannot be arbitrary, too. For instance, the backward integration of the N-body problem can yield reliable results in the Solar system on time scales $\approx 10^5$ years, whereas on scales 10^{10} years it produces definitely wrong results, since the Solar system did not exist 5 Gyr ago. Therefore, we also cannot predict, e.g., how many inhabitants will be on Earth after 1 Myr or to predict any reliable weather forecast after 1 yr. Note that any climatological model is much more simple than Einstein's equations (cf. (5.3)).

7.2 The Origin of Excessive Extrapolations

In 1922, cosmologists had no idea about the actual size of the physical universe, since other galaxies were discovered later by Edwin Powell Hubble. In spite of that Alexander Friedmann [108] implicitly assumed that Einstein's equations describe exactly the behavior of gravity on cosmological scales. However, this hidden assumption is highly speculative, since Einstein's equations (more precisely only some simple formulae derived from the parametrized post-Newtonian formalism) "are tested" on the scales of astronomical units and excessive extrapolations are made to the entire universe that is at least $1{,}000{,}000{,}000{,}000{,}000 = 10^{15}$ times larger than 1 au.

Remark 7.2 The well-known relationship $s = vt$ from Newtonian mechanics is scale invariant in Euclidean space. However, it is not scale invariant in the physical universe, since any velocity v never exceeds c. Einstein's equations also cannot be scale invariant on an expanding sphere $\mathbb{S}^3_{a(t)}$, since it is bounded. Furthermore, they do not trustworthily describe phenomena at the atomic level which also implies that they are not scale invariant, cf. Remark 1.12. In this case, the modeling error is very large (see the right part of Fig. 7.3). However, it is tacitly assumed that Einstein's equations describe reality very well on arbitrarily large scale. Einstein's equations also cannot be used to model an extreme state just after the Big Bang [122] when matter and antimatter collided, highly energetic photons appeared, and the temperature thus briefly increased [37, p. 113]. Additionally, they contain fixed fundamental physical constants c and G, are fully deterministic and nonlinear.

Example 7.4 To see why nonlinearities break scale invariance, consider a simple nonlinear relation between the magnetic intensity H and the magnetic induction B as sketched in Fig. 7.2. This represents a typical behavior of ferromagnetic materials contained in electromagnets, transformers, rotary machines, electron microscopes, etc., see [191, p. 133] for details. We observe that this relation is linear on very small scales near the origin. On large scales it is also linear, but the slope is different. Hence, this relation cannot be scale invariant globally. Similar considerations can also be done for other kinds of nonlinearities describing, e.g., elasto-plastic material (with hardening), nonlinear heat conduction, or Einstein's equations.

Remark 7.3 We present one more argument indicating why Einstein's equations should not be applied to the entire universe. The present state of our universe depends on its history, i.e., its evolution is surely not reversible. On the other hand,

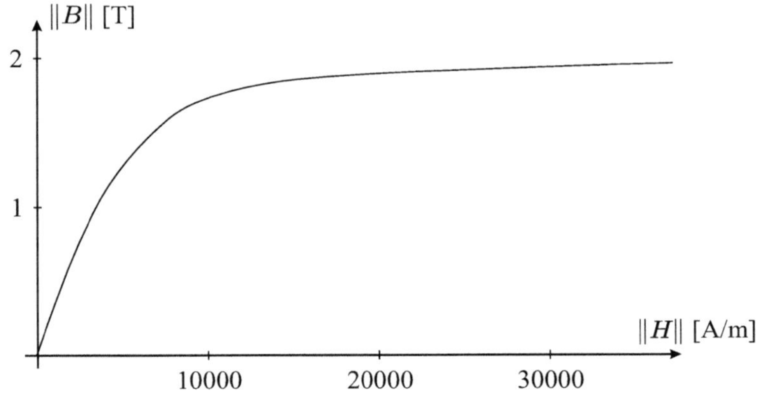

Fig. 7.2 A typical nonlinear dependence of magnetic induction B on magnetic intensity H

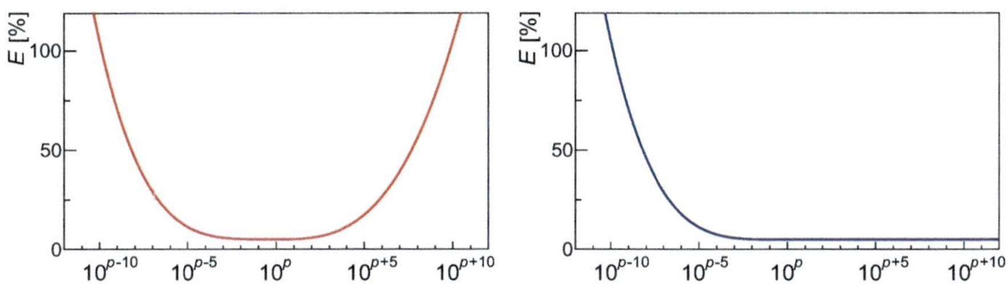

Fig. 7.3 Left: A schematic illustration of a general behavior of the relative modeling error E for equations of mathematical physics. The horizontal axis has a logarithmic scale and p is the exponent for which the modeling error is the smallest. Right: Behavior of the modeling error for Einstein's equations proposed by the current standard cosmological model

the Friedmann equation is completely reversible. That is its solution does not depend on the history, but only on one prescribed value of the expansion function at some given time instant (e.g. the present time), since it is an ordinary differential equation of the first order, see (7.21) below.

From Fig. 1.9 we know that no equation of mathematical physics describes reality absolutely exactly on any scale. The reason is that the laws of physics are not unchanged under a change of scale, in general. Each mathematical model has, in general, a different behavior of the modeling error. In the left part of Fig. 7.3, we see a typical behavior of this error for different scales. Therefore, it has no sense to look for a model that would perfectly describe reality on any scale (micro and macro together).

Therefore, we have to study models that are independent of the Friedmann equation (see e.g. [183–185], [271, p. 299]) to check whether dark matter exists or not. If dark matter does not exist, then the relation (6.30) derived from the Friedmann equation can hardly yield some reliable distances. We will return to this important problem in Sects. 7.3 and 7.4. The standard cosmological model thus resembles the situation of article with the title *Lemma 1*, see [151], which can be characterized as follows:

> Suppose that Lemma 1 implies Lemma 2, from which we further derive Lemma 3. These auxiliary results imply a certain important mathematical theorem, which is a basis of a new fascinating and beautiful theory. But after some time we find that Lemma 1 is wrong, and therefore the theorem need not hold.

In contemporary cosmology such an assertion as Lemma 1 is the statement that Einstein's equations describe the evolution of the entire universe very well. This hidden assumption is considered to be obvious. However, it is just a null hypothesis which has not been verified by any experiment and was obtained by excessive extrapolations by many orders of magnitude in the articles [94] by Einstein and [108] by Friedmann. Since the twenties, this fatal mistake stretches throughout the entire cosmology. Therefore, from a similar reason it also makes no sense to look for a theory of everything.

7.3 Distances in Time

In this section, we derive a relation between the redshift z and the so-called look-back time (see (7.9) below) by means of the Friedmann equation, i.e., how much time has passed before the photons from the observed object reached us. We shall follow the idea of Carroll et al. [59] from 1992 in more detail.

For $z \geq 0$ we introduce an auxiliary function

$$X(t) := \frac{a(t)}{a(t_0)} = \frac{1}{z+1}, \tag{7.1}$$

where t_0 is the age of the universe. By the law of conservation of mass and energy we find for zero pressure that

$$\rho(t)a^3(t) = \rho(t_0)a^3(t_0) \tag{7.2}$$

is a time independent constant for nonrelativistic matter in matter dominated universe with $z \lesssim 1000$. Thus we have

$$\frac{\rho(t)a^2(t)}{a^2(t_0)} = \frac{\rho(t)a^3(t)}{a^2(t_0)a(t)} = \frac{\rho(t_0)a^3(t_0)}{a^2(t_0)a(t)} = \frac{\rho(t_0)a(t_0)}{a(t)} = \frac{\rho(t_0)}{X(t)}.$$

From this and the Friedmann equation (6.32) divided by $a^2(t_0)$, we get

$$\dot{X}^2(t) = \frac{8\pi G\rho a^2}{3a^2(t_0)} + \frac{\Lambda c^2 X^2(t)}{3} - \frac{kc^2}{a^2(t_0)} = \frac{8\pi G\rho(t_0)}{3X(t)} + \frac{\Lambda c^2 X^2(t)}{3} - \frac{kc^2}{a^2(t_0)}.$$

Dividing this equation by the Hubble-Lemaître constant

$$H_0 = H(t_0) = \frac{\dot{a}(t)}{a(t)}\bigg|_{t=t_0} > 0,$$

we obtain

$$\frac{1}{H_0^2}\dot{X}^2(t) = \frac{\Omega_\mathrm{M}}{X(t)} + \Omega_\Lambda X^2(t) + \Omega_\mathrm{K}, \tag{7.3}$$

where for brevity

$$\Omega_\mathrm{M} = \Omega_\mathrm{M}(t_0), \quad \Omega_\Lambda = \Omega_\Lambda(t_0), \quad \text{and} \quad \Omega_\mathrm{K} = \Omega_\mathrm{K}(t_0) \tag{7.4}$$

denote the present values of the cosmological parameters (6.29). The left-hand side of (7.3) is nonnegative and thus

$$\frac{1}{H_0}\dot{X}(t) = \pm\sqrt{\frac{\Omega_M}{X(t)} + \Omega_\Lambda X^2(t) + \Omega_K}. \tag{7.5}$$

Since by (7.1),

$$\dot{X}(t) = \frac{\mathrm{d}}{\mathrm{d}t}\Big(\frac{1}{z+1}\Big) = -\frac{1}{(z+1)^2}\frac{\mathrm{d}z}{\mathrm{d}t}$$

and since the derivative $\mathrm{d}z/\mathrm{d}t$ is negative, we find from (7.5) that

$$\frac{1}{H_0}\dot{X}(t) = -\frac{1}{H_0(z+1)^2}\frac{\mathrm{d}z}{\mathrm{d}t} = \sqrt{\Omega_M(z+1) + \frac{\Omega_\Lambda}{(z+1)^2} + \Omega_K} \tag{7.6}$$

and thus,

$$\frac{\mathrm{d}z}{H_0(z+1)\sqrt{\Omega_M(z+1)^3 + \Omega_K(z+1)^2 + \Omega_\Lambda}} = -\mathrm{d}t. \tag{7.7}$$

By (6.28) we can exclude Ω_K from the denominator using the identity

$$\Omega_M(z+1)^3 + (1-\Omega_M-\Omega_\Lambda)(z+1)^2 + \Omega_\Lambda = (\Omega_M z+1)(z+1)^2 - \Omega_\Lambda z(z+2). \tag{7.8}$$

Now by integration of (7.7) we get the relation between the redshift z and the *look-back time* t,

$$t_0 - t = \frac{1}{H_0}\int_0^z \frac{\mathrm{d}\overline{z}}{(\overline{z}+1)\sqrt{(\Omega_M\overline{z}+1)(\overline{z}+1)^2 - \Omega_\Lambda\overline{z}(\overline{z}+2)}}. \tag{7.9}$$

We see that the actual look-back time essentially depends on the fact whether the Friedmann equation describes well the evolution of the universe. Numerical relations between the redshift and the corresponding look-back time are given in [284] for $k = 0$.

Remark 7.4 If

$$\Omega_M \geq 0 \quad \text{and} \quad \Omega_\Lambda \leq 1, \tag{7.10}$$

then the integral in (7.9) exists and is finite. These additional conditions are essential, because then the difference under the square root in (7.9)

$$(\Omega_M \overline{z} + 1)(\overline{z} + 1)^2 - \Omega_\Lambda \overline{z}(\overline{z} + 2) \geq (\overline{z} + 1)^2 - \overline{z}(\overline{z} + 2) = 1$$

is positive (cf. Remark 7.15 and Example 4.1).

Remark 7.5 According to [59, p. 509] and also [299, Sect. 4.3], the formula (7.9) for $z \to \infty$ gives the age of the universe only by means of the present values of the cosmological parameters Ω_M, Ω_Λ, and H_0. However, this need not be true, since (7.9) was derived under the assumption (7.2) which is not valid for very large $z \geq 1000$. According to [383, Chapt. 2], for ultrarelativistic matter we have $\rho(t)a^4(t) \equiv$ const. for $t \in (0, \tau)$ which corresponds to $z \gtrsim 1000$ in a radiation dominated universe. The use of (6.32) for this case has to be done with special care.

Remark 7.6 In applying the current standard cosmological model various "delicate" limits are performed: $a \to 0$, $a \to \infty$, $t \to 0$, $t \to \infty$, $z \to \infty, \ldots$ (see, for instance, [249, 369]). At present it is believed that $a(t) \to \infty$ for $t \to \infty$. By (6.28) and (6.29) for $k \leq 0$ it follows that $1 > \Omega_\Lambda(t)$, and thus $\frac{1}{3}\Lambda c^2 < H^2(t)$ for arbitrary time. From this and (6.27) we observe that also the time derivative of the expansion function grows beyond all bounds if Λ is a positive constant. Hence, the derivative $\dot{a}(t) \to \infty$ for $t \to \infty$ in an infinite universe (Euclidean or hyperbolic).

Remark 7.7 The present measured value of the Hubble-Lemaître constant is

$$H_0 \approx 70 \text{ km s}^{-1} \text{ Mpc}^{-1} \approx 2.27 \cdot 10^{-18} \text{ s}^{-1}, \tag{7.11}$$

where 1 pc $= 3.086 \cdot 10^{16}$ m, $\Omega_M \approx 0.3$ and $\Omega_\Lambda \approx 0.7$. According to [290],

$$t_0 = 13.82 \text{ Gyr},$$

i.e., this age was derived up to four significant digits by formula (7.9). Nevertheless, from such a simple calculation we should not make any definite predictions or categorical conclusions about the actual age of the physical universe as is often done. The Friedmann equation was derived from Einstein's equations by excessive extrapolations to cosmological scales. Its solution for $t > t_0$ depends only on the value of the expansion function at the present time t_0 and not on the history. Since the terminal condition $\dot{a}(t_0)/a(t_0) = H_0$ is known, we may evaluate $a(t_0)$ from equation (6.32) and then integrate (6.32) backward. From this we get the above value $t_0 = 13.82$ Gyr, see [198, p. 274]. Other models that are not based on the Friedmann equation may yield a different age of our universe, e.g. an infinite age for $a(t) = C_1 e^{C_2 t} + C_3$ and some suitable constants $C_i \geq 0$, see [184, p. 305]. For instance, if $C_2 = H_0$ and $C_3 = 0$, then $H(t) \equiv H_0$ is by (6.27) constant everywhere. Then no short-time inflation period just after the Big Bang is necessary for the curvature index $k = 1$, because the exchange of information about temperature, density, and pressure does not violate the principle of causality.

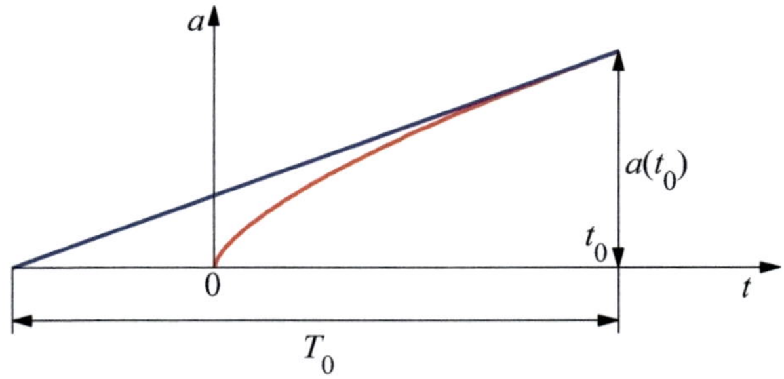

Fig. 7.4 If the expansion function $a = a(t)$ were to be concave, then its graph would be below the tangent line through the point $(t_0, a(t_0))$. Thus for $\dot{a}(t_0) > 0$ by (7.12) the age of the universe t_0 would not exceed the Hubble age $T_0 = 13.97$ billion years

On the other hand, by the inflation theory of Alan Guth [122] the expansion function is firstly strictly convex (exponential), then strictly concave, and then strictly convex again for increasing time t. This seems to be unnatural.

Remark 7.8 The *Hubble time*, sometimes called the *Hubble age* (see Fig. 7.4), is defined by the relation

$$T_0 := \frac{1}{H_0} \approx 13.97 \text{ Gyr} \approx 4.41 \cdot 10^{17} \text{ s},$$

where 1 yr = 31 558 149.54 s and the measured value of H_0 is given in 7.11. The corresponding *Hubble distance* is then defined by

$$d_{\mathrm{H}} := cT_0 = 1.32 \cdot 10^{26} \text{ m}.$$

If the expansion function $a = a(t)$ in its definition domain were to be concave everywhere (which probably does not correspond to reality) and $\dot{a}(t_0) > 0$, then by (6.27) the following would hold for the true age t_0 of the universe (see Fig. 7.4):

$$t_0 \le T_0 = \frac{a(t_0)}{\dot{a}(t_0)} = \frac{1}{H(t_0)} \approx 13.97 \text{ Gyr}. \tag{7.12}$$

Remark 7.9 If $0 \le z \ll 1$ then by (7.9)

$$z = \int_0^z \mathrm{d}\bar{z} = H_0(t_0 - t),$$

i.e., for a small z the look-back time t is independent of Ω_{M} and Ω_Λ. Multiplying this equation by c and using the classical *Hubble law*

$$v = H_0 d,$$

where v the recession speed of galaxies at distance d, we get

$$v = cz.$$

Hence, the distance can be calculated from the measured redshift $z \ll 1$ as follows (cf. [301])

$$d = \frac{cz}{H_0}.$$

This relation is a very good approximation for $z \lesssim 0.1$, i.e., for distances not greater than 400 Mpc. Notice that the Hubble law predicts recession speeds faster than c of very distant galaxies. However, this does not violate STR, since they do not belong to our local inertial systems, see [79].

Remark 7.10 If $\Omega_K = \Omega_\Lambda = 0$ then by (6.28) and (7.9)

$$\int_0^z \frac{d\overline{z}}{(\overline{z}+1)^{5/2}} = \left[\frac{-2}{3(\overline{z}+1)^{3/2}}\right]_0^z = \frac{2}{3}\left(1 - \frac{1}{(z+1)^{3/2}}\right) = H_0(t_0 - t).$$

From this and $z \to \infty$ we get $H_0 = 2/(3t_0)$, i.e.,

$$H_0(t_0 - t) = \frac{2}{3} - \frac{2t}{3t_0}.$$

For instance, if $z = 8$ then $(z+1)^{3/2} = 27$ and the corresponding age of the universe would be $t = \frac{1}{27}t_0$. In fact, this represents an upper estimate of the age $\overline{t}$ for nonzero cosmological parameters Ω_K and Ω_Λ. To see this, assume that (7.10) is valid. Then

$$(\Omega_M\overline{z} + 1)(\overline{z}+1)^2 - \Omega_\Lambda\overline{z}(\overline{z}+2) \leq (\overline{z}+1)^3,$$

and thus by (7.9)

$$\begin{aligned} t_0 - \overline{t} &:= \frac{1}{H_0}\int_0^z \frac{d\overline{z}}{(\overline{z}+1)\sqrt{(\Omega_M\overline{z}+1)(\overline{z}+1)^2 - \Omega_\Lambda\overline{z}(\overline{z}+2)}} \\ &\geq \frac{1}{H_0}\int_0^z \frac{d\overline{z}}{(\overline{z}+1)^{5/2}} = t_0 - t, \end{aligned}$$

that is $\overline{t} \leq t = \frac{1}{27}t_0$.

Remark 7.11 Using (6.27) and (7.1), we get

$$H(t) = \frac{a(t_0)}{a(t)} \cdot \frac{\dot{a}(t)}{a(t_0)} = (z+1)\dot{X}(t).$$

From this, (7.6), and (7.8) we can express the behavior of the Hubble parameter under the assumption (7.10) as a function of the redshift z as follows

$$\mathcal{H}(z) = (z+1)H_0\sqrt{\Omega_{\mathrm{M}}(z+1) + \frac{\Omega_\Lambda}{(z+1)^2} + \Omega_{\mathrm{K}}}$$

$$= H_0\sqrt{(\Omega_{\mathrm{M}} z + 1)(z+1)^2 - \Omega_\Lambda z(z+2)},$$

where $\mathcal{H}(z) = H(t)$ and $z = a(t_0)/a(t) - 1$ by (7.1). The inverse relation between z and t can be derived from (7.9).

Remark 7.12 Consider the linear equation of the second order

$$\ddot{a} = -\frac{4\pi G}{3}\Big(\rho + \frac{3p}{c^2}\Big)a + \frac{\Lambda c^2}{3}a \tag{7.13}$$

which is a direct consequence of the Einstein equations, see Remark 6.11. For $\dot{a} \neq 0$ define the dimensionless *deceleration parameter* by

$$q = q(t) := -\frac{\ddot{a}a}{(\dot{a})^2} = -1 - \frac{\dot{H}}{H^2},$$

where the last equality follows from (6.27). In particular, if the pressure $p \equiv 0$, then q can be expressed by means of the cosmological parameters (6.29) as

$$q(t) = \frac{4\pi G\rho}{3H^2} - \frac{\Lambda c^2}{3H^2} = \frac{1}{2}\Omega_{\mathrm{M}}(t) - \Omega_\Lambda(t)$$

due to (7.13). For $\Omega_{\mathrm{M}} \approx 0.3$ and $\Omega_\Lambda \approx 0.7$ the present value of the deceleration parameter $q(t_0) \approx -0.55$ is negative, and thus the expansion of the universe is accelerating. Thus, the energy of the entire universe is not conserved.

Remark 7.13 The current standard cosmological model assumes that time flows completely uniformly from the Big Bang on (see Fig. 7.4). Therefore, we often hear the question: *What was before the Big Bang?*

However, it is important to realize that in the observable universe we actually look in any direction into the vast spacetime singularity. The more distant the observed objects are, the more it seems to us that time passes slower. If there were a huge clock placed at $z = 1$ from the Earth, then we would see that it goes twice as slow. The largest currently observed distance corresponds to the CMB with redshift $z = 1089$ which appeared 380,000 years after the Big Bang. At that time the universe was 1090 times smaller and the radius of the corresponding CMB sphere was about 10^{24} m. The most typical diameter of fluctuations in the CMB is about one angular degree which is about 10^{21} m in diameter, i.e. greater than our Galaxy.

Note that this is too large an object for the well-known method of baryonic acoustic oscillations, cf. [97].

During one earthly hour we would observe that the CMB clock had moved only about $3.3 = \frac{3600}{1090}$ s. If we would ever see objects more distant than $z = 1089$ (using e.g. detectors of relict neutrinos), we would find that time there passes even more slowly, and so on. To avoid this drawback, a conformal time is usually applied, see (7.17) below. Thus the Big Bang singularity deforms not only space but also time. This is a similar phenomenon to time dilatation in a gravitational potential hole, since photons reduce their frequency.

7.4 Classification of Cosmological Distances in Space

In cosmology many different distances are defined: angular, comoving, Euclidean, Hubble, light-year, luminosity, Minkowski, parallax, proper motion, redshift, ...distance, see [189]. There is no single natural distance, since the universe is expanding, curved, and we look back in time. In this section we will concentrate on geometrical interpretations of the above-mentioned distances.

By [369] the *angular distance* (sometimes called the *angular-size distance*) is in Euclidean space defined by the relation

$$d_{\mathrm{A}} = \frac{D}{\delta}, \tag{7.14}$$

where D is a known (or assumed) proper size of an object and $\delta \in (0, \pi)$ is its apparent angular size (see e.g. Fig. 1.7, where $D = 2r \sin(\delta/2) \approx r\delta$). Since δ is usually very small, we have $\tan \delta \approx \sin \delta \approx \delta$ in radians.

Remark 7.14 Let an observer be located at the North Pole of the sphere $\mathbb{S}^2$ in Fig. 6.3. The arcs of the parallel $\pm 60°$ between the meridians $0°$ and $30°$ are obviously viewed under the same angle. A similar nonuniqueness may cause some troubles when using the angular distance (7.14) in $\mathbb{S}^3$. On the other hand, a very long and distant segment in $\mathbb{H}^3$ can be viewed under a very small angle. Since the actual curvature index k is not known up to now, one has to be careful when using the angular distance in cosmology for large z.

The *proper motion distance* (sometimes called only the *proper distance* or the *comoving transverse distance*) is defined by the relation

$$d_{\mathrm{M}} = a(t_0) f(\chi), \tag{7.15}$$

where f depends on the curvature index k as follows

$$f(\chi) = \begin{cases} \sin\chi & \text{if } k = 1, \\ \chi & \text{if } k = 0, \\ \sinh\chi & \text{if } k = -1. \end{cases}$$

Note that for $k = 0$ the proper motion distance d_M is the usual *Euclidean distance*. Only in this case the space manifold at any fixed time is not curved. We see that d_M is always bounded for $z \to \infty$, since the maximum possible coordinate χ is bounded. When $k = 1$ the formula (7.15) should be used by Remark 7.14 only for $\chi \le \pi/2$. For $\pi/2 \le \chi \le \pi$ and $k = 1$ the proper motion distance d_M is maximal for $\chi = \pi/2$, and $d_\mathrm{M} = 0$ for $\chi = \pi$, i.e., the proper motion distance does not satisfy the standard mathematical requirements attributed to the distance, see [383].

Let L be the bolometric luminosity (i.e. total—integrated over all frequencies) of an object. It is measured in Watts. Then the *luminosity distance* d_L is related with the flux ℓ measured in W/m^2 by

$$\ell = \frac{L}{4\pi d_\mathrm{L}^2}, \quad \text{i.e.} \quad d_\mathrm{L} = \sqrt{\frac{L}{4\pi\ell}}.$$

The luminosity distance in Mpc can be established from the observed apparent magnitude μ and the absolute magnitude μ_abs of an object by the formula

$$\mu - \mu_\mathrm{abs} = 25 + 5\log_{10} d_\mathrm{L}.$$

This is a key formula in cosmology [59], since one can use the observed flux to establish d_L. The main idea is based on the known density of light flux, the distance and flux of some reference galaxy, and Pogson's equation (for details see [184, p. 110] and [369]).

According to [241, Eq. (6)], the luminosity distance is given by the formula

$$d_\mathrm{L} = a(t_0)f(\chi)(1+z).$$

Hence, the formula for the flux ℓ has to be modified as follows (see [369]):

$$\ell = \frac{L}{4\pi a^2(t_0)f^2(\chi)(z+1)^2}.$$

Since photons lose their energy during the expansion of the universe, the factor $(z+1)$ appears in the denominator. Another factor $(z+1)$ is due to the fact that one second lasted longer when a photon was emitted than when it was received. According to [369, p. 485],

$$d_\mathrm{L}(z) = (z+1)d_\mathrm{M}. \tag{7.16}$$

Now we will derive the dependence of the proper motion and luminosity distances, respectively, on the redshift z. To this end, we will consider the *conformal time* η given by

$$d\eta = \frac{c\,dt}{a(t)} \tag{7.17}$$

instead of the usual time t (i.e., η is linearly proportional to t if $a(t) \equiv$ const.). Then from (7.7) and (7.1) we find that

$$\frac{dz}{a(t_0)H_0(z+1)\sqrt{\Omega_M(z+1)^3 + \Omega_K(z+1)^2 + \Omega_\Lambda}} = -\frac{a(t)d\eta}{c\,a(t_0)} = -\frac{d\eta}{c(z+1)}.$$

Multiplying both sides by $z + 1$, we get by (7.8) and integration as in (7.9) that

$$\chi = \eta_0 - \eta = \frac{c}{a(t_0)H_0} \int_0^z \frac{d\overline{z}}{\sqrt{(\Omega_M\overline{z} + 1)(\overline{z} + 1)^2 - \Omega_\Lambda\overline{z}(\overline{z} + 2)}}. \tag{7.18}$$

This is a dimensionless number that represents the distance in the comoving coordinates. The largest possible distance in χ is thus $\chi_{\text{horizon}} = \eta_0$. Therefore, the observable universe is bounded.

Note that the conformal time enables us to replace the term c^2dt^2 by $a^2(t)d\eta^2$ which makes the relation (6.5) more elegant.

Remark 7.15 For admissible values of cosmological parameters $\Omega_M = 0.01$, $\Omega_\Lambda = 1.1$ (cf. Fig. 6.10), and $\overline{z} = 3$, the difference under the square root in (7.18) is negative, because

$$(\Omega_M\overline{z} + 1)(\overline{z} + 1)^2 - \Omega_\Lambda\overline{z}(\overline{z} + 2) = 1.03 \cdot 4^2 - 1.1 \cdot 3 \cdot 5 = -0.02,$$

which would yield complex values. In spite of that the standard Ned Wright's Cosmological Calculator [396] gives a quite acceptable value $d_L = 164.1$ Gly. Such a value is sometimes also called the *light-year distance*.

From now on we shall assume that conditions (7.10) hold. These conditions are only sufficient (not necessary) for the existence of the integral in (7.18).

Remark 7.16 Let $k = 0$. Then by (7.15), (7.16), and (7.18) we get

$$d_L(z) = (z+1)a(t_0)\chi = \frac{(z+1)c}{H_0} \int_0^z \frac{d\overline{z}}{\sqrt{(\Omega_M\overline{z} + 1)(\overline{z} + 1)^2 - \Omega_\Lambda\overline{z}(\overline{z} + 2)}}.$$

Remark 7.17 By (7.4) and (6.29) we see that

$$\Omega_K = -\frac{kc^2}{H_0^2 a^2(t_0)}.$$

Hence, by (7.15), (7.16), and (7.18) we get for $k = 1$ that

$$\begin{aligned} d_{\mathrm{M}}(z) &= a(t_0)\sin\chi \\ &= \frac{c}{H_0\sqrt{|\Omega_{\mathrm{K}}|}}\sin\Big(\sqrt{|\Omega_{\mathrm{K}}|}\int_0^z \frac{\mathrm{d}\overline{z}}{\sqrt{(\Omega_{\mathrm{M}}\overline{z}+1)(\overline{z}+1)^2-\Omega_\Lambda\overline{z}(\overline{z}+2)}}\Big). \end{aligned}$$

For $k = -1$ we only replace sin above by sinh and for $k = 0$ the factors $\sqrt{|\Omega_{\mathrm{K}}|}$ are shortened, see Remark 7.16. These formulae derived in [59] were also used by Perlmutter et al. in [280, p. 566]. Moreover, note that the observed flux essentially depends on the used measuring instruments.

Remark 7.18 By [369, p. 485] the *angular distance* for any $k \in \{-1, 0, 1\}$ is defined as follows

$$d_{\mathrm{A}}(z) = \frac{d_{\mathrm{M}}(z)}{z+1},$$

where the proper motion distance d_{M} is given by (7.15). Hence, by (7.16),

$$\frac{1}{z+1}d_{\mathrm{L}}(z) = d_{\mathrm{M}}(z) = (z+1)d_{\mathrm{A}}(z). \tag{7.19}$$

For $k = 0$ the above angular distance coincides with (7.14) and on very short distances one may also define the standard parallax distance by means of a parallactic angle (see e.g. [369, pp. 420 and 485]).

From (7.15) we see that $d_{\mathrm{L}}(z)$ is unbounded for $z \to \infty$, $d_{\mathrm{M}}(z)$ remains bounded for $z \to \infty$, and $d_{\mathrm{A}}(z)$ converges to zero for $z \to \infty$. Here one has to realize that the light of distant objects with redshift $z > 0$ propagates into a larger volume than if the universe were not to expand. This gives the factor $(z+1)^{-1}$ on the left-hand side of (7.19). A valid criticism of formula (7.19) is given in [363].

Moreover, we must consistently distinguish between "at that time" and "actual dimensions", i.e. the size of the universe when observed photons left a distant object and the universe was much smaller, and "today's" dimensions when "ancient" photons arrived at our ground-based telescopes. Roughly speaking, the younger the objects which are observed, the larger the magnification appears. Therefore, by angular measurements we paradoxically see a very distant object as being larger. We call this phenomenon the *time lens*. The magnification is given by the factor

$$(z+1),$$

which appears on the right-hand side of (7.19). In [184, p. 314], the functioning of the time lens is illustrated by three real-world examples.

From (7.19) and Remark 7.9 we also see that

$$d_{\mathrm{L}}(z) \approx d_{\mathrm{M}}(z) \approx d_{\mathrm{A}}(z) \approx \frac{cz}{H_0} \approx c(t_0 - t)$$

when z is small, i.e. $0 \leq z \ll 1$. The corresponding error is only a few percent for $z \lesssim 0.1$.

7.5 A Simple Form of the Friedmann Equation

Throughout this section we shall suppose for simplicity that

$$\boxed{\Lambda > 0.}$$

The case $\Lambda \leq 0$ can be investigated similarly, see e.g. [348]. The inequality $\Lambda < 0$ leads to an oscillating universe, while for $\Lambda = 0$ the behavior of the expansion function depends on the so-called *critical density* $\rho = 3H_0^2/(8\pi G) \approx 10^{-26}$ kg/m^3, see e.g. [184, p. 285], [231, p. 75].

The time instant τ will again correspond to the decoupling of the cosmic microwave background (6.33). We shall assume that the expansion function $a = a(t)$ satisfies the following condition

$$\dot{a}(\tau) > 0, \tag{7.20}$$

because the universe was expanding for $t = \tau$.

Since the product $\rho(t)a^3(t)$ is constant (see (7.2)) during the time period when matter dominates over radiation, the Friedmann equation (6.32) takes the equivalent form

$$\dot{a}^2 = Aa^2 + B + \frac{C}{a} \tag{7.21}$$

with time independent constant coefficients

$$A = \frac{\Lambda c^2}{3}, \quad B = -kc^2, \quad C = \frac{8\pi G\rho a^3}{3} > 0 \tag{7.22}$$

which are not exactly known. This simplified notation will be useful in further analysis.

Before we analyze the existence and uniqueness of the solution of (7.21) with some initial or terminal condition, we present several remarks, examples, and definitions.

Remark 7.19 Since the ratio

$$\frac{\dot{a}(t_0)}{a(t_0)} = H_0 = H(t_0) \approx 70 \text{ km/(s Mpc)}$$

for the present time t_0 is known, we may calculate from (7.21) and (7.22) the terminal condition $a_0 = a(t_0) = c/\sqrt{\Lambda c^2/3 - kH_0^2 + 8\pi G\rho_0/3} \approx 10^{26}$ m. Thus we can solve equation (7.21) backward and also forward in time for given cosmological parameters (7.4).

Remark 7.20 Since the expansion function $a = a(t) > 0$ is continuous, its time derivative $\dot{a} = \dot{a}(t)$ is by Eq. (7.21) also a continuous function. Anyway, from such a simple ordinary differential equation we should not make any definitive conclusions about the deep past and the distant future of the universe as is often done. Equation (7.21) does not contain the term corresponding to radiation. The reason is that the present measured value of relative density of radiation is less than $\approx 5 \cdot 10^{-5}$, compare with Fig. 6.9. However, for $t \leq \tau$, i.e., for the time period when radiation dominated over matter (see Remark 7.5), another term D/a^2 is added to the right-hand side of Eq. (7.21).

Remark 7.21 If $a > 0$ is sufficiently small, then C/a is much larger than the terms Aa^2 and B on the right-hand side of (7.21). Hence, $\dot{a}^2 a \approx C$ and the corresponding density parameter $\Omega_M \simeq 1$ also dominates the other parameters from (6.29). In this case the expansion function behaves as $t^{2/3}$ up to some multiplicative positive constant (as in the Einstein-de Sitter parabolic model with $\Omega_M = 1$, $\Omega_\Lambda = 0$, and $\Omega_K = 0$).

A function $f : \mathbb{E}^1 \to \mathbb{E}^1$ is said to be *Lipschitz continuous* if there exists a constant L such that

$$|f(x_1) - f(x_2)| \leq L|x_1 - x_2| \quad \forall x_1, x_2 \in \mathbb{E}^1.$$

Theorem 7.1 *Let $f : \mathbb{E}^1 \to \mathbb{E}^1$ be a Lipschitz continuous function and let $y_0 \in \mathbb{E}^1$ be given. Then the problem*

$$\dot{y} = f(y), \quad y(0) = y_0$$

has exactly one continuously differentiable solution on $\mathbb{E}^1$.

For the proof see e.g. [68, Chapt. 1] and [297]. The next two examples illustrate that if f is not Lipschitz continuous, then the global existence and the uniqueness of the solution in Theorem 7.1 are not guaranteed.

Example 7.5 The function $y(t) = \tan y$ for $t \in [0, \pi/2)$ solves the Riccati equation $\dot{y} = y^2 + 1$ whose right-hand side is not Lipschitz continuous for $y \in (-\infty, \infty)$. Thus we see that the existence of a global solution over $[0, \infty)$ need not be guaranteed for right-hand sides that are not Lipschitz continuous.

Example 7.6 The equation $\dot{y} = \sqrt{y}$ with the initial condition $y(0) = 0$ has two solutions $y(t) = \frac{1}{4}t^2$ and $y \equiv 0$. The right-hand side $\sqrt{y}$ is not Lipschitz continuous at 0.

Now we shall investigate the solvability of the Friedmann equation (7.21). Since this equation is autonomous, it does not matter in which point the initial condition is prescribed.

Theorem 7.2 *If the curvature index $k \leq 0$ and (7.20) holds, then (7.21) with an arbitrary initial condition $a(\tau) > 0$ has exactly one continuously differentiable solution which is increasing over the interval $[\tau, \infty)$.*

Proof The right-hand side of (7.21)

$$F(a) = Aa^2 + B + \frac{C}{a}, \quad a \in (0, \infty), \tag{7.23}$$

is a strictly convex function, since its second derivative $F''(a) = 2A + 2C/a^3$ is positive for $a > 0$. From the equation

$$F'(a) = 2Aa - \frac{C}{a^2} = 0 \tag{7.24}$$

we find that F attains its unique minimum at the point

$$a_{\min} = \sqrt[3]{\frac{C}{2A}} = \sqrt[3]{\frac{3C}{2\Lambda c^2}} > 0. \tag{7.25}$$

From (7.22) we see that the right-hand side of (7.21) is positive for $k \leq 0$ and, therefore, $\dot{a}$ does not change its sign. By Theorem 7.1 the expansion function $a = a(t)$ exists over the whole interval $[\tau, \infty)$ and is continuously differentiable there, since by (7.20) the lower bound $\dot{a} = \dot{a}(t) > 0$ holds for all $t \geq \tau$ and the following linear upper bound

$$\dot{a} = \sqrt{Aa^2 + B + \frac{C}{a}} < 2\sqrt{A}a$$

holds for sufficiently large a. □

Let us still note that the function $b(t) = \exp(2\sqrt{A}t)$ is the solution of the upper bound equation $\dot{b} = 2\sqrt{A}b$ and thus $0 < a(t) < b(t)$ for $t \to \infty$.

By Theorem 7.1 solutions $a = a(t)$ of (7.21) for various initial conditions $a(\tau) > 0$ do not intersect and they also do not bifurcate, since the right-hand side of the equation $\dot{a} = \sqrt{F(a)}$ is a Lipschitz continuous. In the next theorem, we will investigate the case $k = 1$ which is more complicated and quite technical.

Theorem 7.3 *If the curvature index* $k = 1$ *and (7.20) holds, then for the solution of (7.21) with an arbitrary initial condition* $a(\tau) > 0$ *there exists at most one point* $t_2 \in [\tau, \infty)$ *such that* $\dot{a}(t_2) = 0$.

Proof Let F and $a_{\min}$ be defined again by (7.23) and (7.25), respectively. For $k = 1$ consider the following three cases:

1. Let $F(a_{\min}) > 0$. Then by (7.20) and (7.21) we have $\dot{a}(t) \geq \sqrt{F(a_{\min})} > 0$ for all $t \geq \tau$, where τ is given by (6.33). Therefore, the expansion function is again increasing over the whole interval $[\tau, \infty)$ due to the following linear upper bound $\dot{a} \leq 2\sqrt{A}a$ for sufficiently large a. Note that this important case is often forgotten in the cosmological literature.
2. Let $F(a_{\min}) = 0$. Then $F(a) > 0$ for all $a \neq a_{\min}$ and by (7.21) the second derivative is

$$\ddot{a} = \pm \frac{2Aa - C/a^2}{2\sqrt{Aa^2 + B + C/a}} \dot{a} = \pm \Big(Aa - \frac{C}{2a^2} \Big). \tag{7.26}$$

Comparing this equation with (7.13), we observe that the sign $+$ should only be considered. Moreover, we see that its value for $a = a_{\min}$ is

$$Aa_{\min} - \frac{C}{2a_{\min}^2} = 2^{-1/3} A^{2/3} C^{1/3} - 2^{-1/3} A^{2/3} C^{1/3} = 0. \tag{7.27}$$

If $a(\tau) > a_{\min}$ then

$$\ddot{a}(t) > Aa_{\min} - \frac{C}{2a_{\min}^2} = 0$$

and $a = a(t)$ is strictly convex and increasing over the interval $[\tau, \infty)$ due to (7.21), see Fig. 7.5.

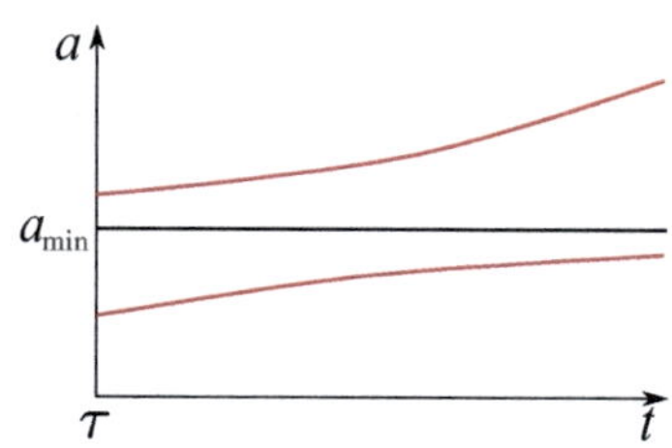

Fig. 7.5 Eddington-Lemaître solutions $a = a(t)$ of the Friedmann equation (7.21) for the case $k = 1$ and $F(a_{\min}) = 0$. Moreover, if (7.20) holds, then all solutions are increasing. For $0 < a(\tau) < a_{\min}$ the corresponding solution is strictly concave and for $a(\tau) > a_{\min}$ it is strictly convex. The constant Einstein solution $a(t) \equiv a_{\min}$ does not satisfy (7.20)

If $a(\tau) = a_{\min}$ then $a = a(t)$ is a constant function everywhere due to (7.21), (7.26), and (7.27) which contradicts (7.20). Hence, this case cannot happen. It corresponds to the Einstein static universe, which will be treated in detail in Theorem 7.4.

If the initial condition satisfies the inequalities $0 < a(\tau) < a_{\min}$, then

$$\ddot{a}(t) = Aa(t) - \frac{C}{2a^2(t)} < Aa_{\min} - \frac{C}{2a_{\min}^2} = 0$$

and $a = a(t)$ is strictly concave and increasing for all $t \in [\tau, \infty)$. The sharp inequality in the above relation follows from the uniqueness, since no solution can cross the line $a(t) \equiv a_{\min}$.

3. Finally, let $F(a_{\min}) < 0$. Then there exist two roots $a_1 < a_2$ such that $F(a_1) = F(a_2) = 0$ and F is negative on the interval (a_1, a_2). However, the left-hand side of (7.21) is nonnegative everywhere. Hence, the case $a_1 < a(t) < a_2$ for $t \geq \tau$ cannot happen.

If $a(\tau) = a_1$ or $a(\tau) = a_2$, then by (7.21) we have $\dot{a}(\tau) = 0$ which contradicts (7.20), i.e., these cases can be excluded, too.

If the initial condition satisfies $a(\tau) > a_2$, then $F(a) > 0$ for all $a > a_2$ and by (7.20) the expansion function is again increasing.

We will not investigate the case $a_1 \leq 0$, since the radius of $\mathbb{S}_a^3$ is positive. So let the initial condition $a(\tau)$ satisfy the inequalities $0 < a(\tau) < a_1$. Then F is positive on the interval $(0, a_1)$. Set

$$K := Aa_1 - \frac{C}{2a_1^2}.$$

Since $0 < a_1 < a_{\min}$, we obtain by (7.26) and (7.27) that

$$\ddot{a}(t) = Aa(t) - \frac{C}{2a^2(t)} \leq K < Aa_{\min} - \frac{C}{2a_{\min}^2} = 0$$

for all $t > \tau$ such that $a(t) > 0$, i.e.

$$\ddot{a}(t) \leq K < 0$$

which means that $a = a(t)$ is strictly concave. By integration we find that

$$\dot{a}(t) - \dot{a}(\tau) = \int_\tau^t \ddot{a}(\sigma)\,\mathrm{d}\sigma \leq K(t - \tau) < 0.$$

Consequently, $Kt \to -\infty$ as $t \to \infty$ and since $\dot{a}(\tau)$ and $K\tau$ are fixed numbers, there exists exactly one time instant $t_2 > \tau$ such that

$$\dot{a}(t_2) = 0 = F(a_1).$$

Then the cosmological density parameters are not well defined due to the division by zero in (6.29). This case corresponds to a universe of finite duration. The unique solution of (7.21) is given by the equation $\dot{a} = \sqrt{F(a)}$ on the interval $[\tau, t_2]$ and by the equation $\dot{a} = -\sqrt{F(a)}$ after the point t_2 until $a = a(t)$ reaches the zero value.

□

7.6 Einstein's Static Universe

Friedmann described the dynamical behavior of the universe as an alternative to Einstein's static universe. Now we show that the expansion function associated to Einstein's static universe also satisfies the Friedmann equation. Instead of (7.20) we shall assume that $\dot{a} \equiv 0$ and moreover, let $k = 1$ and let again

$$\Lambda > 0. \tag{7.28}$$

This special case corresponds to *Einstein's static universe* from [94]. By (7.23) and (7.25) we have

$$\begin{aligned} 0 = F(a_{\min}) &= Aa_{\min}^2 - c^2 + \frac{C}{a_{\min}} = A\sqrt[3]{\frac{C^2}{4A^2}} - c^2 + C\sqrt[3]{\frac{2A}{C}} \\ &= \sqrt[3]{\frac{AC^2}{4}} - c^2 + 2\sqrt[3]{\frac{AC^2}{4}}. \end{aligned}$$

Hence, by (7.22) and (7.28),

$$C = c^3\sqrt{\frac{4}{27A}} = \frac{2c^2}{3\sqrt{\Lambda}}. \tag{7.29}$$

Substituting (7.29) into (7.25), we get the formula for the radius of Einstein's static universe, namely

$$a = \frac{1}{\sqrt{\Lambda}}. \tag{7.30}$$

Notice that this solution does not require any initial and boundary conditions. However, Einstein in [94] did not ask a natural question whether stars could shine for an infinitely long time. The constant solution of Einstein's equations (7.30) is thus unphysical.

Theorem 7.4 *The total mass of Einstein's static universe M and its mean density ρ can be expressed by the following fundamental constants π, c, G, and Λ,*

$$M = \frac{\pi c^2}{2G\sqrt{\Lambda}}, \qquad \rho = \frac{\Lambda c^2}{4\pi G}.$$

Proof By (6.2) the volume of the sphere $\mathbb{S}_a^3$ with radius a is

$$V = 2\pi^2 a^3.$$

From this, (7.29), and (7.22) we obtain

$$\frac{2c^2}{\sqrt{\Lambda}} = 3C = 8\pi G\rho a^3 = \frac{4G\rho V}{\pi} = \frac{4GM}{\pi},$$

which yields the proposed relation for M. From this and (7.30) we get the formula for the density ρ. This formula can also be obtained from (7.13) for $\ddot{a} \equiv 0$ and $p \equiv 0$. □

Remark 7.22 The mass of our Galaxy is about 10^{42} kg. For comparison let us evaluate the mass of Einstein's static universe by Theorem 7.4 for $\Lambda \approx 10^{-52}$ m^{-2},

$$\boxed{M = \frac{\pi c^2}{2G\sqrt{\Lambda}} \approx 2.12 \cdot 10^{53} \text{ kg.}} \tag{7.31}$$

Hence, Einstein's static universe is tuned very finely. However, the resulting stationary solution

$$a = \Lambda^{-1/2} \approx 10^{26} \text{ m} \tag{7.32}$$

of Eq. (6.32) is not stable, i.e., any small deviation from the constant $a = a(t) = \Lambda^{-1/2}$ will cause either a gravitational collapse, or expansion (see [249, p. 746] or [348, p. 405]). We also see that the constant C in (7.29) is independent of the total mass of the universe, while in a more general case (7.22) the constant C depends on the total mass. Furthermore, we observe that the density parameter $\Omega_{\mathrm{M}}(t)$ defined in (6.29) is infinity even though nothing dramatic happens. The mean density of baryonic matter is of course finite, namely

$$\rho = \frac{M}{V} = \frac{\pi c^2}{2G\sqrt{\Lambda} \cdot 2\pi^2 a^3} = \frac{\Lambda c^2}{4\pi G} \approx 10^{-26} \text{ kg/m}^3. \tag{7.33}$$

This density differs from that stated in Remark 7.27 below due to a special choice of $\Lambda \approx 10^{-52}$ m^{-2}.

Remark 7.23 The numbers appearing in the Friedmann equation (6.32) are very different in size in the SI units, e.g., $\Lambda \approx 10^{-52}\ \text{m}^{-2}$, $\rho(t_0) \approx 10^{-27}\ \text{kg/m}^3$, $G = 6.674 \cdot 10^{-11}\ \text{m}^3\ \text{kg}^{-1}\ \text{s}^{-2}$, $c = 3 \cdot 10^8$ m/s, and $a(t_0) \approx 10^{26}$ m. This might produce various numerical troubles in finite arithmetic precision. For $k = 1$ the radius $a(t_0)$ is usually estimated as $a(t_0) \approx c/H(t_0) \approx 10^{26}$ m, but several times larger values can also be found in the literature.

Theorem 7.5 *The pressure $p = 0$ for Einstein's static universe.*

Proof Since $\ddot{a} = 0$, we find from the linear Friedmann equation (7.13) and (7.30) that

$$\frac{4\pi G p}{c^2} a = -\frac{4\pi G}{3} \rho a + \frac{\Lambda c^2}{3} a = -\frac{c^2\sqrt{\Lambda}}{3} + \frac{c^2\sqrt{\Lambda}}{3} = 0.$$

The proof is completed. □

7.7 Some Consequences of the Friedmann Equations

Remark 7.24 Now let us calculate the inflection point of the expansion function $a = a(t)$, i.e., when the strictly concave expansion function changes into the strictly convex function. Note that the inflection point need not exist in the interval $[\tau, \infty)$ under some special conditions, see e.g. Fig. 7.5. From (7.21) we find that

$$\ddot{a} = \frac{2Aa - C/a^2}{2\sqrt{Aa^2 + B + C/a}} = 0.$$

Hence, $\frac{2}{3}\Lambda c^2 a^3 = 2Aa^3 = C$ and thus for the inflection point we obtain surprisingly the same relation as (7.25), i.e. the radius of Einstein's static universe (see (7.29) and (7.30)), namely

$$a = \sqrt[3]{\frac{3C}{2\Lambda c^2}}$$

which is independent of the curvature index k. From this and (7.22) we get $2\Lambda c^2 = 8\pi G\rho$ which for $\Lambda \approx 10^{-52}\ \text{m}^{-2}$ yields the corresponding mass density

$$\rho = \frac{\Lambda c^2}{4\pi G} \approx 10^{-26}\ \text{kg/m}^3.$$

Interestingly, its value is again the same as that in (7.33). Since the measured mean density in our neighborhood is smaller, the expansion function seems to be strictly convex at present. This fact was predicted already in 1931 by Georges Lemaître

[219, p. 422] (see also [249, p. 746], [272, 345], [348, p. 411]). Moreover, from Remark 7.12 we immediately find that if

$$\Omega_{\rm M}(t_3) = 2\Omega_\Lambda(t_3)$$

for some time t_3 and $\dot a(t_3) \neq 0$, then $\ddot a(t_3) = 0$.

Theorem 7.6 *The Friedmann equations (7.21) and (7.13) imply that the pressure* $p = p(t) = 0$ *for all* $t > \tau$.

Proof By differentiation of (7.21) we find that

$$2\dot a\ddot a = 2Aa\dot a - \frac{C}{a^2}\dot a.$$

If $k = 1$ then by Theorem 7.3 there exists at most one point $t_2 \in [\tau, \infty)$ such that $\dot a(t_2) = 0$. If $\dot a \neq 0$, then we get

$$\ddot a = Aa - \frac{C}{2a^2},$$

i.e., this equation is valid except for at most one point t_2. However, (7.13) can be rewritten as

$$\ddot a = Aa - \frac{C}{2a^2} - \frac{4\pi Gpa}{c^2}$$

implying that $p \equiv 0$, since $p = p(t)$ is supposed to be continuous (including t_2).

□

Note that the zero pressure corresponds to incoherent dust (cf. (6.21)). This fits to astronomical observations for $t \geq \tau$.

Remark 7.25 If $\Lambda = 0$ and $k = 1$, then from (7.21) we get $\dot a^2 a^2 = c^2(\alpha a - a^2)$, where $\alpha = C/c^2$ is a positive constant. Using the conformal time η defined by (7.17), we find that $\dot a = \frac{c}{a}a'$, where $a' = \mathrm{d}a/\mathrm{d}\eta$. Therefore, $(a')^2 = \alpha a - a^2$. Employing the substitution $x = \frac{2a}{\alpha} - 1$, we obtain

$$\int \frac{\mathrm{d}a}{\sqrt{\alpha a - a^2}} = \int \frac{2}{\alpha}\frac{\mathrm{d}a}{\sqrt{1 - \big(\frac{2a}{\alpha} - 1\big)^2}} = \int \frac{\mathrm{d}x}{\sqrt{1 - x^2}} = \arcsin\Big(\frac{2a}{\alpha} - 1\Big).$$

Hence, $\eta - \eta_0 = \arcsin\big(\frac{2a}{\alpha} - 1\big)$ and $\sin(\eta - \eta_0) = \frac{2a}{\alpha} - 1$, which leads to $-\sin\eta_0 = -1$ and $\eta_0 = \pi/2$. Thus, $a(\eta) = \frac{\alpha}{2}(1 - \cos\eta)$ for $\eta \in [0, 2\pi]$ which leads to an oscillating universe, see also [348, p. 412].

Remark 7.26 Consider again the situation sketched in Fig. 6.2. Gravity (e.g. of our Earth or Galaxy) does not act on bodies which are in the observable universe, but it acts on these bodies in their future cones (see also Remark 11.12 below). This fact is not described by the Friedmann equation (7.21). Hence, there is another problem of the standard cosmological model.

Remark 7.27 Concerning the baryonic mass density $\Omega_{\rm B} = 0.049$ (see Fig. 6.9), we find by (6.29) and (6.31) that

$$\rho(t_0) = \frac{3\Omega_{\rm B} H_0^2}{8\pi G} = \frac{3 \cdot 0.049 \cdot (2.27)^2 \cdot 10^{-36}}{8\pi \cdot 6.674 \cdot 10^{-11}} \ {\rm kg/m^3} = 4.5 \cdot 10^{-28} \ {\rm kg/m^3}$$

for the present age of the universe t_0. This is less than one proton in three cubic meters, since $m_{\rm p} = 1.67 \cdot 10^{-27}$ kg. However, it is usually said that there are 6 protons per $\rm m^3$ on averge.

7.8 The Age of the Universe

Expressing the expansion function $a = a(t)$ as a Taylor expansion in time $t_0 = 0$, which corresponds to the present time, we have by (6.27) and (6.29)

$$a(t) = a(0) + \dot{a}(0)t + \frac{1}{2}\ddot{a}(0)t^2 + \cdots = a(0)\left(1 + H_0 t - \frac{1}{2}q_0 H_0^2 t^2 + \ldots\right) \tag{7.34}$$

$$= a(0)\left(1 + H_0 t - \frac{1}{2}\Lambda q_0 \frac{c^2}{3\Omega_\Lambda(0)} t^2 + \ldots\right).$$

Here $H_0 = H(0)$ and $q_0 = q(0) = -0.6$ is the usually accepted present value of the deceleration parameter (see [300, p. 110]) which is negative, since the expansion of the universe accelerates. This means that the expansion function $a = a(t)$ is strictly convex on a given time interval, see Fig. 7.6.

M. Carrera and D. Giulini [58, p. 175] correctly derive that at the distance of Pluto (i.e. about 40 au) the acceleration of the expansion of the universe is only $2 \cdot 10^{-23}$ m/s^2 which is indeed an entirely negligible quantity. A similar tiny value was derived by B. Mashhoon et al. [236, p. 5041]. However, all these authors concentrated only on the single quadratic term in expansion (7.34) and did not consider the large value of the Hubble constant (6.31) which stands at the linear term in (7.34). In other words, an accelerated expansion does not manifest itself on scales of the Solar system, but the expansion itself is observable. In particular, we have

$$|H_0 t| \gg \frac{1}{2}|q_0| H_0^2 t^2 = \frac{1}{2}|q_0| \frac{\Lambda c^2}{3\Omega_\Lambda(0)} t^2$$

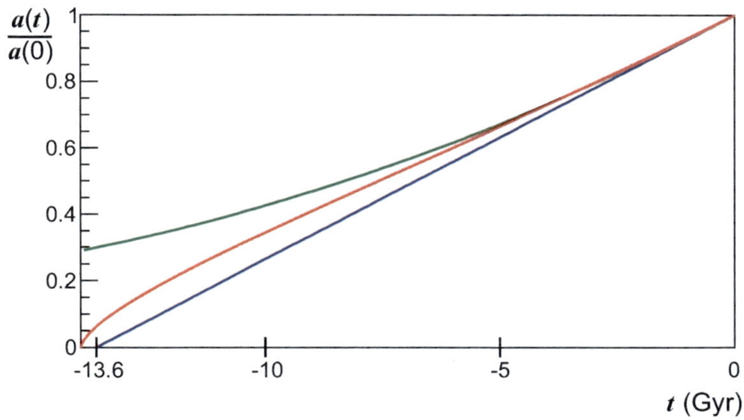

Fig. 7.6 The assumed behavior of the normalized expansion function $a(t)/a(0)$. The time variable is shifted for simplicity such that $t_0 = 0$ corresponds to the present time. The lower blue graph corresponds to the linear function $1 + H_0 t$ from (7.34) on the interval $[-1/H_0, 0]$, where $1/H_0 = 13.6$ Gyr is the Hubble time. The upper green graph shows the quadratic function $1 + H_0 t - \frac{1}{2} q_0 H_0^2 t^2$ with $q_0 = -0.6$. The middle red graph illustrates the behavior of the normalized expansion function calculated numerically for $k = 1$, $\Omega_M = 0.317$, $\Omega_\Lambda = 0.683$, and $H_0 = 67.15$ km/(s Mpc), see [284]. We observe that the accelerated expansion differs only very little from the linear expansion during the last few Gyr

for t close to 0. Consequently, the accelerated expansion given by the quadratic term only appears at cosmological distances. In spite of that, the single quadratic term is so small that the linear term $|H_0 t|$ from (7.34) essentially dominates not only in the neighborhood of 0, but for all t in the whole interval $(-1/H_0, 0)$, since

$$0.3 \cdot |H_0 t| > \frac{1}{2} |q_0| H_0^2 t^2,$$

where $\frac{1}{2}|q_0| = 0.3$. Recall that the physical dimension of the constant Λ is m^{-2}. On the other hand, the physical dimension of H_0 is s^{-1} which does not depend on space variables. Therefore, we can detect the Hubble expansion also on small scales, where the effect of Λ is negligible.

We cannot reliably estimate the remainder of the Taylor series on the whole domain of definition, since the first derivatives of the expansion function $a = a(t)$ were extremely large just after the Big Bang. To see this, denote by

$$\boxed{t_0 = 13.82 \text{ Gyr}}$$

the estimated age of the universe by the standard cosmological model, see [290]. The microwave background radiation appeared $t_1 = 380{,}000$ years after the Big

Bang, see (6.33). Then for the measured cosmological redshift $z = 1089$ we find that

$$\frac{t_1 a(t_0)}{t_0 a(t_1)} = \frac{t_1}{t_0}(z+1) = \frac{380{,}000 \cdot 1090}{13.82 \cdot 10^9} = \frac{1}{33.3},$$

where the first equality follows directly from (see [249, p. 730], [272, p. 96]) the definition of redshift. From this we get

$$33.3 \cdot \frac{a(t_0)}{t_0} = \frac{a(t_1)}{t_1},$$

and thus the mean expansion speed of the universe on the interval $(0, t_1)$ was 33.3 times larger than on the interval $(0, t_0)$. This indicates that the expansion function had a much larger derivative after the Big Bang than today and that it was strictly concave on some subinterval (cf. Figs. 7.6 and 6.2).

Remark 7.28 The oldest known star is HD 140283. Its age is estimated to 14.46 ± 0.8 billion years independently of cosmological models. Since it is in our close neighborhood (only 190 ly away from us), it is very probable that there are even older stars in the entire universe. This age would possibly conflict with the calculated age 13.82 billion years of the universe as determined from the Friedmann equation by the Planck satellite data.

7.9 Flight Around the Universe

Consider a homogeneous and isotropic universe which can be modeled by three maximally symmetric manifolds $\mathbb{S}^3$, $\mathbb{E}^3$, and $\mathbb{H}^3$ up to scaling (see Theorem 6.2). Let us ask a purely theoretical question:

Could we orbit the universe in finite time?

If the curvature index $k \leq 0$, then the answer is NO, since the space is unbounded. So let $k = 1$. In the next example we show that for Einstein's static universe the answer is obviously YES.

Example 7.7 Assume that the universe is modeled by the three-dimensional sphere $\mathbb{S}_a^3$ with fixed radius $a = \Lambda^{-1/2}$, see (7.32). Further, assume that a space shuttle travels along a given great circle (geodesic) with constant velocity $v = 0.999\,c$. Then its total cosmic travel time T can be estimated by the relation $1 \text{ yr} \approx \pi \cdot 10^7$ s as follows

$$T = \frac{2\pi a}{v} \approx \frac{2\pi \cdot 10^{26}}{3 \cdot 10^8} \text{ s} \approx \frac{2\pi \cdot 10^{26}}{3 \cdot 10^8 \cdot \pi \cdot 10^7} \text{ yr} = 6.666 \cdot 10^{10} \text{ yr} \qquad (7.35)$$

(which holds also for the speed of light). This is five times the present age of the universe $t_0 = 13.82$ Gyr according to the standard cosmological model. Note that

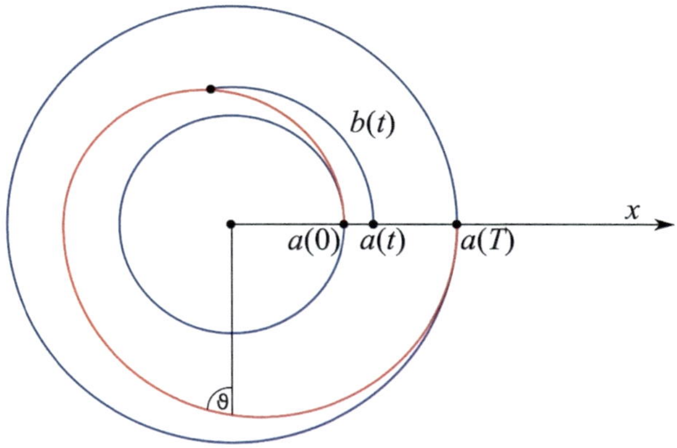

Fig. 7.7 Trajectory of a space shuttle in an expanding universe modeled by the sphere $\mathbb{S}^3_{a(t)}$ for $a(T) = 2a(0)$. Its equation in polar coordinates reads $r(\vartheta) = a(0)\exp(\vartheta V/v)$ and its length is $(a(T) - a(0))/\cos\arctan(v/V)$

Λ (if it exists) is known only very roughly, see (7.32). However, the proper time of the traveler would be much smaller than T due to time dilation (2.5).

Further, we shall consider a variable radius $a = a(t)$ of the sphere $\mathbb{S}^3_a$ and for simplicity, we assume that it increases linearly with constant velocity $V > 0$ (cf. Fig. 7.6). The expansion velocity V can be even faster than light and there is no contradiction with STR outside the observer's inertial system, see [79]. Moreover, assume that the space shuttle has also a constant velocity $v > 0$ with respect to its local neighborhood (i.e. intergalactic dust). In this particular case, its trajectory will be one turn of the logarithmic spiral instead of the great circle, see Fig. 7.7. Recall that the *logarithmic spiral* is a self-similar curve making a constant angle $\vartheta \in (0, \pi/2]$ with the polar radii at all its points [296, p. 139], (for an approximately logarithmic double spiral pattern see also Fig. 8.7).

The following surprising theorem was inspired by Luboš Pick and Mirko Rokyta (see [283, p. 228]). It states that if the space shuttle has an arbitrarily small constant velocity $v > 0$ with respect to V, then it always needs finite time to travel around a linearly expanding universe. Therefore, we will deal with the nonrelativistic addition of velocities in Theorem 7.7 below. The spiraling trajectory can be easily unfolded into a straight line. Thus, in fact, we shall investigate only a one-dimensional problem. For brevity, we also make a linear shift of cosmic time so that $t = 0$ corresponds to the start time.

Theorem 7.7 (Pick–Rokyta) *Let $\mathbb{S}^3_{a(t)}$ expand with constant velocity $V > 0$, i.e. $\dot{a}(t) = V$ for all $t \geq 0$, and let $v > 0$ be a constant velocity of the space shuttle. Then its total cosmic travel time around the universe is*

$$T = a(0)\frac{\exp(2\pi V/v) - 1}{V}. \tag{7.36}$$

Proof Assume that the space shuttle will launch from the axis x at time $t = 0$. The circumference of a great circle (geodesic) of $\mathbb{S}_a^3$ at time $t \geq 0$ is clearly given by

$$2\pi a(t) = 2\pi(a_0 + Vt), \tag{7.37}$$

where $a_0 = a(0)$ is the initial radius. Let $b(t)$ be the instant distance of the space shuttle at time $t \geq 0$ along the great circle with radius $a(t)$ from the axis x, i.e. from the launch point in an expanding universe (see Fig. 7.7). It satisfies the following initial problem

$$\dot{b}(t) = v + w(t), \quad b(0) = 0, \tag{7.38}$$

where $w = w(t)$ is the drifting velocity of a linearly expanding space. We see that the ratio between the velocity $w(t)$ and the expansion velocity $2\pi V$ of the great circle is the same as the ratio between the distance $b(t)$ and the circumference $2\pi a(t)$, namely,

$$\frac{w(t)}{2\pi V} = \frac{b(t)}{2\pi a(t)}.$$

From this, (7.37), and (7.38) we get the initial problem

$$\dot{b}(t) = v + V\frac{b(t)}{a_0 + Vt} \quad \text{and} \quad b(0) = 0. \tag{7.39}$$

The right-hand side of this first order differential equation contains the velocity v which is increased by the drifting velocity $w(t)$ of space expansion. By the substitution $t = 0$ and differentiation, we can easily verify that

$$b(t) = (a_0 + Vt)\frac{v}{V}\ln\frac{a_0 + Vt}{a_0} \tag{7.40}$$

is the solution to problem (7.39). The space shuttle will return to the launch point from the opposite direction at time T when

$$2\pi a(T) = b(T).$$

From this, (7.37), and (7.40) we obtain the equation

$$\frac{v}{V}\ln\frac{a_0 + VT}{a_0} = 2\pi. \tag{7.41}$$

Consequently,

$$a_0 + VT = a_0 \exp \frac{2\pi V}{v} \tag{7.42}$$

and thus (7.36) follows. □

Remark 7.29 The assumption that $V > 0$ is constant in Theorem 7.7 is not too restrictive, since the expansion function $a = a(t)$ in Fig. 7.6 is almost linear during the last 10 Gyr. Moreover, for the linear function $a(t) = a_0 + Vt$ the associated Hubble parameter

$$H(t) = \frac{\dot{a}(t)}{a(t)} = \frac{V}{a_0 + Vt}$$

has a similar behavior as that in Fig. 10.4. By l'Hospital's rule, (7.36) converges for $V \to 0$ to $T = 2\pi a(0)/v$ corresponding to Einstein's static universe (7.35).

The assumption that $v > 0$ is constant in Theorem 7.7 is also not restrictive, since for each variable velocity of the space shuttle greater than v we clearly get a smaller (i.e. finite) travel time.

Remark 7.30 For simplicity, we have chosen $a(T) = 2a(0)$ in Fig. 7.7. From this and (7.41) it follows that $v/V = 2\pi/\ln 2$. Hence, the constant slope angle ϑ between the polar radii and the tangent line to the spiral is only slightly less than $\pi/2$, where $\vartheta = \arctan(v/V)$. Nevertheless, Theorem 7.7 covers also the case $0 < v \ll c < V$ which is more realistic.

Now we show that the CMB radiation (the cosmological horizon) might be just the image of the antipodal point of our neighborhood $\approx$ 14 Gyr ago.

Example 7.8 To illustrate the main idea of the above hypothesis, consider the trajectory of a photon which is moving at the velocity $v = c$ with respect to its local space. Then for the trajectory from the antipodal point of $\mathbb{S}^3_{a(0)}$ to our present location we find by (7.37) and similarly to (7.42) that

$$a(t) = a(0) + Vt = a(0) \exp \frac{\pi V}{c},$$

where t is the travel time. From this and (7.1) the cosmological redshift of the photon is given by

$$z = \frac{a(t)}{a(0)} - 1 = \exp(2.22\,\pi) - 1 \approx 1075 \quad \text{for } V = 2.22c.$$

This value is quite close to the observed redshift $z = 1089$ of the CMB (see Fig. 6.1), even though the actual expansion velocity of the universe is variable and not a fixed constant V.

7.10 Conclusions

The current ΛCDM cosmological model supposes that time flows completely uniformly from the Big Bang on. By [361, 362] the proper time should be replaced by the conformal time (7.17). The main objections against the Friedmann equation can be summarized as follows:

1. It implies the existence of dark matter and dark energy that have not been confirmed by measurements, yet, see Sect. 6.7.
2. It was derived by excessive extrapolations from Einstein's equations applied to the entire universe, see Sect. 7.2. For instance, the Hubble-Lemaître constant was calculated by extrapolating 13.82 Gyr old CMB data to the present time, see Remark 6.23. This is a typical ill-conditioned problem.
3. It is too simple model described by the first order ordinary differential equation with constant coefficients, see Sect. 7.5.
4. It yields a smaller age of the universe than the oldest stars in our neighborhood, see Remark 7.28.
5. It is fully deterministic and reversible, see Remarks 5.6 and 7.3.
6. It is used to calculate the time and space distances, see Sects. 7.3 and 7.4. The difference in the denominator of (7.9) can produce numerical instabilities, see Remarks 7.4 and 7.15.
7. The cosmological parameters allow division by zero, see Sect. 6.7.
8. The cosmological parameters have a strange behavior in time, see Remark 6.21.
9. No significant digit of the cosmological constant is known, see Remark 3.15. Let us emphasize that this constant stands at the leading and most dominant quadratic term in (7.21).
10. The emergence and evolution of a homogeneous and isotropic hyperbolic (or Euclidean) universe cannot be well described by the Friedmann equation, since this would require an infinite speed of information transfer, see Remark 6.27.

Chapter 8
Arguments Against the Proclaimed Amount of Dark Matter

8.1 Problem of Missing Matter

The term *dark matter* was first briefly mentioned in Jan Oort's paper [266, p. 285] for mass that is not visible. One year later this term (in German *dunkle Materie*) was independently used by Fritz Zwicky in his 1933 paper [385] on page 125. This paper essentially changed the development of astronomy and cosmology for many decades. Using the Virial Theorem, Zwicky found that in order to explain rapid movements of only eight galaxies in the giant galaxy cluster Abell 1656 in Coma Berenices (called *Coma cluster*), he had to assume the existence of a 400 times larger amount of nonluminous matter than luminous to keep the cluster together. In 1936, Sinclair Smith [326, p. 27] independently reached a similar conclusion for the Virgo cluster. The huge factor 400 was reduced to 150 in another paper [386] by Zwicky. None of the above-mentioned authors claimed that dark matter must be nonbaryonic other than just being optically invisible. But in modern terms dark matter stands for hypothetical nonbaryonic dark matter of unknown composition, i.e., it does not include baryonic matter that is not luminous (i.e. also black holes). Moreover, it does not interact electromagnetically with "normal" matter. At present, it is supposed that

$$\text{the ratio of masses of dark matter to baryonic matter} \approx \mathbf{6:1}. \tag{8.1}$$

The aim of this chapter is to collect the main arguments from our papers [165, 183, 185] showing that this ratio is exaggerated. We do not claim that dark matter does not exist. However, we do claim that the ratio on the left-hand side of (8.1) is smaller and it can even be zero. We present several facts and phenomena which should be taken into account to estimate the amount of dark matter.

In Sect. 8.2, we briefly recall the Virial Theorem which establishes the equilibrium of the total kinetic and potential energy of gravitationally interacting bodies of a stable system. Then we present Zwicky's method and comment on it. We also

M. Křížek, L. Somer, *Mathematical Aspects of Paradoxes in Cosmology*,
https://doi.org/10.1007/978-3-031-31768-2_8

point out several effects that were not taken into account by Zwicky and that may essentially reduce the calculated mass of the whole Coma cluster.

In Sect. 8.3, we present two "back of the envelope" calculations illustrating whether it is necessary to assume extragalactic dark matter in the Coma cluster. In Sect. 8.4, we analyze rotational curves proposed by Vera Rubin and in the last section we give some concluding remarks on the existence of dark matter.

8.2 Analysis of Zwicky's Method

Zwicky in [385, 386] investigated large velocities of galaxies in the Coma galaxy cluster which is located near the north Galactic pole. Therefore, the recession speed of the Coma cluster from the Sun is practically equal to the recession speed of the Coma cluster from the Milky Way, even though the orbital speed $v_\odot \approx 230$ km/s of the Sun about the Galactic center is relatively high. Using the Virial Theorem

$$V + 2T = 0, \tag{8.2}$$

Zwicky estimated its mass M by the following simple formula

$$M = \frac{5R\overline{v}^2}{3G}, \tag{8.3}$$

where $V = -\frac{3}{5}GM^2/R$ is the total potential energy of the cluster, $T = \frac{1}{2}M\overline{v}^2$ is the kinetic energy, R is the radius of the cluster, and $\overline{v}$ is the mean quadratic speed with respect to the center of gravity of the cluster. For detailed derivations of the Virial Theorem and formula (8.3) we refer to [183]. Note that Smith [326, p. 29] presents a similar formula to (8.3), where the coefficient $\frac{5}{3}$ is replaced by $\frac{1}{2}$. This is a quite big difference.

To explain that

$$\overline{v} > 1000 \text{ km/s}$$

in the Coma cluster, Zwicky had to assume the existence of a large amount of nonluminous matter to keep the cluster gravitationally bound together. However, he used very imprecise data and made many simplifying assumptions. For instance:

1. Zwicky assumed that the distance to the Coma cluster is 14 Mpc which was measured by Milton Humason. However, the actual distance is about 100 Mpc. Therefore, also Hubble in [134] got the following large value of the Hubble constant $H_0 = 500$ km s^{-1}Mpc^{-1}, see Remark 6.17.
2. Zwicky was mistaken by two orders when calculating masses of galaxies in the Coma cluster from their luminosity.

3. He used Newtonian mechanics with an infinite speed of gravitational interaction, while the actual speed is finite.
4. He replaced galaxies of diameter $\sim 10^{10}$ au by point masses. However, points cannot rotate. Hence, this does not allow one to consider angular moments of rotating galaxies which also contribute to the total angular momentum.
5. Tidal forces that influence the dynamics of the system were not considered. For instance, an isolated binary system of two galaxies that orbit closely to each other is not stable, since galaxies will merge due to tidal friction, whereas the classical two-body problem has a periodic solution.
6. Zwicky substituted the spacetime curved by a thousand of galaxies by Euclidean space.
7. The gravitational redshift was ignored. Photons have to overcome the potential well of each star, the potential well of the corresponding galaxy, and also the potential well of the whole Coma cluster.
8. Zwicky assumed that the Virial Theorem (8.2) holds absolutely exactly and that the Coma cluster is in equilibrium.
9. He supposed a uniform distribution of galaxies inside the Coma cluster, see [386, p. 229]. However, the central region is much denser than the region around the boundary and larger galaxies are closer to the center.
10. In Zwicky's papers [385, 386], an isotropic distribution of velocities was assumed and relativistic effects of high velocities were ignored.

There are many other nonnegligible phenomena and facts (e.g., the existence of nonluminous intergalactic baryonic matter, local Hubble expansion, thousands of dwarf galaxies of the Coma cluster that were unknown to Zwicky, gravitational lensing, see Fig. 8.1) which were not accounted for in Zwicky's method, see [183] for details. A large amount of errors of various origins may essentially distort the resulting calculation of the mass. Thus we cannot make any reliable estimates of the actual mass of the Coma cluster by means of (8.3). The discrepancy between the virial mass and luminosity mass can be essentially reduced by means of actual data (see Fig. 8.2).

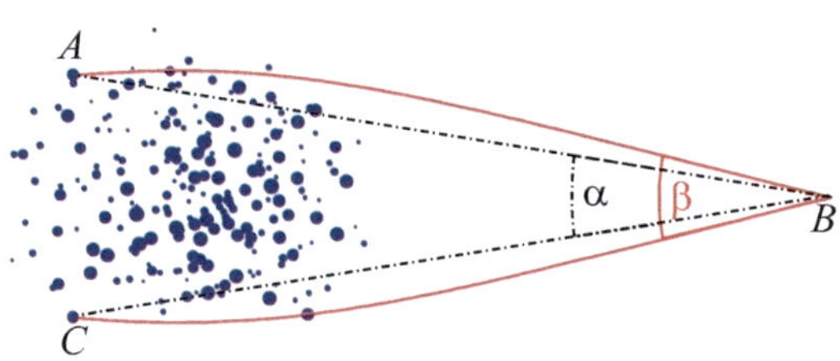

Fig. 8.1 A schematic illustration of the selflensing effect. The observation angle $\beta = \widetilde{\sphericalangle} ABC$ is slightly larger than the angle $\alpha = \sphericalangle ABC$ due to the bending of light caused by the gravitation of a galaxy cluster itself, see [184, p. 117]

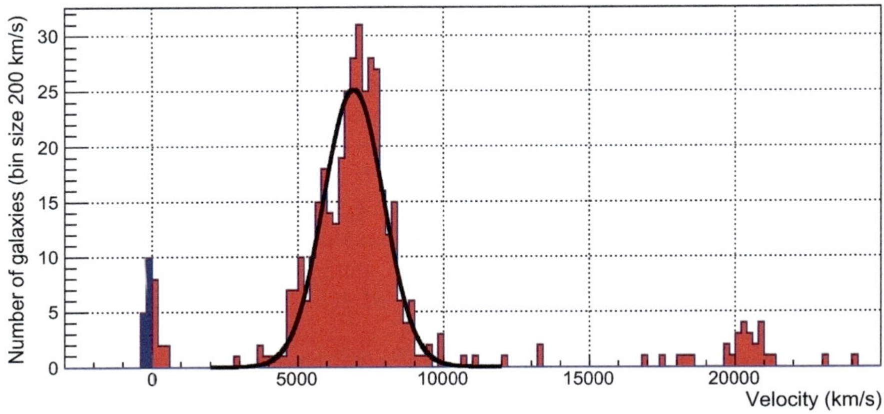

Fig. 8.2 Histogram of radial velocities less than 25,000 km/s of galaxies which are projected to the Coma cluster region. Here we consider only those galaxies whose magnitudes do not exceed 20. These data are available in [10], [31], and [69]. Galaxies possessing blueshift are on the left. The dark line represents the Gaussian curve fitted to those data that correspond to galaxies belonging to the Coma cluster. The other galaxies, which are contained in the same region of the sky, are not contained in this cluster, since their velocities with respect to the center of gravity of the cluster are too large. They should not be used to calculate the virial mass (8.3), because they do not form a bound system

8.3 Two Simple Examples

Now we present two examples showing that (8.1) is exaggerated. They can be convincingly verified by simple hand calculations.

Example 8.1 The two largest galaxies in the Coma cluster are supergiant elliptic galaxies NGC 4889 and NGC 4874 which have almost the same luminosity (see Fig. 8.3). For simplicity assume that they have the same mass m and that they orbit along a circular trajectory with center O, radius r, and velocity v (see Fig. 8.4). If one of these two galaxies were to be smaller, it would orbit the larger one by a higher velocity and along a longer path. Then it would meet and then absorb more neighboring galaxies than the larger one. By this mechanism the masses of both supergiant galaxies are balanced. By [397] they are ten times more massive than our Galaxy, i.e.

$$m = 10M_G = 10^{13}M_\odot = 2 \cdot 10^{43} \text{ kg}, \tag{8.4}$$

where the total mass of our Galaxy $M_G = 10^{12}M_\odot$ is given in [214, p. 127].

Since the gravitational potential inside a homogeneous spherical layer is constant, external galaxies and possible dark matter outside the sphere with center

Fig. 8.3 Giant galaxy cluster Abell 1656 in the constellation Coma Berenices. In the middle there are two supergiant elliptic galaxies NGC 4889 and NGC 4874 (photo NASA)

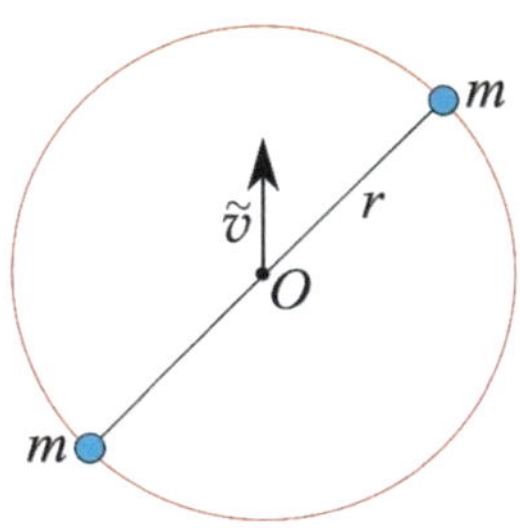

Fig. 8.4 An idealized circular trajectory of the two largest galaxies in the Coma cluster. The recession velocity of the center O from us is $\tilde{v}$

O and radius r have almost no influence on this motion, see Theorem 1.2. From Newton's laws we get

$$\frac{Gm^2}{4r^2} = \frac{mv^2}{r}. \tag{8.5}$$

The distance of both galaxies on the celestial sphere is $8.15'$ which in projection on the distance 100 Mpc gives $7.32 \cdot 10^{21}$ m. Thus for the radius r we have

$$r \geq 3.66 \cdot 10^{21}\ \text{m}. \tag{8.6}$$

According to [10, p. 19], the measured radial velocities of both supergiant galaxies are 6472 km/s and 7189 km/s, respectively. Their average velocity

$$\tilde{v} = 6830.5 \text{ km/s}$$

nicely corresponds to the mean recession speed of the whole cluster 6877 km/s, see [183, p. 53]. For the radial velocity v_{radial} with respect to $\tilde{v}$ we get by (8.4), (8.5), and (8.6)

$$3.585 \cdot 10^5 = \frac{7,189,000 - 6,472,000}{2} = v_{\text{radial}} \leq v = \sqrt{\frac{Gm}{4r}}$$

$$\leq \sqrt{\frac{6.674 \cdot 10^{-11} \cdot 2 \cdot 10^{43}}{4 \cdot 3.66 \cdot 10^{21}}} \text{ m/s} = 3.02 \cdot 10^5 \text{ m/s}. \tag{8.7}$$

Comparing the left-hand and the right-hand sides, we find a small discrepancy. This simplified example shows that Newtonian mechanics does not describe reality correctly or the masses or radial velocities of both the elliptic galaxies are wrongly estimated or we have to assume the existence of invisible matter between galaxies.

Remark 8.1 Slightly different velocities 6505 and 7108 km/s of the two giant galaxies as given in [31] would even yield a smaller value $3.015 \cdot 10^5$ m/s on the left-hand side of (8.7). Moreover, considering the gravitational influence of other matter that is inside the sphere with center O and of radius r, the right-hand side of (8.7) would even be larger, i.e., no discrepancy would appear.

Remark 8.2 By Tutukov and Fedorova [353] the intracluster medium contains 30–50% of stars from all the stars of the cluster. According to [14, 35, 366], clusters of galaxies contain five times more baryonic matter in the form of hot gas producing X-rays than baryonic matter in galaxies themselves. Also [35] and [135] claim that the intracluster medium contains more baryonic matter than the matter of all stars in a given galaxy cluster. Moreover, the lower bound in (8.6) is slightly smaller, since the cluster itself magnifies the angular distances due to lensing effects (see e.g. Fig. 8.1). Hence, the right-hand side of (8.7) should be larger. Therefore, it does not seem that there would be six times more dark matter than baryonic matter as proposed in (8.1).

Example 8.2 Assume, for simplicity, that the additional baryonic matter has a spherically symmetric distribution and denote its mass by M. Concentrating the mass M to the center O, the speed v in (8.7) can be, using Newton's First Theorem 1.1, replaced by

$$\overline{v} = \sqrt{\frac{G(m + 4M)}{4r}} \gg v.$$

To see this, it is enough to consider the relation

$$\frac{Gm^2}{4r^2} + \frac{4}{4} \cdot \frac{GmM}{r^2} = \frac{m\overline{v}^2}{r}$$

instead of (8.7). Zwicky's paradox of large observed velocities thus disappears, since it has a quite natural explanation without dark matter.

8.4 Analysis of Rotation Curves by Vera Rubin

It is said that Vera Rubin's greatest discovery was the fact that spiral galaxies have "flat" rotation curves [310]. On that basis, in the '70s of the last century she developed her own theory of rotation curves of galaxies. From the high orbital speed of stars she concluded that galaxies should contain much more nonluminous than luminous matter to be kept together by gravity—see e.g. her review articles [307, 309] on dark matter.

In nearby spiral galaxies Rubin found, see [309] and also [311, p. 480], that all stars of these galaxies move at almost the same constant speed of order $v \approx 200$ km/s for $r > r_0$, where $r_0 > 0$ approximately corresponds to the radius of the central bulge and is typically equal to a few kpc (see Fig. 8.5). On the other hand, she observed that for $r \leq r_0$ the inside of the spiral galaxy (including a possible bar) rotates with roughly constant angular speed in a manner like that of a DVD record, i.e., the speeds of these stars are approximately linearly proportional to their distance from the center (see Fig. 8.5). An exception is a very close neighborhood of the central black hole, see Remark 8.3 below.

In this section we shall look more closely at Rubin's hypothesis. Consider a test particle of mass m (typically this will be a star) and let $M \gg m$ be the mass of some body generating the central force field. Assume that the test particle revolves about the center along a circular orbit with radius r and speed v. Then from Newton's law

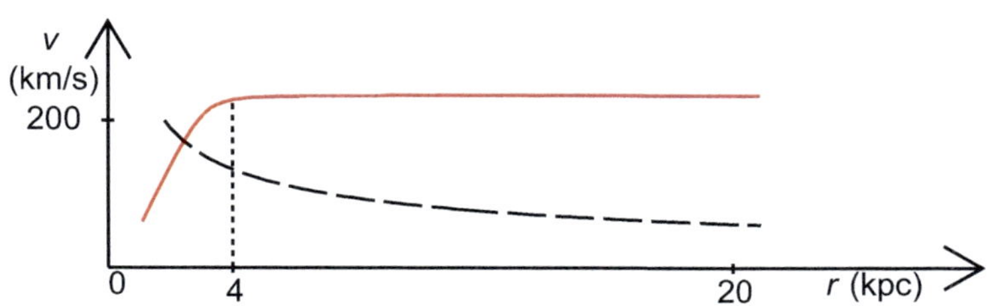

Fig. 8.5 The solid line shows an idealized rotation curve whose shape was derived by Rubin by means of measurements. It is flat for $r > r_0 \approx 4$ kpc. The dashed line shows the rate of decrease of velocities for Keplerian orbits that depend on the distance r from the center of a spiral galaxy

of gravitation and the relation for centripetal force Rubin easily obtained that, see [307],

$$G\frac{Mm}{r^2} = \frac{mv^2}{r}, \tag{8.8}$$

that is

$$v = \sqrt{\frac{GM}{r}}. \tag{8.9}$$

The velocity v of a particle on a circular orbit is thus proportional to $r^{-1/2}$, see Fig. 8.5. Such orbits are called *Keplerian*.

In [310, p. 491] Rubin states that

> the stellar curve does not decrease as is expected for Keplerian orbits.

Remark 8.3 To explain this discrepancy, it is important to realize that spiral galaxies do not have a central force field except within a close neighborhood of the center, where e.g. in the Milky Way the stars S1, S2, ... orbit a central black hole according to Kepler's laws with velocities up to 7000 km/s. The mass of this black hole is roughly 4 million solar masses, which is less than 0.01% of the total mass of our Galaxy $M_G \approx 10^{12} M_\odot = 2 \cdot 10^{42}$ kg, see (8.4).

In the Solar system, on the contrary, 99.85% of the mass is concentrated at the Sun. The planets barely interact gravitationally among themselves and their movements are determined mainly by the central force of the Sun. On the other hand, trajectories of stars in a galactic disk are substantially influenced mainly by neighboring stars, because the central bulge contains only about 10% of all mass of a galaxy.

Remark 8.4 It is noteworthy to mention the following big mathematical paradox of Rubin's observations. The stars of spiral galaxies are measured to move at an almost constant speed on different radii (see e.g. [309, p. 7]) which causes a differential rotation. However, these galaxies are not winding up and surprisingly do not show an expected tightening of arms as shown in Fig. 8.6. The idealized constant orbital velocity in Fig. 8.5 for $r \in [4, 20]$ kpc produces special spirals of the form $(r\cos(\alpha + b/r), r\sin(\alpha + b/r))$ for some suitable constants α and b. Therefore, their shape cannot be stable if they have gone through so many revolutions. This is known as the *winding problem* or *wind-up problem* [165, 362]. To see this we present the following example.

Example 8.3 Consider a spiral galaxy as marked in the upper left part of Fig. 8.6. Assume that the outer radius is 20 kpc and the inner radius $r_0 = 4$ kpc (see Fig. 8.5). Then after one revolution of a star on the outer orbit a star on the inner orbit makes $20 : 4 = 5$ revolutions if it has the same speed. This contradicts observations (see e.g. the galaxy M51 in Fig. 8.7). So there is has to be some systematic error.

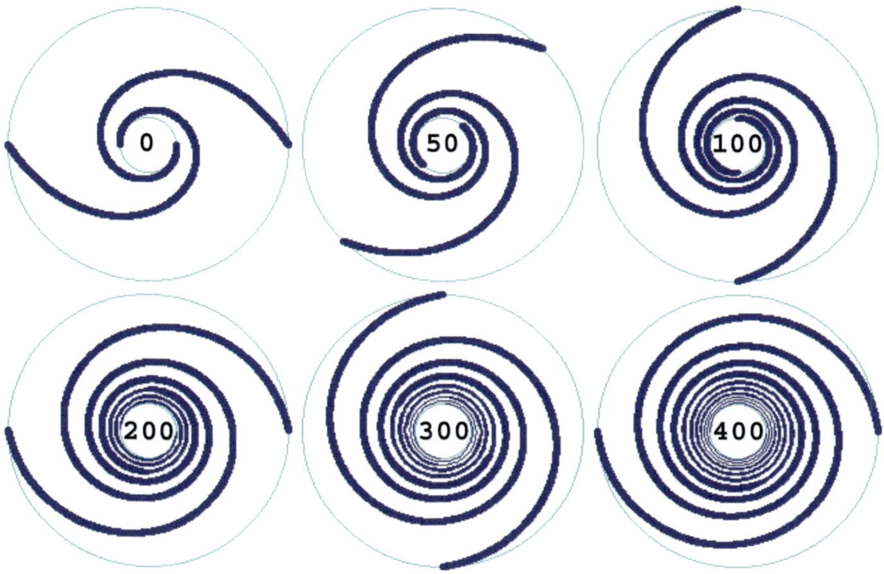

Fig. 8.6 The flat rotation curve from Fig. 8.5 causes a very quick tightening of spiral arms after one revolution of the most external stars. This takes approximately 400 Myr for a typical spiral galaxy (note that the age of the Sun is more than 10 times larger). Here stars on the galactic edge revolved about the angles: 0 (an idealized initial state), $\pi/4$, $\pi/2$, π, $3\pi/2$, and 2π. The corresponding time in given in Myr. After a relatively short time period of 100 Myr we would get a shape similar to Fig. 8.7. After a few hundred million years the spiral arms would be highly twisted. After 1 Gyr they would be completely wrapped up. However, observed spiral galaxies exist for many Gyr

With difficulty it can also be assumed that the arms of e.g. type Sbc galaxies are formed by some kind of density waves as suggested in [29, p. 544]. Such barred spiral galaxies resemble an open letter S. Therefore, some astronomers did not become convinced of the need for dark matter halos in spiral galaxies, see e.g. [206, 238, 246, 316].

Now we shall investigate orbital speeds of stars in our Galaxy to show that no dark matter is necessary.

Example 8.4 Our Sun orbits the center of the Milky Way with the speed

$$v_\odot = 230 \text{ km/s} \tag{8.10}$$

(most sources give the range of 220–240 km/s) on a path of radius $r_\odot = 8$ kpc. The radius of the visible part of the disk of our Galaxy is estimated by

$$r_\mathrm{G} = 16 \text{ kpc} = 4.938 \cdot 10^{20} \text{ m}, \tag{8.11}$$

Fig. 8.7 Spiral galaxies could not have such a large symmetry if there were to be six times more uniformly distributed dark matter than structured baryonic matter (photo NASA)

i.e., the Sun stays about halfway out from the center O of the Galaxy. Stars orbiting the center of our Galaxy at any distance $r > r_0 \approx 4$ kpc should have a speed similar to $v_\odot$ due to the expected flat rotation curve (see Fig. 8.5).

Denote the mass of all baryonic matter contained inside the ball of radius (8.11) centered at O by $\mathcal{M}(r_G)$. By Mikulášek and Krtička [245, p. 353], we have the following guaranteed lower bound

$$\mathcal{M}(r_G) \geq 3.85 \cdot 10^{41} \text{ kg.} \tag{8.12}$$

We shall proceed in two steps:

1. First, let us concentrate all baryonic matter contained inside this ball to the central point O. Then from relations (8.9), (8.11), and (8.12) the velocity of a test particle on the orbit of radius r_G is

$$v = \sqrt{\frac{G\mathcal{M}(r_G)}{r_G}} \geq \sqrt{\frac{6.674 \cdot 10^{-11} \cdot 3.85 \cdot 10^{41}}{4.938 \cdot 10^{20}}} \text{ m/s} = 228 \cdot 10^3 \text{ m/s.} \tag{8.13}$$

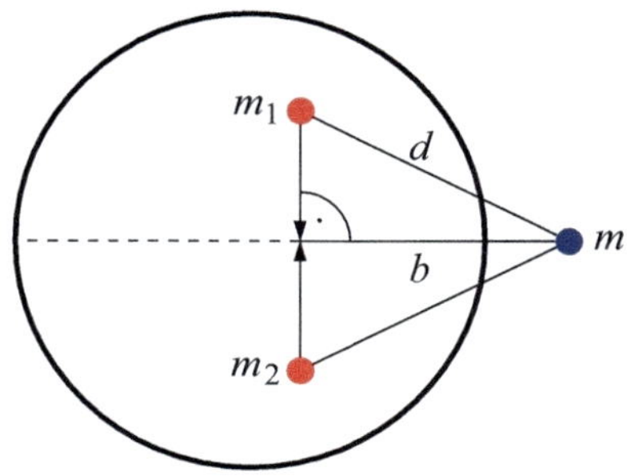

Fig. 8.8 A ball with symmetrically distributed mass with respect to the horizontal plane acts on a test particle by a smaller force than the mass projected perpendicularly to the horizontal plane of the disk – dashed

By this simple "back of the envelope" calculation, we see that v is indeed comparable to the measured speed (8.10). Although these relations are only approximate, to postulate the existence of six times more dark matter than baryonic matter (see (8.1)) to hold the Galaxy together by gravity seems to be highly overestimated.

2. Second, we claim that the orbital velocity around a flat disk of the same mass is even higher than that in (8.13). This results from the following theorem proved in [184, pp. 131–134].

Theorem 8.1 *A particle orbiting a central point mass along a circular trajectory of radius R has a smaller speed than if it were to orbit a flat disk of radius R and the same mass with an arbitrary rotationally symmetric density distribution.*

To explain the main idea of the proof, consider the situation in Fig. 8.8. Let two point masses $m_1 = m_2$ be located inside a ball placed symmetrically with respect to the horizontal plane. Let a test particle with mass m be in this plane. Then the total force F of both point masses acting on the test particle of mass m will be less than the force $\overline{F}$ of both point masses projected perpendicularly to the disk and acting on m. Let d be the distance between m_1 and m. Denoting by b its orthogonal projection on the horizontal plane, we find that

$$F = G\frac{2m_1 m}{d^2} \cdot \frac{b}{d} \quad \text{and} \quad \overline{F} = G\frac{2m_1 m}{b^2}.$$

Thus we see that the ratio of forces $\overline{F}$ and F is equal to the third power of the fraction d/b:

$$\overline{F} = \left(\frac{d}{b}\right)^3 F \geq F.$$

This cubic nonlinearity causes a greater attractive gravitational force by the disk than by the ball (cf. (8.8)), and thus also a higher orbital speed around the disk. For instance $\left(\frac{d}{b}\right)^3 = \left(\frac{5}{4}\right)^3 \approx 2$. The above Theorem 8.1 thus explains the large orbital velocities of stars in our Galaxy, even though spiral galaxies do not have rotationally symmetric distribution of mass. It also indicates why (8.1) is overestimated.

Remark 8.5 Rubin in [306, p. 1340] and [308, p. 29] assumed a spherical symmetry of the whole galaxy and that the density $\mathcal{M}(r)/r^3$ decreases as r^{-2}. Then from the relation $G\mathcal{M}(r)m/r^2 = mv^2(r)/r$ (cf. (8.8)) she got $v(r) \equiv$ const. This is indeed only a very rough calculation.

Remark 8.6 Orbital speed v satisfies the well-known baryonic Tully-Fisher law $v^4 \sim M$, where M is the mass of a spiral galaxy, see [351, 380] and also [178, p. 75].

In [184, p. 131], we prove why the force of a disk-shaped galaxy acting on a test particle is much larger than it would be if its whole mass were to be concentrated at one central point (cf. Newton's First Theorem 1.1). Therefore, the speed v of stars on circular orbits in a spiral galaxy should be higher than for Keplerian orbits (see Fig. 8.5).

Remark 8.7 The average thickness of the disk (except for the bulge) of spiral galaxies varies from 300 pc to 1 kpc. It is therefore about 30–100 times thinner than the diameter of the visible part of the galaxy. This is easily seen when galaxies are observed edge-on. Moreover, the gas and dust are mainly found close to the plane of symmetry of the disk [291]. Consequently, in [185] we treat the disk just as a two-dimensional body, which is obviously a better approximation than a central point mass. The gravitational field of spiral galaxies will therefore be approximated by the gravitational field of a flat disk with rotationally symmetric mass density distribution.

8.5 Final Remarks

In previous sections we introduced a number of arguments showing that the amount of dark matter as given in (8.1) seems to be considerably overestimated. Consequently, the amount of dark matter proposed by the standard ΛCDM (i.e. Lambda-Cold Dark Matter) cosmological model is mistakenly determined, or that dark matter does not exist at all. This model based on the Friedmann ordinary differential equation (7.21) of the first order is too simple and thus cannot yield correct conclusions.

Let us briefly mention several further arguments:

1. The influence of dark matter in the Solar system has not been observed (see [251]), even though our Sun is a large gravitational attractor.
2. At the end of the twentieth century it was thought that red dwarfs of spectral class M form only 3% of the total number of stars, see e.g. [29, p. 93]. Nevertheless, according to the current Harvard spectral classification of stars, red dwarfs are in the vast majority—over 76% (cf. Table 8.1). The main reason is that they stay on the main sequence a very long time and they are not visible due to the selection effect. However, they essentially contribute to nonvisible baryonic matter at large distances. The same is true for the other spectral classes L (red-brown dwarfs), T

Table 8.1 The Harvard spectral classification of stars by [399]. The first column shows the main spectral classes of stars, T is their surface temperature, and M is their mass. The last two columns contain their maximum age on the main sequence of the Hertzsprung-Russell diagram and the corresponding fraction of all main-sequence stars

Class	Color	T (K)	$M/M_{\odot}$	Age (Myr)	Fraction
O	Blue	>30,000	>16	1–10	0.00003%
B	Blue white	10,000–30,000	2.1–16	11–400	0.13%
A	White	7500–10,000	1.4–2.1	400–3000	0.6%
F	Yellow white	6000–7500	1.04–1.4	3000–7000	3%
G	Yellow (Sun)	5200–6000	0.8–1.04	7000–15,000	7.6%
K	Orange	3700–5200	0.45–0.8	~20,000	12.1%
M	Red	2400–3700	0.08–0.45	~50,000	76.45%

(brown dwarfs), and Y (black dwarfs). However, these dwarfs are not considered as ordinary stars.
3. By McLaughlin and van der Marel [239] the ratio between virial mass and luminous mass in globular clusters is less than two.
4. Dark matter should only interact gravitationally (and perhaps also weakly). Particle physics experiments looking for dark matter (i.e. hypothetical particles such as axions, neutralina, wimps, etc.) or dark antimatter found no evidence of their existence.
5. Most of the galaxies have a spiral structure. If these galaxies were to contain six times more uniformly distributed dark matter than baryonic matter, then they could not exhibit such a high symmetry of structured baryonic matter as seen e.g. in Fig. 8.7, since they would be governed by a dark matter halo.

Remark 8.8 It is very probable that Newton's law of gravity and the General Theory of Relativity on large cosmological scales approximate reality only very roughly and thus the proclaimed dark matter is nothing else than a modeling error. Several modifications of Newtonian theory, e.g. MOND (Modified Newtonian Dynamics) [238, 246, 316] and its relativistic generalization TeVeS (Tensor-Vector-Scalar) [27], are at present being developed and studied. Effects that are attributed to dark matter are explained by a different form of the gravitational law on large scales. However, MOND contradicts the principle of causality (a consequence cannot precede its cause), since it assumes an infinite speed of gravitational interaction, i.e., it cannot be a good model on cosmological scales. In [157] and [184, p. 253], we included delays given by a finite speed of gravity into the system (1.14). According to Asher Yahalom [379, 381], any changes inside galaxies may affect other parts tens of millennia afterwards. Also the Tully-Fisher law can be explained by taking retardation effects into account, see [380]. It is a pity that these retardation effects are not included in the present day analysis of galaxy evolution. Finite speed of gravitational interaction contributes to circularization of elliptic orbits, since larger velocities produce larger gravitational aberration angles.

Remark 8.9 Dark matter cannot be hot (having relativistic velocities), since then we would not observe complicated structures of baryonic matter.

Pavel Kroupa (see e.g. [205, 206]) presents a number of observations which rule-out the existence of dark matter particles. He shows that there are no dark matter halos, since then our neighboring dwarf galaxies would slow down due to dynamical friction with dark matter and they would spiral into the center of the Milky Way. This is not observed and too many dwarf galaxies are even flying away from the Local Group too fast. Moreover, several dwarf galaxies orbit our Galaxy and M31 in a few planes (i.e. they are isotropically correlated) which would be impossible with dark matter. The dynamical friction of dark matter halos would also slow down central bars of spiral galaxies, cf. also [304]. Too long tidal tails of several large galaxies are also not consistent with dark matter halos.

At present there are many other papers claiming that it is not necessary to assume the existence of dark matter on scales of galactic disks, that the ratio (8.1) is highly overestimated or that dark matter does not exist at all, see e.g. [101, 111, 140, 165, 183, 185, 251, 285, 400].

On the other hand, Douglas Clowe et al. in [67] assume a collision of two galaxy clusters, where the intergalactic gas is stopped, while the galaxies continue in an unchanged direction together with presumed dark matter. The title of this paper *A direct empirical proof of the existence of dark matter* should imply that dark matter was finally found. Nevertheless, we are not able to measure tangential components of the velocities of these clusters to prove that the collision really happened. The authors neglect dynamical friction of particular galaxies and suppose a quite unrealistic mutual infall velocity (see [67, p. L112])

$$v \approx 4700\ \mathrm{km/s}\ > 0.01c$$

of these clusters to guarantee that the collision does not last more than several billion years. The expected tidal tails were not observed. Moreover, the proposed infall velocity v has the opposite sign to the overall Hubble expansion speed of the universe. How could these two galaxy clusters get such unlikely high velocities several Gyr ago? This would produce an extremely large kinetic energy proportional to v^2 in an almost isotropic and homogeneous universe, where the local speed of galaxies is usually only several hundred km/s. Therefore, any conclusion about the direct proof of a large amount of dark matter is questionable.

The regions with dark matter in [67] are artificially colored in blue by some numerical simulations based on gravitational lensing. In the case of light bending near our Sun during total eclipses, we know exactly the bending angle. However, in the case of hypothetical dark matter regions we have to apply only some inexact heuristic algorithms and interpolation techniques between galaxies, since galaxies are seen projected on the celestial sphere and they are represented only by several pixels in photos.

Chapter 9
Dark Energy and the Local Hubble Expansion

9.1 Local Hubble Expansion

Most cosmologists believe that the universe expands globally, but not locally, see e.g. [249, p. 719]. However, this immediately leads to a mathematical contradiction. We will demonstrate its main idea in the following example.

Example 9.1 Suppose that the universe expands globally and, for simplicity, let us sketch it as an ellipse illustrated in Fig. 9.1. Then it has to expand in its left or right half. If it expands in the right half, for example, then it must expand in the upper or lower quarter. Let us therefore assume that it expands e.g. in the lower quarter. Then we again halve it. In this way, we can proceed as shown in Fig. 9.1 and obtain that the universe must expand at least somewhere locally.

The above example is inconsistent with Kepler's laws, and therefore also with the law of conservation of energy. For instance, all planets of the Solar system are sufficiently isolated from the gravitational influence of nearby stars. By (1.9) the gravitational force between the Earth and Alpha Centauri (whose mass is $1.1M_\odot$ and distance 4.37 ly) is about 1 million times smaller than the maximal gravitational force between the Earth and Venus. To illustrate the sparseness of our Galaxy, let us mention that if we reduce the Sun with diameter $1.391016 \cdot 10^9$ m to the size of a ping-pong ball with a diameter of 3.7 cm, then Alpha Centauri would be

$$d = \frac{0.037 \cdot 4.37 \cdot 9.46 \cdot 10^{15}}{1.391016 \cdot 10^9} \text{ m} = 1.01 \cdot 10^6 \text{ m}$$

from us, i.e. slightly more than 1000 km. This shows how sparse our Galaxy is in our neighborhood. The diameter of the Galaxy in this scale would be 100 times larger than the Earth-Moon distance.

Some authors argue (see e.g. [58, 70]) that dark energy does not manifest itself at all in the Solar system. From Sect. 7.8 we know why their reasoning is erroneous.

M. Křížek, L. Somer, *Mathematical Aspects of Paradoxes in Cosmology*,
https://doi.org/10.1007/978-3-031-31768-2_9

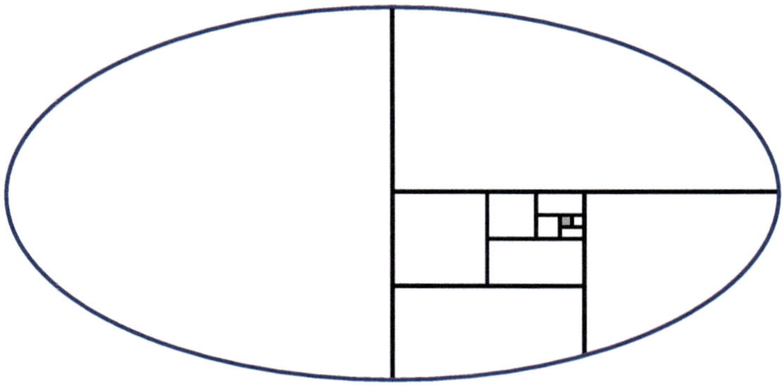

Fig. 9.1 An illustration that the universe has to expand somewhere locally. It can be easily generalized to the three-dimensional space

Example 9.2 Let us rescale the value H_0 to the mean Sun-Earth distance, i.e. one astronomical unit

$$1\,\mathrm{au} = 149{,}597{,}870{,}700\,\mathrm{m} \approx 150 \cdot 10^9\,\mathrm{m}.$$

Taking into account that 1 pc $\approx$ 206,265 au and that one sidereal year has about 31,558,150 s, we get

$$H_0 \approx 70\,\mathrm{km\,s^{-1}Mpc^{-1}} = 70\,\mathrm{m\,s^{-1}kpc^{-1}} = \frac{70 \cdot 31{,}558{,}150}{206{,}265{,}000}\,\mathrm{m\,year^{-1}au^{-1}}.$$

Therefore,

$$\boxed{H_0 \approx 10\,\mathrm{m\,year^{-1}au^{-1}}.}$$

From this we observe that 1 m^3 of space will increase by an average of 0.2 mm^3 per year, since

$$\left(1 + \frac{10}{150 \cdot 10^9}\right)^3 \approx 1 + 3\frac{10}{150 \cdot 10^9} = 1 + 0.2 \cdot 10^{-9}.$$

Hence, each cubic meter of the universe has increased its volume approximately twice on average since the origin of the Solar system $4.6 \cdot 10^9$ year ago.

Michael S. Turner [352] is well known for coining the term *dark energy* in 1998 for the force that is responsible for the expansion of the universe. There is no reason to assume that manifestations of dark energy do somehow avoid our Galaxy or the Solar system. Since the above values in Example 9.2 are relatively large, the influence of dark energy should also be found in the Solar system.

Admitting the manifestation of dark energy in the Solar system, we can easily explain a wide range of puzzles such as the paradox of tidal forces of the Moon, the paradox of the large orbital angular momentum of the Moon, Titan, Triton and Charon, the existence of rivers on Mars and the existence of its satellite Phobos, migration of planets, the faint young Sun paradox [213], the formation of Neptune and the Kuiper belt of comets, the slow rotation of Mercury and the absence of its moons.

In the following sections we present a number of astrobiological, astronomical, geometric, geochronometric, geophysical, heliophysical, climatological, paleontological, and observational arguments to show that the Solar system and also our Galaxy are expanding at a rate comparable to the order of the Hubble-Lemaître constant H_0 (although usually a little bit smaller). From Sect. 7.8 we already know that the influence of the cosmological constant Λ is completely negligible on the local Hubble expansion.

Finally, note that the current location of the American space probe Pioneer 10 launched in 1972 does not conform to the laws of Newtonian mechanics. This is the famous Pioneer anomaly problem [17]. However, it has nothing to do with the local Hubble expansion and dark energy, since:

1. It can be explained by radiation of an asymmetric position of heat sources in the probe.
2. The corresponding time interval is too short compared to cosmological time scales.
3. The additional acceleration has the opposite sign than the local Hubble expansion.

9.2 The Universe Expands Almost as Rapidly as the Moon Recedes From the Earth

In this section we show that the Moon moves away from the Earth more rapidly than would be gathered from classical mechanics. The hypothesis about tiny repulsive forces and the local expansion of the Solar system can be very well tested just by means of a precise measurement of the Earth-Moon distance. The Moon's orbit has a small eccentricity ($e = 0.0554$). Thus the Moon slowly approaches the Earth and then moves away during each period. Current technologies allow us to establish long-term changes in the parameters of its orbit very precisely.

Since the seventies of the last century, changes in the mean Earth-Moon distance

$$D \approx 384{,}402\,\mathrm{km} \tag{9.1}$$

have been carefully measured by laser retroreflectors installed on the Moon by the Apollo missions.

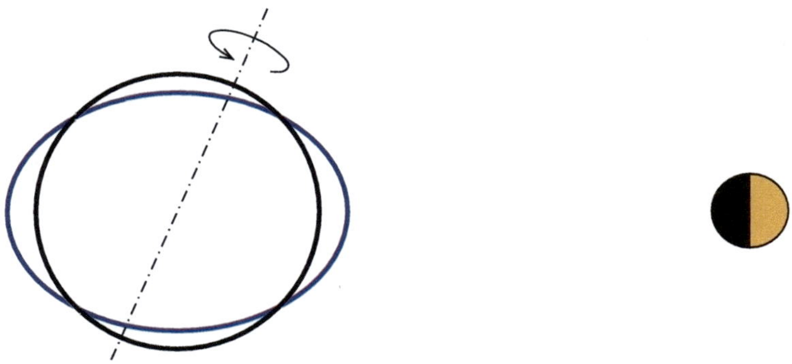

Fig. 9.2 Reduction of the angular velocity of the Earth's rotation is caused by friction when creating tidal bulges, which are formed by the gravitational influence of the Moon. Since the Earth is not an ideal rigid body, its deformations are converted into heat

The nearer side of the Earth to the Moon is attracted by a greater force than the opposite side. As a result of a nonconstant gravitational potential, *tidal forces* are formed. Note that tidal forces played an important role in the deposit of sediments. They provide a record of the lengths of the year and month over time as discovered by G. E. Williams [378]. Due to the Earth's rotation, they give rise not only to ebbs and flows of the seas and oceans, but also to the lifting of the Earth's mantle (see Fig. 9.2). Continual deformations of the Earth cause its rotation to slow down, and thus its spin (rotational) angular momentum decreases over time. Then from the law of conservation of total angular momentum it follows that the orbital angular momentum of the Earth-Moon system has to increase.

Already in 1975, T. C. van Flandern [354] (cf. also [84], [88, Chapt. 6], [232], [335, p. 155]) noticed that the Moon has a somewhat anomalous orbit, because it moves away from the Earth faster than can be explained by tidal forces and Newtonian mechanics. He examined whether it could be due to a slight decrease in the value of the gravitational constant. Long-term measurements show (see [82]) that the mean distance D continually increases on average

$$v = 3.84\,\text{cm per year}. \tag{9.2}$$

Hence, the trajectory of the Moon is not an ellipse, but a very dense spiral with respect to the Earth-Moon center of gravity. However, the action of tidal forces on the Earth's crust, hydrosphere, atmosphere, etc., can explain a much smaller expansion velocity – only about 55% of the value given in (9.2), see (9.20) and also [264, p. 67], [273, Sec. 9.10.4]. This is usually called the *Paradox of tidal forces of the Moon*, see [354]. Verbunt [365] even writes about the *tidal catastrophe*.

Therefore, the Moon is receding from us probably not only due to tidal forces. For the remaining increment of almost the same order of about 1.71 cm per year we should look for other explanations. For instance, one possibility is that some apparent repulsive forces also contribute to the observed recession of the Moon.

The current expansion rate of the universe is characterized by the Hubble-Lemaître constant, see Example 9.2 and (7.11),

$$H_0 \approx 10\,\text{m}\,\text{year}^{-1}\text{au}^{-1} \approx 2.3 \cdot 10^{-18}\,\text{s}^{-1}.$$

It is not difficult to recalculate the speed of this expansion only to the changing distance between the Earth and Moon. Denoting year to be the unit of time corresponding to one sidereal year and $R = 1$ au, then by (9.1) we have

$$H_0 \approx 10\,\text{m}\,\text{year}^{-1}\text{au}^{-1} = 10\frac{D}{R}\,\text{m}\,\text{year}^{-1}D^{-1} = 2.57\,\text{cm}\,\text{year}^{-1}D^{-1}. \tag{9.3}$$

We see that this value is surprisingly quite close to the mean value (9.2) of the recession speed of the Moon from the Earth. This is a remarkable coincidence. At the same time, the tiny repulsive forces can explain a larger measured recession speed (9.2) than that resulting from the tidal forces.

Further, we show that from the law of conservation of the total angular momentum one can derive a recession speed of the Moon from the Earth due to tidal forces of only about 2.13 cm per year (cf. (9.20)) assuming a constant moment of inertia of the Earth. First, we will estimate in detail the contribution of tidal forces to the speed (9.2) by the method introduced in [160].

Consider the isolated binary system Earth-Moon with masses

$$m_1 = 5.97219 \cdot 10^{24}\,\text{kg}, \quad m_2 = 7.3477 \cdot 10^{22}\,\text{kg} \tag{9.4}$$

and assume, for simplicity, that their orbits are circular (estimation of the tidal influence of the Sun on the change in the Earth's rotation will be discussed later). Then the corresponding distance (see (9.1)) can be expressed as follows:

$$D = R_1 + R_2, \tag{9.5}$$

where

$$R_1 = \frac{Dm_2}{m_1 + m_2} \quad \text{and} \quad R_2 = \frac{Dm_1}{m_1 + m_2} \tag{9.6}$$

are the distances of the Earth and the Moon from their Newtonian centre of gravity, respectively.

By the conservation of the total angular momentum of this system (i.e. the sum of the spin angular momenta and the orbital angular momenta), the value

$$J = I_1\omega_1 + I_2\omega_2 + m_1R_1v_1 + m_2R_2v_2 \tag{9.7}$$

has to be constant, i.e., its time derivative is zero,

$$\dot{J} = \frac{\mathrm{d}J}{\mathrm{d}t}.$$

Here v_1 and v_2 are the orbital speeds of the Earth and Moon, respectively, relative to their center of gravity, I_1 and I_2 are the momenta of inertia of the Earth and Moon,

$$\omega_1 = \frac{2\pi}{T_1} = 7.292 \cdot 10^{-5}\,\mathrm{s}^{-1}, \quad \omega_2 = \frac{2\pi}{T_2} = 2.669 \cdot 10^{-6}\,\mathrm{s}^{-1} \tag{9.8}$$

are the angular frequencies of the Earth's and Moon's rotations about their axes, $T_1 = 86{,}164.1$ s is the sidereal day, and $T_2 = 27.32166\,T_1$. By Burša and Peč [53] we have

$$I_1 = (8.036 \pm 0.008) \cdot 10^{37}\,\mathrm{kg\,m}^2. \tag{9.9}$$

We shall examine the sizes of the terms $I_1\dot{\omega}_1$ and $I_2\dot{\omega}_2$ appearing in the equation $\dot{J} = 0$ (see (9.14) and (9.17) below). Consider the following three steps:

1. Using the formula for the momentum of inertia of a homogeneous ball [296, p. 109], we find for the momentum of inertia of the Moon (whose density $\rho(r)$ increases towards its center) the inequality

$$I_2 < \frac{4}{3}\pi r_2^3 \cdot \rho_2 \cdot \frac{2}{5} r_2^2 = 8.87 \cdot 10^{34}\,\mathrm{kg\,m}^2, \tag{9.10}$$

where

$$\rho_2 = 3348\,\mathrm{kg/m}^3 \tag{9.11}$$

denotes the mean density of the Moon and $r_2 = 1737\,\mathrm{km}$ is its radius. Note that the term $I_2\omega_2 < 2.37 \cdot 10^{29}\,\mathrm{kg\,m^2s^{-1}}$ corresponding to the Moon is much smaller than $I_1\omega_1 = 5.86 \cdot 10^{33}\,\mathrm{kg\,m^2s^{-1}}$. However, we have to compare their time derivatives.

According to [315], the Earth's rotation has slowed down so much during the last 2700 years that the length of the day increased on average about 1.7 ms per century, i.e.

$$\dot{T}_1 = 1.7 \cdot 10^{-5}\,\mathrm{s\ per\ year}. \tag{9.12}$$

Estimations of the size of these long-term changes in Earth's rotation caused by tidal forces were obtained by a thorough analysis of ancient Babylonian (see e.g. [264, p. 62], [315, p. 270]), Arab et al. [273] records concerning the angular heights of the Sun during observed eclipses. This is in agreement with today's precise radio measurements of the Earth's rotation slow-down using distant quasars. Contemporary increased melting of glaciers, internal processes in the Earth, mass transfer in the atmosphere, hydrosphere, etc., cannot explain such a large value as given in (9.12). For simplicity, we will first assume that the momentum of inertia of the Earth I_1 is independent of time. In the next section we shall allow the time dependence $I_1 = I_1(t)$.

The Sun and Moon have practically the same angular diameter. Since tidal forces decrease with the third power of the distance from the Earth and the volume increases also with the third power, the tidal influence of each of the two bodies on the Earth is directly proportional to its mean density. Since $\rho_\odot = 1409\,\mathrm{kg/m^3}$ due to (1.13), we find by (9.11) that the ratio of the densities of the Moon and Sun is about $3348 : 1409 = 2.38$. The lunar tidal effect on the Earth is thus larger than the solar tidal effect. The actual ratio is slightly smaller than 2.38, since the mean angular size of the Sun is $31.98'$, whereas for the Moon it is $31.07'$.

According to [282], the Earth's rotation slows down about 68.5% due to the influence of the Moon's tidal forces, and 31.5% from those of the Sun. Hence, the increase $\mathcal{T}_1 = 0.685\mathcal{T}$ corresponds to the Moon and $0.315\mathcal{T}$ to the Sun. After 1 year the angular frequency of the Earth's rotation changes as follows

$$\varpi_1 = \frac{2\pi}{T_1 + \mathcal{T}_1}. \tag{9.13}$$

From relation (9.12) and (9.13) we see that the corresponding time change of the Earth's angular frequency is[1]

$$\dot{\omega}_1 = \frac{\varpi_1 - \omega_1}{T} = -\frac{2\pi}{T}\frac{\mathcal{T}_1}{T_1(T_1 + \mathcal{T}_1)} = -3.123 \cdot 10^{-22}\,\mathrm{s}^{-2},$$

where

$$T = 31{,}558{,}149.54\,\mathrm{s}$$

is the sidereal year. By (9.9) the corresponding change of the spin angular momentum of the Earth is equal to

$$I_1\dot{\omega}_1 = -2.509 \cdot 10^{16}\,\mathrm{kg\,m^2\,s^{-2}}. \tag{9.14}$$

2. The Moon also reduces its angular frequency about its rotational axis due to the recession speed (9.2) and the $1:1$ resonance between the Moon's rotation and the orbital period T_2 of the Moon around the Earth. The Moon has fallen into the so-called *tidal trap*.

[1] The correct calculation of the derivative $\dot{\omega}_1$ requires the use of the measured value (9.12) of Earth's rotation slowdown, and not a theoretically derived value $\mathcal{T}$ using the conservation law of the total angular momentum and relation (9.2) as done in [264, p. 65].

Now, we show that $I_2\dot{\omega}_2$ is several orders of magnitude smaller than the value given in (9.14). Taking into account (9.5), we may apply Kepler's third law which approximates the actual situation on short time scales quite well. By (1.10) the ratio D^3/T_2^2 is constant. Hence, the product $\omega_2^2 D^3$ is also constant by (9.8). Differentiating $\omega_2^2 D^3$ with respect to time, we obtain the differential equation

$$2\omega_2\dot{\omega}_2 D^3 + 3\omega_2^2 D^2 \dot{D} = 0.$$

Hence,

$$\dot{\omega}_2 = -\frac{3}{2}\omega_2\frac{\dot{D}}{D} \quad \text{with} \quad \frac{\dot{D}}{D} = \frac{3.84\,\text{cm year}^{-1}}{384{,}402\,\text{km}} = 0.999 \cdot 10^{-8}\text{cy}^{-1}. \tag{9.15}$$

By (9.2) the mean long-term observed time change of D is given by

$$\dot{D}_{\text{observed}} = \frac{3.84\,\text{cm}}{T} = 1.2 \cdot 10^{-9}\,\text{m/s}. \tag{9.16}$$

From this, (9.1), (9.8), (9.10), and (9.15) we obtain

$$|I_2\dot{\omega}_2| < 1.1 \cdot 10^8\,\text{kg}\,\text{m}^2\,\text{s}^{-2}. \tag{9.17}$$

Hence, the decrease of the Moon's spin angular momentum is negligible in comparison to the value given in (9.14).

3. The decrease of the spin angular momentum of the Earth in (9.14) has to be compensated by the increase of the orbital angular momentum $m_1R_1v_1 + m_2R_2v_2$ in (9.7). We will again employ the $1:1$ resonance between the angular frequency of the Moon about its rotational axis and the angular speed of the Earth about their common center of gravity, i.e., $\omega_2 = v_2/R_2 = v_1/R_1$. Consider further the Earth-Moon system whose center of gravity is at rest. Then by the momentum conservation law $m_1v_1 = m_2v_2$, (9.5), and (9.6), we get

$$\begin{aligned} m_1R_1v_1 + m_2R_2v_2 &= (R_1 + R_2)m_1v_1 = D\,m_1v_1 \\ &= Dm_1R_1\omega_2 = D^2\frac{m_1m_2}{m_1+m_2}\omega_2. \end{aligned}$$

From this we get by differentiating (9.7) with respect to time and neglecting $I_2\dot{\omega}_2$ (cf. (9.17)) that

$$\begin{aligned} I_1\dot{\omega}_1 &= -\frac{m_1m_2}{m_1+m_2}\frac{\mathrm{d}(D^2\omega_2)}{\mathrm{d}t} = -\frac{m_1m_2}{m_1+m_2}(\dot{\omega}_2D^2 + 2\omega_2D\dot{D}) \\ &= -\frac{m_1m_2}{m_1+m_2}\frac{\omega_2 D}{2}\dot{D}, \end{aligned} \tag{9.18}$$

where the last equality follows from (9.15). Substituting from (9.1), (9.4), (9.8), and (9.14), we finally obtain

$$\dot{D} = 0.674 \cdot 10^{-9}\,\text{m/s}, \tag{9.19}$$

which is only slightly more than one half of the measured value (9.16).

Multiplying (9.19) by the length of the sidereal year T, we find that the Earth-Moon distance should increase by only about the value

$$v_{\text{tides}} \approx 2.13\,\text{cm per year}. \tag{9.20}$$

Denote the difference between the measured value from (9.2) and the value (9.20) by

$$v_{\text{remainder}} \approx 1.71\,\text{cm per year}, \tag{9.21}$$

that is,

$$v = v_{\text{tides}} + v_{\text{remainder}}.$$

The origin of the term $v_{\text{remainder}}$ is unknown, but its magnitude is only a little bit smaller than the current value of the Hubble-Lemaître constant in (9.3). This supports the hypothesis concerning the local influence of dark energy. More precisely, the speed 1.71 cm per year from (9.21) equals 67% of the value 2.57 cm per year of the Hubble-Lemaître constant, as given in (9.3), i.e., the local Hubble expansion is

$$\boxed{H_0^{(\text{loc})} = 0.67\,H_0.} \tag{9.22}$$

Such a large recession does not seem to be caused by tidal forces nor other nongravitational effects, as we shall see below and in further chapters.

In 2003, Yurii Dumin (see [84, p. 2463]) derived from astrometric measurements corrected to ancient eclipses a similar local expansion rate of $H_0^{(\text{loc})} \approx 0.5\,H_0$ for the Earth-Moon system. In his paper [85] from 2008 this value was increased to $H_0^{(\text{loc})} \approx 0.85\,H_0$ for the data from the last three centuries.

Remark 9.1 By Kokubo et al. [149] our Moon was formed at a distance of only $d = 20{,}000$ km from the Earth approximately 4.5 billion years ago. In this case, solar and lunar eclipses would have been much more frequent than at present, since the angular diameter of the Moon was about $10^\circ \approx 2\arcsin(r_2/d)$. The Moon was shining extremely brightly at that time, since its area on the celestial sphere was 400 times larger than now. Its orbital period was less than 6 h by Kepler's third law (1.10). Moreover, tidal forces were huge. This corresponds to the mean recession speed

$$\overline{v} = 8\,\text{cm per year}$$

to reach the current distance (9.1). Because tidal forces decrease with the third power of the distance, tidal forces played a significant role especially at the beginning. However, apparent repulsive forces could slightly but permanently increase the recession speed of the Moon from the Earth within the entire 4.5 Gyr long interval.

9.3 Mars Was Much Closer to the Sun When There Were Rivers

The total solar power incident per unit area of 1 m^2 perpendicular to the Sun's rays at the distance of 1 au is equal to the *solar constant*

$$L_0 = 1.361\,\text{kWm}^{-2}. \tag{9.23}$$

The value of L_0 currently varies by less than 0.1% depending mainly on the number and size of sunspots. Hence, the total solar power is about

$$L_\odot = 4\pi R^2 L_0 = 3.828 \cdot 10^{26}\,\text{W}, \tag{9.24}$$

where $R = 1$ au.

Since the Sun is a star on the main sequence of the Hertzsprung-Russell diagram, its position slowly rises to the left upper corner. However, in doing so, the Sun travels only a very short distance on the main sequence. According to [314, p. 461], the Sun had a surface temperature of 5586 K about 4.5 Gyr ago and its luminosity was only 70% of the current value (corresponding to $\approx$ 1 kW/m^2 at the distance 1 au from the Sun). Its luminosity increased approximately linearly with time up to the present (see Fig. 9.3). The current effective surface temperature is 5770 K and after 3 Gyr it will rise to $\approx$ 5843 K, see [314, p. 461]. At that time the Sun will already produce 133% of its current power. Thus we see that the Sun's surface temperature increases relatively slowly, while its power grows much faster. The Sun will become a red giant after 12 Gyr from its formation.

Hydrologists estimated from the number of craters in dry river valleys (cf. Fig. 9.4) that Mars had liquid water on its surface 3–4 Gyr ago. At that time the luminosity of the Sun was only 75% of its present value. The solar power decreases with the square of the distance from the Sun. Therefore, the corresponding solar constant for Mars would only be

$$\boxed{L_{\text{Mars}} = 0.75 L_0 \left(\frac{150}{225}\right)^2 = \frac{L_0}{3}} \tag{9.25}$$

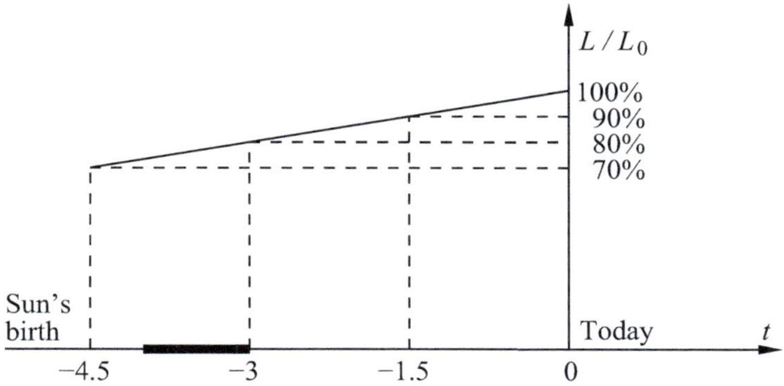

Fig. 9.3 Relative luminosity L/L_0 of the Sun from the origin of the Solar system up to the present. The time t is given in Gyr. The bold interval on the time axis indicates the period when Mars had liquid water on its surface

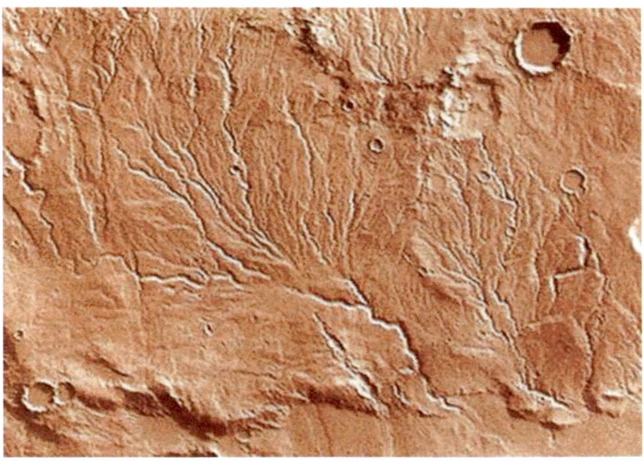

Fig. 9.4 Dry riverbeds on Mars tending to the ancient sea at the bottom right. The center of the image (whose dimensions are $175 \times 125\,\mathrm{km}^2$) is at 42.3° south Martian latitude and 92.7° west longitude (photo NASA)

provided that Mars had been on average

$$r = 225 \cdot 10^9\,\mathrm{m} \tag{9.26}$$

away from the Sun as it is now. In fact, the luminosity (9.25) corresponds to a 67%-decrease of the solar constant L_0 given in (9.23). In this case the existence of rivers on Mars would be impossible.

Note that a decrease of L_0 by only 2% causes ice ages on the Earth, even though there is the greenhouse effect. An ancient atmosphere on Mars 3–4 Gyr ago had one-third to two-thirds of the surface atmospheric pressure that Earth has today (see [124]). Higher concentration of CO_2 (as suggested by Bertotti et al. [28, p. 177]) surely contributed to a higher surface temperature on Mars, but cannot fully explain liquid water there because of the huge 67%-decrease of the luminosity given in (9.25). Therefore, Mars must have been much closer to the Sun to account for liquid water.

There is a possibility that Mars had escaped to freezing in the past due to a strong greenhouse effect, which could have compensated for the lower solar luminosity (9.25). However, the absence of carbonates in lake sediments according to analysis performed by the Curiosity Rover in Gale Crater of Mars [48] is casting serious doubts on this interpretation. For a rapid degassing of the Martian atmosphere during the first 400 million years we refer to [318, p. 16].

Mars also had a larger albedo than now, since there were water clouds feeding many lakes and hundreds of large rivers whose dry riverbeds are now between $-50°$ and $50°$ of Martian latitude (see Fig. 9.4). Due to measurements of the missions Viking I and II, Pathfinder, Spirit, Opportunity, Curiosity, Mars Orbiter, Perseverance, etc., we know that the current annual average temperature (about $-63°$C neglecting the greenhouse effect) on Mars is very much below the freezing point of water.

By the Stefan-Boltzmann law (see [184, p. 173]), the equilibrium temperature T_{eq} at the distance r from the Sun satisfies $\sigma T_{eq}^4 = L_\odot/(\pi r^2)$, where $L_\odot = 3.846 \cdot 10^{26}$ W is the present value of the total Solar luminosity and $\sigma = 5.669 \cdot 10^{-8}$ $Wm^{-2}K^{-4}$ is the Stefan-Boltzmann constant. Since Mars' surface area is four times larger than the area of its maximal cross-section, the current overall average surface temperature can be estimated by the Stefan-Boltzmann law as follows:

$$T_{Mars} = \frac{1}{2}\Big(\frac{(1-A)L_\odot}{\pi\sigma r^2}\Big)^{1/4} = 211\,\text{K}, \tag{9.27}$$

where $A = 0.25$ is the present value of the Bond albedo. We see that T_{Mars} is really very close to the yearly planet-wide mean measured temperature $\approx -60°$C. When the Sun's luminosity was 75% of the present value (see Fig. 9.3), we get from (9.27) that $T_{Mars} = 196$ K only. Note that if the temperature of water is 273.16 K and the pressure 611.7 Pa, then water can exist simultaneously in gaseous, liquid, and solid state. This is called a *triple point* (see Fig. 9.5). As a consequence we must conclude that Mars has been moving away from the Sun with an average speed of at least several meters per year (cf. (9.3)).

When the Sun's luminosity was only $0.75L_\odot$, then similarly we would get $r = 116.782 \cdot 10^6$ km to reach the freezing point of water $T_{Mars} = 273.15$ K. However, this distance is more than 100 million km smaller than the current radius $r = 225 \cdot 10^6$ km. The infrared emissivity does not change these values too much.

The above arguments show that it has no sense to develop some artificially complicated climate models for Mars and that Mars must have been closer to the Sun by several tens of million km when it had liquid water (Fig. 9.6).

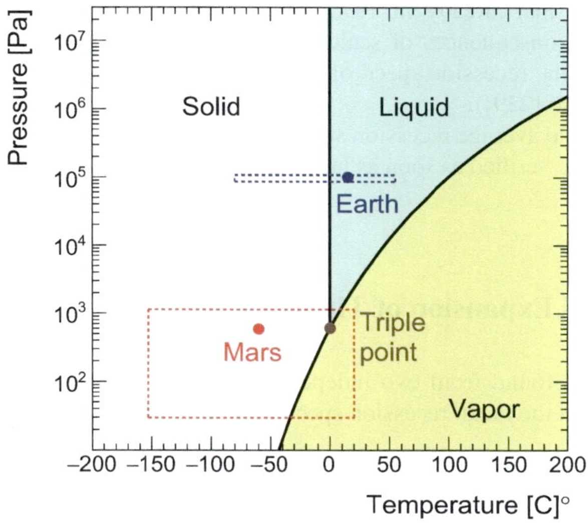

Fig. 9.5 The triple point in the phase diagram of water. The average temperature and pressure on the Earth and Mars is indicated by a small bullet. Boxes show approximate ranges of temperature and pressure that can be achieved on the Earth and Mars. The vertical axis has the logarithmic scale

Fig. 9.6 Pictures of sediments in the Gale crater taken by the Curiosity mission represent a further proof of liquid water on Mars during a long time period. An open problem is why all layers have almost the same thickness and which physical mechanism could produce such a perfect periodicity (photo NASA)

By (9.3) recalculated to the Mars-Sun distance (9.26), we find that Mars could have moved further from the Sun by an amount of about $60 = 4 \cdot 10 \cdot 225/150$ million km during the last 4 Gyr.

Finally note that already in 1979, Maeder and Bouvier [230, p. 88] studied the weak field consequences of scale invariance of gravitation and suggested the possibility that the recession speed of Mars from the Sun should be several meters per year (see also [229]).

The associated average recession speed of cca 10 m/year (cf. (9.3)) of Mars from the Sun could be verified as soon as laser retroreflectors are installed on the Mars or Phobos.

9.4 Orbital Expansion of Titan

Recently it was found from two independent sets of astrometric and radiometric Cassini data that the mean recession speed of Titan from Saturn is

$$v = 11.3 \text{ cm/year}. \tag{9.28}$$

Before these measurements, astronomers assumed that this speed was only 0.1 cm/year, because Saturn is mostly a gaseous planet and thus tidal forces are small.

Lainey et al. [210] claim that this rapid recession speed (9.28) is mainly caused by a resonance locking mechanism of the five inner mid-sized moons of Saturn: Mimas, Enceladus, Tethys, Dione, and Rhea. Nevertheless, their total mass is only 3 % of the mass of Titan. So if the unexplained recession speed of Titan is about 0.1 m/year, the inner moons should approach Saturn by an amount that is on average approximately equal to more than 3 m/year (i.e., 3,000,000 km/Gyr) in order to keep the laws of conservation of energy and momentum valid. This would however contradict the present sizes of their orbits whose semimajor axes are between 185,539–527,108 km. Moreover, the old cratered surfaces on the inner moons (see [259]) contradict their implied young ages.

Therefore, in [181] another explanation is given. It is again based on tiny repulsive forces which could be due to gravitational aberration, cf. [160], [176], [231, p. 25]. Since the Titan-Saturn average distance is $d = 1{,}221{,}870\,\text{km}$, the value (9.28) results in

$$\frac{\dot{d}}{d} = \frac{11.3\,\text{cm/year}}{1{,}221{,}870\,\text{km}} = 0.925 \cdot 10^{-8}\text{cy}^{-1} \tag{9.29}$$

and by means of (9.3) the recalibrated value of the Hubble-Lemaître constant is

$$H_0 = 8.15\,\text{cm year}^{-1}d^{-1}.$$

Therefore, the corresponding expansion rate is

$$H_0^{(\text{loc})} = \frac{11.3}{8.15}H_0 = 1.38\,H_0. \tag{9.30}$$

This value, of course, includes tides which have to be subtracted as in the case of our Moon, see (9.22). Moreover, resonances with mid-sized inner moons should also be taken into account.

9.5 Expansion of the Solar System

In Sects. 10.3 and 10.4 we will present several arguments supporting the hypothesis that repulsive forces also significantly contribute to the increase of the mean Earth-Sun distance by several meters per year. Below we present several other arguments indicating the expansion of the Solar system.

Remark 9.2 The rotation of Mercury is very slow (59 days) which could be caused by a very large impact in the very early stage of its development. However, Mercury has no tectonic activity that would renew its 4.5 billion years old surface which is now uniformly dotted only with relatively small craters. Another possibility is that its slow rotation is due to tiny repulsive forces. We hypothesize that Mercury was once closer to the Sun, and then due to these forces it slowly moved to a higher orbit with a semimajor axis of about 57.9 million km.

Because tidal forces are decreasing with the cube of the distance from the Sun, they act on Mercury by a $(149.6/57.9)^3 \approx 17$ times larger value per unit mass than on the Earth. Moreover, if Mercury were e.g. only 40 million km from the Sun at the time of its creation, which is in accordance with (9.3), then the tidal forces from the Sun would be $(57.9/40)^3 \approx 3$ times greater than today. Altogether, we get a $3 \cdot 17 = 51$ times larger tidal influence per unit mass than that acting on the Earth at present. This would significantly slow the rotation of Mercury, which has about a 100 times smaller moment of inertia than the Earth.

Mercury could also be closer to the Sun due to the absence of its moons whose orbits would be unstable there.

Remark 9.3 There is also an open problem concerning how Neptune could be formed as far away as 30 au from the Sun, where the original protoplanetary disc was relatively sparse and all motions would be very slow (by Kepler's third law its mean speed is only 5.4 km/s). Assuming that the average Neptune-Sun distance increases roughly at the rate of H_0, we find by (9.3) that Neptune could have been formed many astronomical units closer to the Sun 4.5 Gyr ago.

Remark 9.4 There is a well-known puzzle, how the large moon Triton with a diameter of 2705 km was captured so far from Neptune. Triton is a retrograde moon, since it orbits in the direction opposite to the direction in which Neptune rotates. Thus the spin angular momentum of Neptune has the opposite sign than the orbital angular momentum of the Neptune-Triton system. Hence, Triton should continually approach Neptune to keep the law of conservation of angular momentum. Triton has probably orbited Neptune for a very long time, because its present generally recognized orbital eccentricity is incredibly small, namely,

$$e = 0.000{,}016. \tag{9.31}$$

This is the smallest eccentricity of all known bodies in the Solar system. When Triton was captured during some N-body collision, its orbit was almost certainly an elongated ellipse and it took billions of years for Triton to reach an almost circular orbit due to tides. Its present radius $a = 354{,}759$ km is very large.

Using (9.31), we find that

$$\sqrt{1-e^2} = \sqrt{1 - 256 \cdot 10^{-12}} = 0.999{,}999{,}999{,}872.$$

From this and (1.2) we get for the difference of the semimajor and semiminor axes that

$$a - b = a(1 - \sqrt{1-e^2}) = 354{,}759\,\text{km} \times 0.000{,}000{,}000{,}128 = 4.5\,\text{cm}.$$

Is it really possible that (9.31) would be some measured value without any error estimate?

We claim that Triton was captured closer and then migrated from Neptune, since the apparent repulsive force are bigger than tidal forces that act in opposite direction. This could explain the current very large orbital angular momentum of Triton.

Remark 9.5 Similar arguments can be applied to the Pluto-Charon system. Since this system is completely locked into the 1:1:1 resonance, tidal forces are almost negligible and its large orbital angular momentum can again be due to tiny repulsive forces.

Remark 9.6 We know 11 rapidly-moving satellites that orbit Uranus below its stationary orbit with radius 82,684 km. The average difference between radii of their neighboring orbits is only 2663 km. The distances of such neighbors above the stationary orbit are much bigger (cf. Fig. 9.7). The orbit of the satellite Puck has a radius of 86,010 km, followed by Mab with 97,700 km and Miranda with 129,390 km. This is because for satellites below the stationary orbit, the tidal forces and apparent repulsive forces are mutually subtracted, while for the satellites above the stationary orbit they are summed. A similar statement holds also for rapidly-moving satellites of Neptune. For more details concerning satellites below and above the stationary orbit of Uranus and Neptune, see [184, p. 226].

In [184] we present several other examples suggesting that gradual expansion can also be observed in other smaller subsystems of the Solar system. This apparently contradicts the classical Newtonian law of conservation of energy.

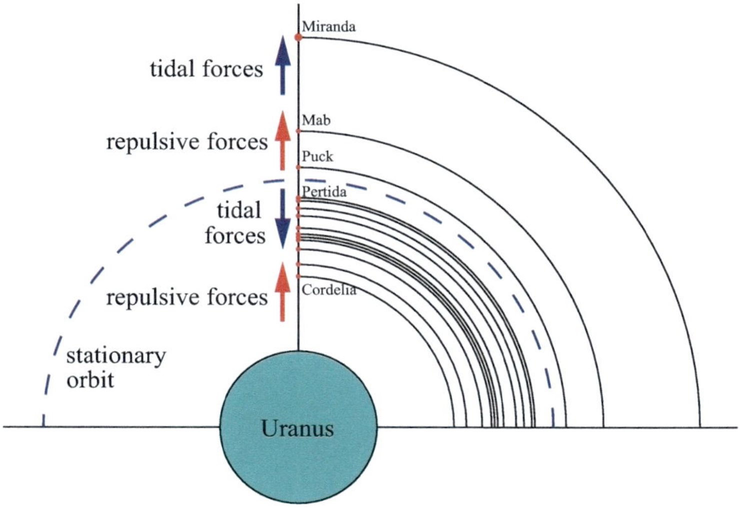

Fig. 9.7 Rapidly-moving satellites of Uranus. Below the stationary orbit the tidal forces and apparent repulsive forces are subtracted, since they have opposite directions, whereas above this orbit they are summed up coherently. The distances between neighboring satellites above the stationary orbit are substantially greater than under it

Emmy Noether was a leading German mathematician who proved the following theorem [263], which completely changed the face of physics:

Theorem 9.1 (Noether) *The energy of each isolated system is conserved if it possesses symmetry with respect to time translations.*

If the trajectories of the moons of planets in the Solar system develop in time along spirals, then this may indicate the existence of apparent repulsive forces that would violate the law of conservation of energy. The symmetry with respect to time translations is not true for spiraling trajectories. Similarly, another theorem of Noether states that spatial symmetry under rotation, or rotational invariance, leads to the conservation of angular momentum. This symmetry is also slightly violated in our case. Finally, she also proved that spatial translation symmetry ensures that momentum is conserved.

9.6 Gravithermal Catastrophe

Almost all open and globular clusters seem to be unstable, as was shown by Pavel Kroupa [204]. Smaller stars tend to move away from the center of a cluster, while the massive ones gradually accumulate around the center. This leads to mass segregation. During multiple collisions light-mass stars are sometimes even ejected

Fig. 9.8 Globular cluster M13 in the constellation Hercules (photo NASA)

from the cluster which is usually called *evaporation of stars* in the clusters. The surrounding, heavier stars drift closer to the center due to loss of potential energy. This process ends by the so-called *gravithermal catastrophe* (see [115]).

However, the local Hubble expansion should cause all the stars on average to move away from the center. Hence, it would act against the gravithermal catastrophe inside the cluster and it would slow the entire process.

There exist about 150 globular clusters in the Milky Way, but some other galaxies contain thousands of globular clusters. They are very old systems and contain hundreds of thousands or up to millions of stars, see Fig. 9.8. Some globular clusters have existed for more than 13 billion years and their gravithermal catastrophe has not been fully manifested, yet.

9.7 Do Single Galaxies Expand?

There is no reason to assume that apparent repulsive forces would somehow not be present in the interior of galaxies, since their manifestations are observed not only at large cosmological distances, but also inside the Solar system as shown in previous sections, see also [335, p. 154]. In spite of that it is generally claimed that galaxies do not expand, because they are gravitationally bound and that only the space between them expands. However, galaxies are usually included in clusters that should also not expand, because they are gravitationally bound as well. Galaxy clusters are gravitationally bound again in superclusters, cf. [262]. So where does the universe expand?

According to [235, Sect. 8], the observed conservative expansion rate of the Milky Way is 0.6–1 kpc/Gyr, which is approximately 0.6–1 km/s, since 1 kpc $= 3.086 \cdot 10^{16}$ km and 1 Gyr $= 3.156 \cdot 10^{16}$ s. This expansion rate nicely fits to the Hubble-Lemaître constant recalculated by (6.31) on the radius

$$R = 50{,}000\,\text{ly} = 15{,}328\,\text{pc} \approx \frac{1}{70}\,\text{Mpc}$$

of our Galaxy, namely,

$$H_0 \approx 1\,\text{km/(s}\,R)$$

and

$$\frac{\dot{R}}{R} = \frac{1\,\text{kpc/Gyr}}{15{,}328\,\text{pc}} = 0.652 \cdot 10^{-8}\,\text{cy}^{-1}.$$

This value is again comparable with similar ratios (9.15) and (9.29) which include effects of tidal forces.

Many other astronomical observations document the expansion of galaxies themselves. For instance, Bouwens et al. [39] found that galaxies grow slowly. Also Ignacio Trujillo with his team [350] discovered that the size of massive galaxies increases with time. This increase can, of course, be partially explained by intergalactic gas and dust that falls into galaxies due to gravity, as well as galactic cannibalism. However, galaxies at cosmological distances have on average more stars per unit volume when compared with the present situation, i.e., this is in line with the proposed local Hubble expansion on galactic scales.

By Ferré-Mateu and Trujillo [102] superdense galaxies were quite common in the early universe with redshift $z > 1.5$. At present they are quite sparse. Papers [357, 362, 376] also suggest that early galaxies were smaller and denser just after their formation. According to [51], the mass density of some galaxies for $z > 1$ is even comparable with the density of globular clusters themselves, i.e., on average several stars per pc^3. On the basis of this enumeration (see also [75, 119, 349], etc.) we may suppose that repulsive forces essentially contributed to the above-mentioned expansion and thus led the interiors of the galaxies themselves to become more tenuous. Similarly as in the Solar system, the rate of expansion of galaxies themselves may be smaller than the Hubble expansion, but it may also have the same order. This would support the picture that galaxies "swell up" very slowly like sea sponges or bread rising in the oven.

Further, we introduce a geometrical argument from [184, p. 243] which is based on proof by contradiction. Suppose for a moment that galaxies have constant volume (i.e., they do not expand with time) and that they do not collide. The right part of

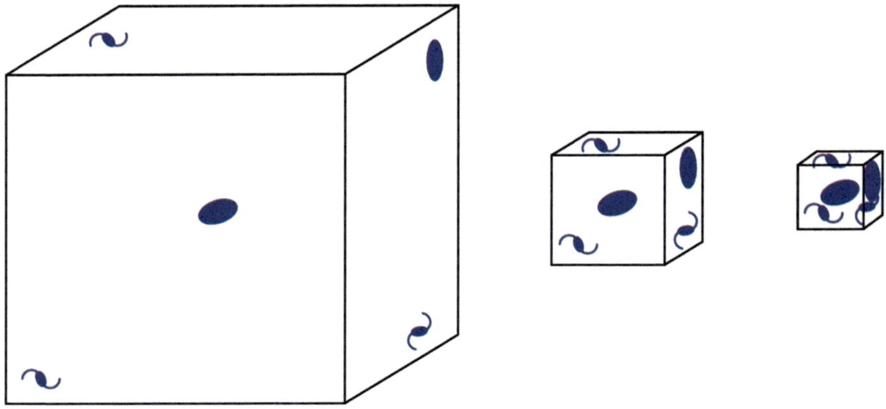

Fig. 9.9 A schematic illustration of nongrowing galaxies of constant size over time in an expanding universe. The unit cube is on the left. It contains several galaxies in our neighborhood for the redshift $z = 0$. The distribution of galaxies at cosmological distance $z = 2$ is in the middle and for $z = 4$ on the right. Such a picture of tightly crowded galaxies has not been observed by astronomers

Fig. 9.9 illustrates what we would observe at cosmological distances, if galaxies would have the same size at all times and no new galaxies were formed. On the left, five galaxies are schematically depicted in the unit cube for $z = 0$. Thus for $z = 2$ there would be on average $5(z + 1)^3 = 5 \cdot 3^3 = 135$ crowded galaxies in the unit cube, since the middle cube from Fig. 9.9 can be placed into the left unit cube $3^3 = 27$ times. Analogously we find that for $z = 4$ there would be $5 \cdot 5^3 = 625$ tightly packed galaxies in the unit cube. They could touch or even penetrate each other, if their sizes would be constant st all times. However, such a tight arrangement is not observed in various Hubble Deep Fields, since galaxies were smaller at that time.

We may also argue in the following way: The volume of our Galaxy is about $10^5 \times 10^5 \times 10^3\,\mathrm{ly}^3$. The total volume of all 10^{12} observed galaxies can thus be approximated as $10^{25}\,\mathrm{ly}^3$. Hence, if the radius of this part of the universe was smaller than 10^8 ly and protogalaxies were to have the same size as today's galaxies, they would necessarily interpenetrate each other, which would lead to a contradiction.

The above arguments show that galaxies themselves grow (see Fig. 9.10) although probably by a somewhat lower rate than the Hubble parameter. The repulsive forces (sometimes called *dark forces*) slightly, but continually generate energy in any system of free bodies that interact gravitationally [198, p. 276]. They increase its total (i.e. kinetic + potential) energy [265]. These forces thus act locally in our Galaxy and in other star systems.

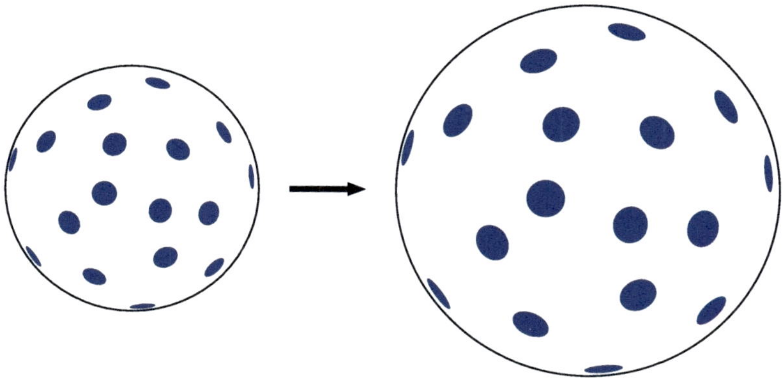

Fig. 9.10 A two-dimensional model $\mathbb{S}^2_{a(t)}$ of an expanding universe with comoving and swelling galaxies. On the other hand, in [78, 79], and [249, p. 719] galaxies have constant sizes over time. The center of this "inflating rubber balloon" corresponds to the Big Bang at an initial time

Remark 9.7 The angular momentum of spiral galaxies is also not conserved, see e.g. [217]. This is naturally expressed by the galactic angular momentum paradox:

How is it possible that spiral galaxies (originating from small random fluctuations in a hot homogeneous and isotropic universe) rotate so fast?

Danail Obreschkow even writes about the angular momentum catastrophe, see [265, p. 4]. How do galaxies gain their extremely large angular momentum? The answer that comes easily reads: From the local Hubble expansion. The kinetic energy hidden in each of the individual instances of the billions of rotating spiral galaxies is enormous. Local as well as global expansion increases the total potential energy. This is similar to the increase in potential energy of a comet moving away from the Sun.

Remark 9.8 By Hulse and Taylor [136] the orbital period of close binary pulsars is shortened. In this case, the binary system creates a strong and rapidly changing gravitational field in a curved spacetime. According to the GTR, the system loses energy in the form of gravitational waves. Since pulsars have extremely strong magnetic fields, the total energy can partly decrease also in the form of electromagnetic waves. These phenomena predominate over the local Hubble expansion.

Finally, note that the gravitational potential behaves like $C_1 r^{-1}$ while the magnetic potential like $C_2 r^{-2}$. Therefore, for sufficiently small r we have

$$\frac{C_1}{r} < \frac{C_2}{r^2}.$$

This implies that magnetic forces play a crucial role in very close encounters between binary pulsars.

Chapter 10
Anthropic Principle and the Hubble-Lemaître Constant

10.1 Weak Formulation of the Anthropic Principle

According to the *weak formulation of the anthropic principle*, all fundamental physical constants have just such values that enabled the origin of life, see e.g. [26, 57, 60, 61, 184]. This hypothesis is, of course, untestable, because we cannot change values of these constants and observe how the universe would evolve afterwards.

The *strong formulation of the anthropic principle* postulates that evolution necessarily leads to the origin of humans (= *Anthropos* in Greek). However, we may immediately raise the following objection against this formulation. If the asteroid that caused the extinction of the dinosaurs had missed the Earth, then the human race would never have developed, even though it cannot be denied that there would possibly be other intelligent beings. Note that there exist also other definitions of the strong and weak anthropic principles. Nevertheless, it is very surprising that all fundamental physical constants have exactly such values that allow the birth and development of life in the universe.

In this chapter, we demonstrate that also the current value of the Hubble-Lemaître constant essentially contributed to the existence of humankind, even though it is not a fundamental constant [202]. Life on Earth has existed continually for at least 3.5 Gyr and this requires very stable conditions during this quite long time interval. Nevertheless, due to the local Hubble expansion and since the luminosity of the Sun increases, the Earth receded from the Sun by an appropriate speed such that it received an almost constant solar flux during the last 3.5 Gyr.

M. Křížek, L. Somer, *Mathematical Aspects of Paradoxes in Cosmology*,
https://doi.org/10.1007/978-3-031-31768-2_10

Remark 10.1 No physical constant should be considered as a standard mathematical constant. For instance, the irrational numbers

$$\pi = 3.1415926535\ldots \quad \text{(Ludolf number)},$$
$$\mathrm{e} = 2.7182818284\ldots \quad \text{(Euler number)},$$
$$\sqrt{2} = 1.4142135623\ldots$$

have infinitely many digits. On the other hand, the Newton gravitational constant

$$G = 6.674 \pm 0.001 \cdot 10^{-11}\,\mathrm{m^3\,kg^{-1}s^{-2}}.$$

is not a real number with an infinite decimal expansion. Its fourth significant digit is close to four, but the other digits are not known. In the future it will be impossible to find e.g. one thousand digits of G, since physical constants have a completely different character than real numbers. Physical constants should be rather treated as "fuzzy numbers" or "interval arithmetic numbers" or the "density of some probabilistic distribution function". The reason is a special character of all physical measurements and the Heisenberg uncertainty principle which prevents us from obtaining infinitely many true decimal digits.

Anyway, the value of the gravitational constant G has a significant influence on the interior temperature of a star, its age and size, its luminosity, and many other parameters. The reason is that the product GM is proportional to the pressure inside the star, where M is its mass. If G were to be only one part per thousand smaller or larger than its present value, then all stars, galaxies and also their clusters would evolve in a different way, and thus the Earth could not come into being as it is. Unfortunately, we cannot perform any experiments to test the evolution of the universe for a different value of the gravitational constant G.

Similar statements are valid also for other fundamental physical constants like the speed of light c in a vacuum,[1] the mass of the proton m_p, the proton-to-electron mass ratio $m_\mathrm{p}/m_\mathrm{e} \approx 1836.152673$, the elementary charge e (of the electron), Planck's constant h, the vacuum electric permittivity ε_0, the Avogadro constant N_A, the Boltzmann constant k_B, etc. On the other hand, the dimensionless constant of fine structure $\alpha = e^2/(2\varepsilon_0 hc) \approx 1/137$, the reduced Planck constant $\hbar = h/(2\pi)$, the vacuum magnetic permeability constant $\mu_0 = 1/(\varepsilon_0 c^2)$, etc., should not be called fundamental, since the right-hand sides of their definition expressions already contain fundamental constants. Another way of proceeding would be to use a different (but equivalent) set of fundamental constants, where we replace e.g. ε_0 by α.

[1] We can always choose physical units so that $c = 1$, $G = 1$, $\varepsilon_0 = 1$, etc., instead of the standard SI units.

Some physical constants that characterize, for instance, ordinary water also have remarkable values. It is well-known that water has its highest density at 4°C under normal pressure. As a result, in winter, lakes do not freeze from the bottom. Contrary to other substances, ice has a lower density than liquid water. Thus, ice floats on water, which protects the fish and other aquatic animals from freezing. Water also has the highest heat capacity of all substances which significantly contributes to a high stability of the temperature of Earth's surface. Nevertheless, particular values of these physical constants are only side effects of the fundamental physical constants. Hence, the values of c, m_p, e, h, ε_0, etc., are so perfectly tuned that they allow water to have such advantageous properties suitable for the existence of life.

The structure of DNA with covalent bonds in both its strands and with hydrogen bonds between adenine (A) and thymine (T), and cytosine (C) and guanine (G) essentially depends on the interaction diameters of elementary particles and also on the size of the elementary charge e. This value plays a crucial role in replication and transcription processes [186]. The translation process: RNA → PROTEINS in ribosomes depends on e, too. Other fundamental constants allow a proper forming of the helix shape of DNA as well. Note that each cell has about 2 m of DNA strands which do not tangle during replication and transcription. Notice further that the total length of DNA in each human is about 10^{13} m which is more than the Earth-Pluto distance.

The two hydrogen bonds between adenine and thymine (see Fig. 10.1) are contained in the so-called TATA-box which precedes the start codon. Therefore, this box can be more easily split by the enzyme RNA polymerase during transcriptions than the three hydrogen bonds between cytosine and guanine, cf. [368, p. 966].

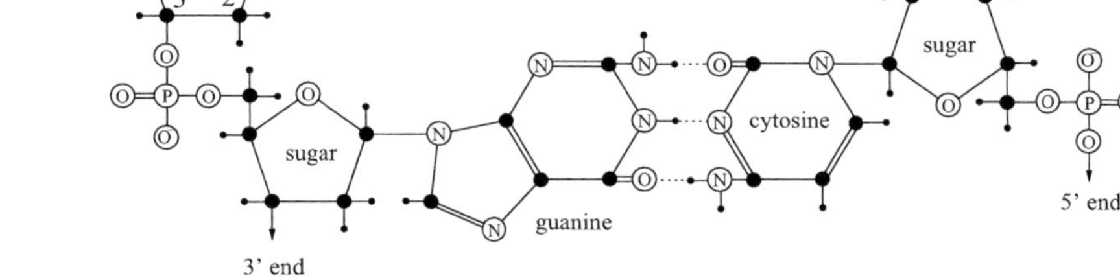

Fig. 10.1 Schematic illustration of the chemical structure of DNA: atoms of carbon are marked by large black dots, atoms of hydrogen by small black dots, covalent bonds are shown as solid lines, and hydrogen bonds as dotted lines. At both edges there is a solid skeleton of sugar phosphates, which protects the genetic information from damage. In these edges the deoxyribose sugar alternates with the phosphate group PO_4

Assigning A = 00, C = 01, G = 10, and T = 11, we observe that genetic information can thus be directly transcribed into binary digital system that was actually first discovered by nature 3.5 Gyr ago, i.e. much earlier than humans [186].

It is also noteworthy that the main components of all organic compounds are the most frequent elements in the universe (except for the inert gas helium). The most frequent element is, of course, hydrogen H, oxygen O is third, carbon C fourth, and nitrogen N fifth. They have small nuclei which supports a large variability of organic molecules. The nucleus of Si is too large in comparison with C to form large complex molecular structures.

A suitable combination of the above-mentioned fundamental constants also guarantees an incredible stability of the Sun for about 10 Gyr. By the formula $E = mc^2$, the Sun's mass losses due to nuclear reactions are 4.26×10^9 kg/s. Since $M_\odot = 1.989 \times 10^{30}$ kg, we find by (9.24) that

$$\frac{L_\odot}{M_\odot} = 0.0002\,\text{W/kg}$$

on average. (For comparison, the human body produces about 1 W/kg.) Consequently, nuclear reactions inside the Sun are incredibly well tuned so that the Sun was shining a very long time ($\sim$ 4.6 Gyr) which allowed the development of life and then the origin of humankind.

This process in the central part of the Sun can be very roughly described as follows. The weak interaction between two colliding protons of the same spin followed by the strong interaction is responsible for the formation of a deuteron. However, this collision is extremely unlikely, since the binding energy of one deuteron is only $\sim$ 1.1 MeV per nucleon. After that the strong interaction between two deuterons causes the formation of helium nuclei (i.e. alpha particles). Their binding energy is much larger, namely, $\sim$ 7.1 MeV per nucleon.

Remark 10.2 The expansion speed of the universe is also a very important quantity in the context of the anthropic principle. If it would be too high, galaxies would not appear. If it would be too small, our universe would gravitationally collapse and life would not have enough time to arise. The current expansion speed essentially depends on the initial conditions during the Big Bang.

There are many different definitions of living organisms. Max Tegmark [344, Chapt. 1] reduces the definition of *life* to a process that can retain its complexity and replicate itself (i.e. a self-replicating information-processing system). It is astonishing that the origin of life on Earth occurs at almost the same time as the end of late heavy bombardment of the Earth by asteroids $\sim$ 3.8 Gyr ago (see Fig. 9.3). Note also that life probably did not arise instantaneously, but it could very slowly evolve over thousands or millions of years in water. During this period one cannot uniquely state if some cluster of molecules is alive or not. We also cannot rule out an extraterrestrial origin of life.

Let us emphasize that H_0 represents only some average value over large structures. The repulsive forces contribute to the expansion of habitable zones (as in the case of the Earth), while the luminosity of the mother star is gradually increasing. In this sense, the habitable zones are then more stable, since they may exist for a longer time period.

10.2 The Faint Young Sun Paradox

Let us first recall what is meant by the faint young Sun paradox. Assume for a moment that the Earth was at its birth about 1 au away from the Sun as it is now. In order to not be completely frozen and to be able to develop life on its surface, the Sun at its origin had to be approximately as hot as it is now. However, this is not in agreement with the fact that the Sun as a star on the main sequence of the Hertzsprung-Russell diagram had a lower luminosity at its origin (cf. Fig. 9.3). This leads to the paradox usually referred as the *faint young Sun paradox* [213]. The mean temperature on the surface of Earth would have been much below the freezing point of water, in contrast with the absence of glaciation in the first 2.7 Gyr of its existence (see [28, p. 177]).

The faint young Sun paradox is, in fact, more severe due to ice-albedo feedback of the frozen ocean. To prevent the Earth from freezing over, a much higher concentration of CO_2 than today is assumed in [145]. This oxide is at present deposited in $CaCO_3$. There is evidence supporting the hypothesis that the Earth during its history was a "snowball" several times (see [114, p. 58]) but, on the other hand, sometimes the temperature of the oceans was much higher than today [224]. Of course, the composition of the atmosphere and also the greenhouse effect played a significant role.

Remark 10.3 If Mars were much closer to the Sun by tens of millions of kilometers (see Sect. 9.3), then our Earth would also have to be closer to the Sun. Otherwise these planets could have close encounters and their orbits would not be stable over time.

In the following sections we present further independent arguments suggesting that the Earth was much closer to the Sun in the past and that their average distance from each other increases by a speed of several meters per year. Assuming that this is due to the repulsive forces, the faint young Sun paradox ceases to be a mystery [86].

10.3 The Expansion of the Ecosphere

The Earth is located inside the so-called *habitable zone*, i.e., the area in which liquid water is present permanently. In this section and mainly in the next Sect. 10.4 we estimate such an expansion speed of the Earth from the Sun to ensure a constant

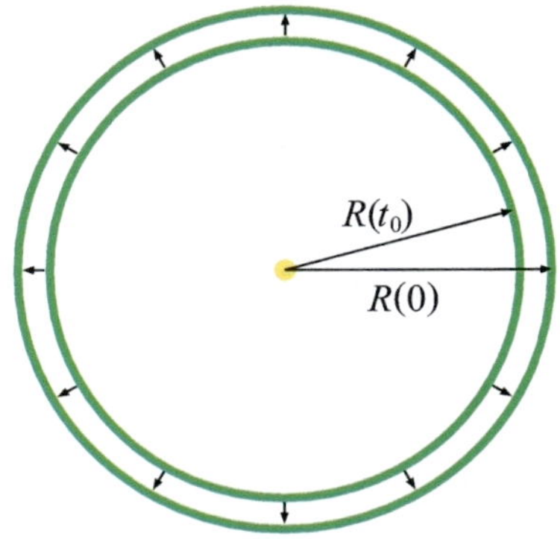

Fig. 10.2 Schematic illustration of the expansion of the ecosphere during the last 3.5 Gyr, where $R(t_0) = 1.3 \cdot 10^{11}$ m, $R(0) = 1.5 \cdot 10^{11}$ m, and $t_0 = -3.5$ Gyr

solar energy flux for 3.5 billion years, which would provide favorable conditions for the development of life on our planet.

So far, unfortunately, we cannot measure the actual average expansion speed of the Earth from the Sun with an accuracy on the order of a meter per year, because the position of the center of gravity of the Solar system shifts every day by about 1000 km primarily as a result of the gravitational influence of the giant planets, see Fig. 4.4. Therefore, we will consider the very long time interval of 3.5 billion years of life on the Earth.

Note that the Earth's elliptic orbit has a very small eccentricity $e = 0.0167$. The semimajor axis $a = 149.598 \cdot 10^6$ km and the semiminor axis $b = a\sqrt{1 - e^2} = 149.577 \cdot 10^6$ km do not differ before the fifth significant digit, i.e., Earth's orbit is almost circular.

To ensure favorable conditions for life on Earth it is necessary at present that the Sun's luminosity be at most 5 % larger or smaller than the solar constant L_0 given by (9.23). Such a ring (resp. spherical layer) is called an *ecosphere*. Since the energy flux from the Sun decreases with the square of the distance, its radii are $(0.95)^{1/2}$ au and $(1.05)^{1/2}$ au, which corresponds to a very narrow interval of 145.8–153.3 million km (see Fig. 10.2). If the Earth's elliptic orbit were to leave this ring for a long time period, it would have disastrous consequences for life on our planet. A permanent reduction of the Sun's luminosity of more than 5 % would cause overall glaciation of the planet. On the other hand, at temperatures over 57°C DNA sequences of mammals decay. Even at temperatures above 50°C mitochondria stop producing adenosine triphosphate (ATP) that supplies chemical energy to eukaryote cells.

When life appeared on Earth (i.e. 3.5 billion years ago), the Sun's luminosity was about 77 % of the present value (see Fig. 9.3). To ensure a favorable climate for the long-term evolution of life, where liquid water is necessary, the Earth was probably tens of millions of kilometers closer to the Sun at that time. This claim is supported, for example, by data on the occurrence of fossil thermophilic bacteria, from which it is believed [224] that the temperature of oceans was about 80°C three and half Gyr ago. Due to higher volcanic activity at that time it is, however, not clear to what extent this may only be a selection effect. On the other hand, formula (9.27) applied to the Earth yields a quite large discrepancy in temperatures, namely, for given a total solar luminosity of $0.77 L_{\odot}$, $r = 1$ au, and a present value for the mean Earth's albedo of $A = 0.306$, we get the mean temperature $-35°$C.

Heat produced by decay of radioactive isotopes also contributed to a higher temperature during the first few hundred million years of the Earth's existence, when elements with a relatively short half-life decayed. According to current measurements, the heat flow from the Earth is less than 0.1 W/m^2. It could not have been much larger 4 Gyr ago, because the half-lives of the contemporary natural radioactive isotopes ^{232}Th, ^{238}U, ^{40}K, and ^{235}U are 13.9, 4.468, 1.248, and 0.704 Gyr, respectively [77]. It is therefore estimated (see [114, p. 58]) that the heat flow from the Earth 4 Gyr ago was no more than 5 times larger than today. Nevertheless, this heat flow (mainly from diverse geothermal sources) is completely negligible when compared to the solar constant (9.23).

Although the Sun's power grew approximately as illustrated in Fig. 9.3 the Earth should have relatively stable conditions that are needed for the development of life during the period lasting 3.5 Gyr.

10.4 Two-Sided Error Estimates

The faint young Sun paradox can well be explained by the tiny repulsive forces that are responsible for the local Hubble expansion, see [161]. First we show that for the constant average speed of the Earth moving away from the Sun, cf. (9.3),

$$\overline{v} = 5.2\,\mathrm{m/year}, \tag{10.1}$$

the Earth would receive an almost constant density of flux of energy. In detail, the velocity $\overline{v}$ is optimal in the sense that the Earth receives practically the same flux of energy from the Sun as the solar constant L_0 (defined by equality (9.23)) over a very long time period of 3.5 billion years. Namely, the optimal flux L_{opt} will correspond to the particular expansion rate (10.1).

In order to prove this (see Theorem 10.1 below), we put

$$t_0 = -3.5\,\mathrm{Gyr}$$

and $t = 0$ will again correspond to the present time. The main results of this section are formulated in the form of mathematical theorems to make it clear what is assumed and what is claimed. The symbol $\forall t$ means "for all t".

Since the luminosity of the Sun increases approximately linearly with time and it was only about 77 % of its present value 3.5 Gyr ago (see Fig. 9.3), we set

$$L(t) = \left(1 - 0.23\frac{t}{t_0}\right) L_0 \quad \forall t \in [t_0, 0], \tag{10.2}$$

i.e. $L(t_0) = 0.77\, L_0$ and $L(0) = L_0$. As the luminosity decreases with the square of the distance, we can state the following quite surprising theorem.

Theorem 10.1 (Optimal Recession Speed of the Earth from the Sun) *Set*

$$L_{\text{opt}}(t) = \frac{L(t)R^2}{(R + \overline{v}t)^2}, \quad t \in [t_0, 0], \tag{10.3}$$

where $R = 1$ au *and* $\overline{v}$ *is given by* (10.1). *Then*

$$|L_{\text{opt}}(t) - L_0| < 0.005\,\text{kW}\,\text{m}^{-2} \quad \forall t \in [t_0, 0]. \tag{10.4}$$

Proof The very small dispersion of luminosity $\pm 0.005\,\text{kW}\,\text{m}^2$ on the right-hand side of formula (10.4) can be easily derived analytically by investigating the rational function $L_{\text{opt}}(t)$. It is concave on the whole interval $[t_0, 0]$, By inspection we find that its minimum is attained in the left end point t_0, $L_{\text{opt}}(t_0) = 1.357387\ldots$ and the maximum of the function L_{opt} is attained in $t^* \approx -1.7 \cdot 10^9$ year, $L_{\text{opt}}(t^*) = 1.364574\ldots$ Hence, the following two-sided estimate holds (cf. Fig. 10.3):

$$1.355 < L_{\text{opt}}(t_0) = \min_{t \in [t_0, 0]} L_{\text{opt}}(t) < \max_{t \in [t_0, 0]} L_{\text{opt}}(t) < 1.365$$

and $L_{\text{opt}}(0) = L_0$. □

We do not claim that the Earth recedes from the Sun just by the speed (10.1), but with a speed whose size has order comparable with $\overline{v}$. The luminosity flux

$$1.36 \pm 0.005\,\text{kW}\,\text{m}^{-2} \quad \forall t \in [t_0, 0]$$

would, of course, guarantee very stable conditions (10.4) for the development of intelligent life on Earth over the very long period of 3.5 Gyr. In particular, the local Hubble expansion seems to be just right for an almost constant influx of solar energy and thus also for the appearance of humankind.

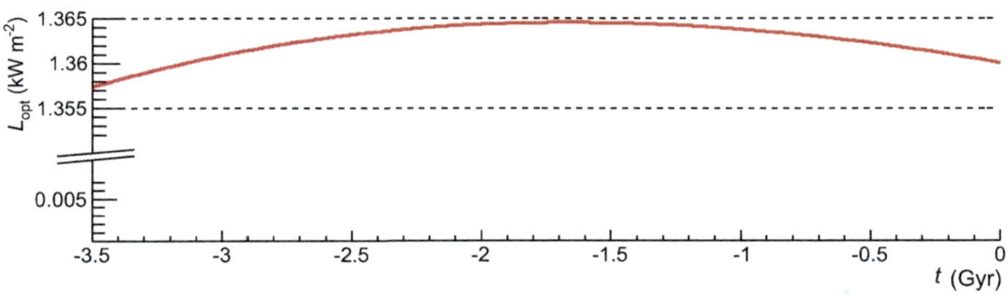

Fig. 10.3 Plot of the almost constant function $t \mapsto L_{\text{opt}}(t)$ over the interval $[t_0, 0]$. The vertical axis is substantially shortened for the clarity of the presentation

The speed (10.1) corresponds to the mean local Hubble expansion

$$\boxed{H_0^{(\mathrm{loc})} = 0.52\, H_0.}$$

Due to a similarity with the order of the Hubble constant (9.3), we propose that the above recession originates from the apparent repulsive forces introduced in the previous chapter. These forces thus cause a secular outward migration of our planet on the order of meters per year, so that it stays within the expanding ecosphere (see Fig. 10.2). If they were not to act, then conditions favorable for the development of life on Earth would exist for only about 1 billion years. Intelligent life would not have had enough time to develop due to a continual rise in temperatures (see Fig. 9.3).

The local Hubble expansion thus represents further support for the (weak) anthropic principle, which states that basic physical constants are favorable to the emergence of life only if they are in very narrow intervals [161]. Moreover, the speed in (10.1) is optimal in the sense that any other slightly different speed would not yield an almost constant flux expressed by the rational function in (10.3) on the time interval 3.5 Gyr. Thus it is probable that the actual average recession speed of the Earth from the Sun oscillates about the value 5.2 m/year (see Theorems 10.2 and 10.3), e.g. due to the influence of other planets.

Now, for a variable continuous recession speed v on the interval $[t_0, 0]$ we define similarly to (10.3) the associated luminosity

$$\mathcal{L}(v,t) = \frac{L(t)R^2}{\left(R - \int_t^0 v(\theta)\mathrm{d}\theta\right)^2}, \qquad t \in [t_0, 0], \tag{10.5}$$

where $t_0 = -3.5$ Gyr, $R = 1$ au, $L(t)$ is given by (10.2), and the integral of the velocity yields the distance. In the case that we are not able to establish a proper value of some quantity, the so-called *two-sided estimates* may be quite useful (see e.g. [193, 364]).

Theorem 10.2 (Two-Sided Estimates) *If the recession speed $v = v(t)$ of the Earth from the Sun lies in the interval from* 4.26 *m/year to* 6.14 *m/year for every $t \in [t_0, 0]$, then the luminosity defined by* (10.5) *changes at most about* 5 % *from L_0, namely,*

$$0.95L_0 \le \mathcal{L}(v,t) \le 1.05L_0 \quad \forall t \in [t_0, 0].$$

Proof By inspection we find that for the constant velocity $v_1 \equiv 4.26$ m/year the rational function $t \mapsto \mathcal{L}(v_1, t)$ is increasing on $[t_0, 0]$ and thus by (10.5),

$$0.95L_0 = \frac{L(t_0)R^2}{(R + v_1 t_0)^2} = \mathcal{L}(v_1, t_0) \le \frac{L(t)R^2}{(R + v_1 t)^2} = \mathcal{L}(v_1, t) \tag{10.6}$$

for any $t \in [t_0, 0]$. Analogously, for the constant velocity $v_2 \equiv 6.14$ m/year, we get that $t \mapsto \mathcal{L}(v_2, t)$ is decreasing, and therefore,

$$\mathcal{L}(v_2, t) \le \mathcal{L}(v_2, t_0) = \frac{L(t_0)R^2}{(R + v_2 t_0)^2} = 1.05 L_0. \tag{10.7}$$

Putting (10.6) and (10.7) together, we obtain by (10.5) and the proposed estimate $v(t) \in [4.26, 6.14]$ m/year that for any $t \in [t_0, 0]$

$$0.95 L_0 \le \frac{L(t)R^2}{(R + 4.26\,t)^2} \le \frac{L(t)R^2}{\left(R - \int_t^0 v(\theta)\mathrm{d}\theta\right)^2} \le \frac{L(t)R^2}{(R + 6.14\,t)^2} \le 1.05 L_0.$$

The proof is completed. □

A more important converse proposition has stronger assumptions on the velocities:

Theorem 10.3 (Additional Two-Sided Estimates) *If the average recession speed* $\overline{v}$ *lies outside the interval* [4.26, 6.14] *m/year, then there exists a nonempty subinterval* $I \subset [t_0, 0]$ *such that the luminosity* $\mathcal{L}(\overline{v}, t)$ *is less than* 95 % *or greater than* 105 % *of* L_0 *for all* $t \in I$.

Proof If $\overline{v} < v_1 \equiv 4.26$ m/year, then similarly to (10.6) we get

$$\mathcal{L}(\overline{v}, t_0) = \frac{L(t_0)R^2}{(R + \overline{v}t_0)^2} < \frac{L(t_0)R^2}{(R + v_1 t_0)^2} = \mathcal{L}(v_1, t_0) = 0.95 L_0.$$

From the continuity of the rational function $t \mapsto \mathcal{L}(\overline{v}, t)$ it follows that there exists a nonempty time interval I_1 such that $\mathcal{L}(\overline{v}, t) < 0.95 L_0$ for all $t \in I_1$.

Analogously to (10.7) we find that for $\overline{v} > v_2 \equiv 6.14$ m/year there exists a nonempty interval $I_2 \subset [t_0, 0]$ such that $\mathcal{L}(\overline{v}, t) > 1.05 L_0$ for all $t \in I_2$. □

An average speed $\overline{v}$ lying outside the interval [4.26, 6.14] m/year corresponds to truly inhospitable conditions for the evolution of life into multicellular forms. The previous two theorems can also be easily modified for other values than just 5 %.

Remark 10.4 In some models the linear function $L(t) = (1 - 0.23t/t_0)L_0$ from (10.2) and Fig. 9.3 is replaced by the rational function (see e.g. [28, p. 177])

$$\hat{L}(t) = \frac{L_0}{1 + 0.3t/\tau_0}, \qquad t \in [t_0, 0],$$

where $\tau_0 = -4.5$ Gyr. In this case the optimal average recession speed (guaranteeing an almost constant energy flux from the Sun) is

$$\overline{v} = 4.36\,\mathrm{m/year}$$

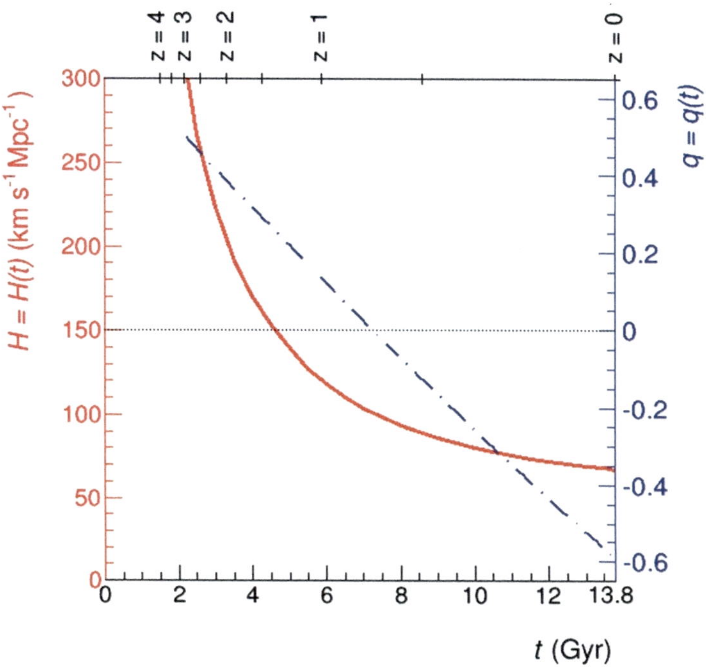

Fig. 10.4 The behavior of the Hubble parameter $H = H(t)$ according to the ΛCDM model is sketched by the solid line. The dashed-dotted line stands for the corresponding deceleration parameter $q = -1 - \dot{H}/H^2$ that was evaluated by means of numerical differentiation of the function $H = H(t)$. The lower horizontal axis shows time in Gyr since the Big Bang. In the upper horizontal axis we see the corresponding redshift z according to [284]

and the mean recession speed should be in the interval [3.27, 5.21] m/year to keep variations of the solar energy flux below 5 % as in Theorems 10.2 and 10.3.

Remark 10.5 It is said that the universe expands exponentially due to dark energy. From Fig. 10.4 we observe that the Hubble parameter $H = H(t)$ has been almost constant during the last 4.5 Gyr when the Solar system was formed. It lies in the interval $[H_0, \frac{5}{4}H_0]$. Assuming that $H(t)$ is constant, i.e., $H(t) \equiv H_0$, we get by (6.27) that Theorem 10.1 can be modified for an exponential expansion as follows. If the average recession speed of the Earth from the Sun is $\tilde{v} = 5.014$ m/year, then $L_{\mathrm{opt}}(t) = 1.36 \pm 0.008\,\mathrm{kW\,m^{-2}}$ for all $t \in [t_0, 0]$, which is analogous to (10.1) and (10.4). Also Theorems 10.2 and 10.3 can correspondingly be slightly modified. Moreover, from Fig. 7.6 we observe that the expansion function $a = a(t)$ is almost linear during the last 4.5 Gyr.

10.5 A Possible Future Development of Earth's Orbit

From Theorems 10.2 and 10.3 we find that the probable secular expansion rate of the radius of Earth's orbit lies in the interval

$$\boxed{H_0^{(\mathrm{loc})} \in [0.426\, H_0,\ \ 0.614\, H_0].}$$

Such a local Hubble expansion is therefore perfectly tuned by the apparent repulsive forces (see [161]). Therefore, not only the fundamental physical constants are finely tuned. For example, the average surface temperature of the Sun $T = 5770\,\mathrm{K}$ is also perfectly balanced. If it were to decrease about 1 %, i.e. a mere 57.7°C, then by the Stefan-Boltzmann law $L_0 = \sigma T^4$, the solar constant L_0 would decrease about 4 % as $0.99^4 L_0 \approx 0.96 L_0$. The total luminosity of the Sun would also decrease by 4 %.

Remark 10.6 It is often argued that after half a billion years, the water in the oceans will evaporate, because the power of the Sun will be too high (see Fig. 9.3). These simplified conclusions do not take into account that the Earth can be inhabited by a technically advanced civilization, which can, for instance, place a giant reflection sheet at the Lagrangian point L_1 to control the incoming flux of variable solar energy. According to [314, p. 461], the Sun's luminosity will increase to $1.33\, L_\odot$ after 3 Gyr. Thus, it may exceed $1.05\, L_\odot$ over half a billion years. However, the recession speed (10.1) would guarantee very stable conditions on the Earth for several Gyr in the future. For instance, in the next 3.5 Gyr the flux density of solar energy will be in the very favorable interval 1.33–1.36 $\mathrm{kW\,m^{-2}}$, if the luminosity would evolve as in (10.3). The function $t \mapsto L_{\mathrm{opt}}(t)$ is in fact decreasing in the interval $[0, |t_0|]$ and from relations (10.2) and (10.3) it follows immediately that

$$L_{\mathrm{opt}}(|t_0|) = 1.33\,\mathrm{kW\,m^{-2}}.$$

Thus there would be pretty good prospects into the far future. Life on the Earth would have had a chance to develop further for a long time period in very stable conditions.

After 5–7 billion years, when reserves of hydrogen in the central region of the Sun will run out, the Sun will begin to change to a red giant. Its radius will extend beyond the current orbit of Venus. At that time the Earth could be about 180 million miles away from the Sun, provided the receding velocity is (10.1). Hence, the tiny repulsive forces could keep the Earth sufficiently far away from the growing Sun.

10.6 Why Does the Law of Conservation of Energy Not Hold in the Physical Universe?

Einstein's equations are assembled so that the law of conservation of energy and momentum is theoretically valid absolutely exactly. The reason is that the covariant divergence of the left-hand and right-hand side of (3.52) is zero. Newtonian mechanics is also designed in such a way that this law holds exactly. Nevertheless, in this and previous chapter we saw that the local Hubble expansion and the corresponding energy non-conservation may explain a number of classical paradoxes in the Solar system, in particular:

1. The faint young Sun paradox.
2. The paradox of the large orbital angular momenta of the Moon, Triton, and Charon.
3. The formation of Uranus and Neptune.
4. The existence of rivers on Mars and the existence of its satellite Phobos.
5. The tidal catastrophe of the Moon.
6. The paradox of the large recession speed of Titan from Saturn.
7. The tidal locking of the moon Iapetus.
8. The slow rotation of Mercury and the absence of its moons.
9. The long existence of rapidly-moving satellites below the stationary orbit of Uranus and Neptune.
10. The long existence of life on Earth.

Thus we claim that Einstein's equations satisfying the law of conservation of energy and momentum cannot well describe the evolution of the Solar system, although no one knows their solution. The main reason it that the energy and momentum in the Solar system are not conserved.

Remark 10.7 The effect of local expansion of the universe has a long history dating back to the 1933 paper [240] by Cunliffe George McVittie. By Carrera and Giulini [58], Cooperstock et al. [70], and Mashhoon et al. [236] the cosmological constant Λ has no effect on the expansion of the Solar system. However, from Sects. 7.8, 9.2, 9.3, 10.4, etc., we already know that we can detect the local Hubble expansion even in the Solar system.

The Solar system but also galaxies are unique astrophysical laboratories for testing whether or not the law of conservation of energy and momentum holds and whether a finite speed of propagation of the gravitational interaction generates as a byproduct the sought energy for general expansion (local and global). To illustrate this hidden energy source consider two bodies A and B of equal masses that orbit symmetrically with respect to their center of gravity. If A attracts B and B also attracts A at their instantaneous positions, then by the Newtonian theory of gravitation these forces are in the same line and in balance.

However, the speed of gravitational interaction is finite. Therefore, the body B is attracted by A towards its previous position A' as depicted in Fig. 10.5. Similarly A

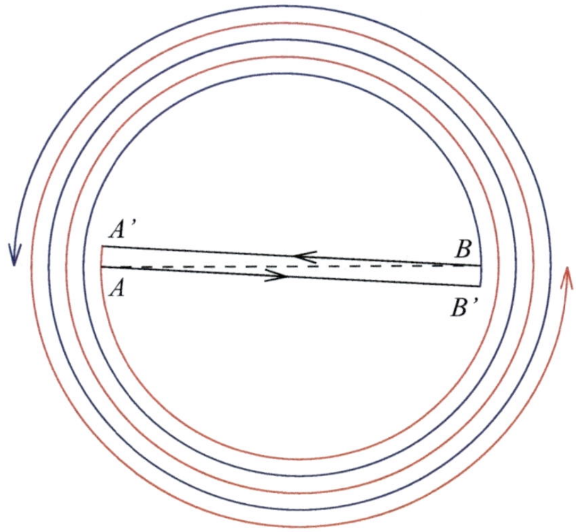

Fig. 10.5 Trajectories of two interacting bodies of equal masses form a double spiral. The distance between two neighboring arms is actually much smaller than in the picture. The gravitational aberration angles ABA' and BAB' are extremely small but positive

is attracted by B towards its previous position B'. Then a couple of non-equilibrium forces arises that permanently acts on this binary system. It increases the total angular momentum and the total energy of the system as well (see [184]). So this system is gradually expanding as a result of apparent repulsive forces.

In the framework of General Relativity Steven Carlip in [55] derived that the gravitational aberration angle γ of a body with speed v is bounded from above by the fraction v^3/c^3 while the angle of light aberration is approximately equal to

$$\alpha = \frac{v}{c}.$$

However, in doing so, he assumes that

(a) the cosmological constant is zero,
(b) some nonlinear terms, which he cannot estimate, are vanishing,
(c) the laws of energy and momentum conservation hold exactly.

Therefore, of course, he cannot obtain spiraling trajectories as schematically depicted in Fig. 10.5. Moreover, since the gravitational interaction has the same speed of propagation as light, the aberration angles α and γ should be the same for weak gravitational fields.

The recession speed and the angle of gravitational aberration depend on masses, velocities, and positions (also in the past) of all free bodies that gravitationally interact among themselves [157]. Numerical simulations show that any N-body system with $N > 1$ expands on average for a finite speed of gravity. Gravitational aberration thus has a nonnegligible influence on the expansion rate of the universe. Note that our Earth is not expanding, because it is not a system of free bodies that gravitationally interact.

Space deformations by the two interacting bodies may also have a non-negligible influence on the local Hubble expansion. Their proper distance is larger than their coordinate distance (see Fig. 5.1) and time dilation should also be taken into account.

Recall (see e.g. [5, 6, 18, 136]) that there exist close binary pulsars whose orbital period, by contrast, decreases over time. In this case, the system loses kinetic energy due to the radiation of gravitational and electromagnetic waves, due to tides, friction by interstellar dust, and so on. The resulting forces then prevail over apparent repulsive forces.

Sometimes we must also take into account various resonances that may cause much larger effects than these apparent repulsive forces whose manifestations on short time scales are indeed negligible. However, they operate continually in any gravitationally bound system of asteroids, moons, planets, stars, galaxies, clusters of galaxies, or orbiting clusters of galaxies. They contribute to migration of planets and their moons over long-term time periods, they cause that star clusters continually "evaporate", they act against gravithermal catastrophe of galactic clusters and star clusters, they reduce the frequency of collisions of stars [157] and galaxies, they slowly expand the "cosmic web", they stabilize the Solar system, and so on. Apparent repulsive forces also helped to create the conditions suitable for the emergence of life on Earth over a period of several billion years, during which solar power was growing.

Chapter 11
Gravitational Waves

11.1 The First Detection of Gravitational Waves

In this chapter we show that masses of binary black hole mergers are overestimated, since a large gravitational redshift is not taken into account (see (11.7) and (11.9) below) which is a direct consequence of time dilation in a strong gravitational field. This fact allows us to explain a high mass gap between observed binary neutron stars and calculated binary black hole mergers. We also present other reasons taken from [199] why masses of black hole mergers are determined incorrectly.

In 2016, the letter [2] about the first detection of gravitational waves was published. They were generated by two merging black holes that had approximately 36 and 29 solar masses, the final black hole had 62 solar masses and 3 solar masses turned into gravitational waves. Now we will analyze this event in detail.

In 1916 (just a century before [2] has appeared), Albert Einstein [93, 95] predicted the existence of gravitational waves. He assumed only very weak gravitational fields which enabled him to linearize his field equations of General Relativity. He considered only small perturbations of Minkowski spacetime and after some further simplifications he got a nonhomogeneous partial differential equation with the d'Alembert operator for plane gravitational waves (see e.g. [32, p. 24]).

Gravitational waves were first detected on September 14, 2015. According to [2], two black holes with masses

$$m_1 = 36^{+5}_{-4} M_\odot \quad \text{and} \quad m_2 = 29^{+4}_{-4} M_\odot, \tag{11.1}$$

merged and the generated gravitational waves GW150914 were independently intercepted by two LIGO detectors.

We show that the masses in (11.1) are not too trustworthy. First of all, we would like to emphasize that our criticism does not concern the LIGO detectors themselves, which are highly sophisticated and remarkable instruments [23]. It concerns the methodology that was used to process and then interpret the measured data. The

M. Křížek, L. Somer, *Mathematical Aspects of Paradoxes in Cosmology*,
https://doi.org/10.1007/978-3-031-31768-2_11

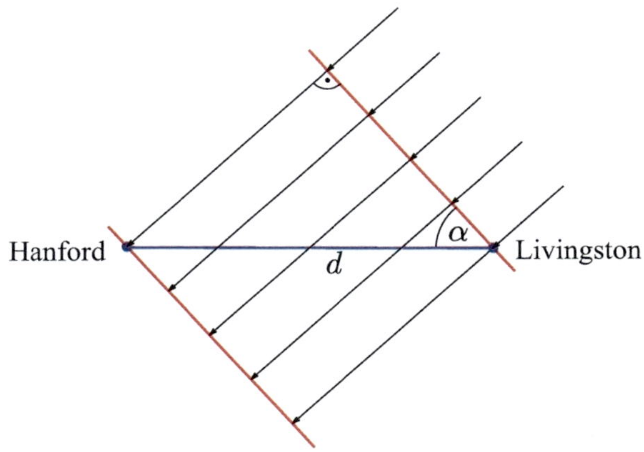

Fig. 11.1 Depiction of the passage of a planar gravitational wave through two detectors LIGO in USA

applied post-Newtonian model assumes an infinite speed of gravity and neglects the infinite gravitational redshift of each single black hole. Therefore, it can only very roughly approximate reality. It is true that it works pretty well in the Solar system. But since this post-Newtonian model is situated in Euclidean space, it does not allow one to consider a highly curved spacetime near black holes. Hence, from such a heuristic model one cannot establish any credible conclusions about the actual masses of these black holes and derive any reliable error estimates.

Remark 11.1 Two LIGO detectors whose distance is $d = 3002\,\mathrm{km}$ recorded the first passage of a gravitational wave. The time delay between these two detections was $\Delta t = 6.9\,\mathrm{ms}$ (see [2, p. 2]). This immediately leads to the following upper bound on the speed of gravitational waves

$$c_{\mathrm{g}} = \frac{d \sin \alpha}{\Delta t} \le \frac{d}{\Delta t} = 435{,}072\,\mathrm{km/s}$$

in a vacuum or in the Earth body, see Fig. 11.1. We observe that this guaranteed upper bound is larger than the speed of light in a vacuum $c = 299{,}792{,}458\,\mathrm{m/s}$. However, since 2017 we know that $c_{\mathrm{g}} \cong c$ with very high accuracy due to observations of a binary neutron star merger, see [5, 6]. The corresponding gravitational waves GW170817 were detected on Earth practically at the same time as electromagnetic waves. Note that the relation $c_{\mathrm{g}} \cong c$ was discussed by Henri Poincaré [294, p. 1507] already in 1905. Anyway, it is surprising that Einstein's equations (3.52) describing gravity do not contain the speed of gravitational interaction c_{g} but the speed of electromagnetic interaction c.

The signal from LIGO and VIRGO detectors of gravitational waves is continually corrupted by white noise due to thermal movement of particles, seismic waves, quantum noise, etc. It is noteworthy that such noise does not seem to be present in Fig. 1 of [2] in a 40 ms long time interval around the maximum amplitude of the detected signal. The white noise can be partially suppressed from the detected signal by means of a wavelet transform which removes components with a small amplitude and high frequencies at the same time (see e.g. [76, 242]).

11.2 Emitted Versus Detected Frequencies

The only relation which is given on a single line in [2] reads:

$$\mathcal{M} := \frac{(m_1 m_2)^{3/5}}{(m_1 + m_2)^{1/5}} = \frac{c^3}{G}\Big[\frac{5}{96}\pi^{-8/3} f^{-11/3} \dot{f}\Big]^{3/5}, \tag{11.2}$$

where $f = f(t)$ and $\dot{f} = \dot{f}(t)$ are the frequency of gravitational waves in time t and its time derivative, m_1 and m_2 are the masses of the components of the binary system for low frequencies, $\mathcal{M}$ is the *chirp mass* in the detector frame, and $G = 6.674 \cdot 10^{-11}\ \mathrm{m^3\,kg^{-1}s^{-2}}$ is the gravitational constant. The chirp mass determines the frequency evolution of the gravitational waves during a binary's inspiral phase.

According to [2, p. 3], f is the detected frequency. However, then the formula (11.2) cannot be true, in general, since f on its right-hand side essentially depends on the total redshift z, whereas the left-hand side of (11.2) is independent of z. The masses m_1 and m_2 cannot depend on z. Therefore, (11.2) can be valid only for $z = 0$. Note that formula (11.2) is based on the paper [33] published in 1995. However, in [33, p. 3516] and [74, p. 2663], the authors correctly consider the emitted frequency f_{em} (or the orbital frequency πf_{em}) in the source frame and not the detected frequency f as in [2]. Therefore, the detected frequency f should be replaced by the emitted frequency

$$f_{\mathrm{em}} = (z + 1) f, \tag{11.3}$$

where z is the total redshift of gravitational waves.

The masses m_1 and m_2 were established by a series of numerical simulations [254] with various input values m_1 and m_2. The numerical methods employed in constructing the templates for the merging black holes can also be found in [63] and [281]. Let us emphasize that no gravitational redshift is taken into account in the cited papers [63, 254, 281] on numerical simulations.

Note also that the corresponding modeling error cannot be established, since the true analytical solution of Einstein's equation for two orbiting bodies with positive masses is not known. Moreover, from [2] one cannot find the method by which the redshift z was estimated and how the derivative $\dot{f}$ from (11.2) was calculated. There is also no relevant reference to this issue.

Now let us look at the equality (11.2) in more detail. This equality was derived from the relation (3) of the 1995 paper [33] in a manner such that several higher order terms were neglected. In spite of that, the authors of [2] kept the equality sign in (11.2).

Example 11.1 Notice that the left-hand side of equality (11.2) does not depend on time t, while the right-hand side is time dependent. Thus the equality (11.2) is satisfied if and only if the product $f^{-11/3}\dot{f}$ is a suitable constant C (e.g. $C = 0.00015142\ldots \text{ s}^{5/3}$ for (11.1)). This leads to the solution of an ordinary differential equation of the first order

$$\dot{f} = Cf^{11/3} \tag{11.4}$$

whose general solution is

$$f(t) = \left(\frac{3}{8}\right)^{3/8} \frac{1}{(K - Ct)^{3/8}}, \tag{11.5}$$

where $K \in (-\infty, \infty)$ is an arbitrary integration constant, see also [74, p. 2663]. By Rektorys et al. [297, p. 14] the function (11.5) is the only solution of the differential Eq. (11.4) for $t < K/C$. Since this equation is autonomous, we may choose $K = 0$, i.e. $t < 0$. Then relation (11.5) determines how the frequency increases with time.

The authors of [2] had only approximate values of the detected frequency f from the interval 35–250 Hz for about 8 "measured orbital periods". Finally note that a numerical computation of the derivatives from smoothed inexact data is an ill-conditioned problem which may produce incorrect results (see e.g. [323, p. 144]). A detailed analysis of the smoothed signal GW150914 can be found in [394, 395].

For the *total mass* $M = m_1 + m_2$ and the *reduced mass* $\mu = m_1 m_2 / M$ we set $\mathcal{M}_1 = \mu^{3/5} M^{2/5}$. We also set $\mathcal{M}_2 = \widetilde{m}(\widetilde{m}/M)^{1/5}$, where $\widetilde{m} = \sqrt{m_1 m_2}$ is the geometric mean. The next theorem states that these formulae represent equivalent definitions of the chirp mass.

Theorem 11.1 *We have* $\mathcal{M} = \mathcal{M}_1 = \mathcal{M}_2$.

Proof From (11.2) we find that

$$\mathcal{M} = \frac{(m_1 m_2)^{3/5}}{(m_1 + m_2)^{1/5}} = \left(\frac{m_1 m_2}{m_1 + m_2}\right)^{3/5} (m_1 + m_2)^{2/5} = \mu^{3/5} M^{2/5} = \mathcal{M}_1$$

and

$$\mathcal{M} = \frac{(m_1 m_2)^{3/5}}{(m_1 + m_2)^{1/5}} = \sqrt{m_1 m_2} \frac{(m_1 m_2)^{1/10}}{(m_1 + m_2)^{1/5}} = \widetilde{m} \left(\frac{\widetilde{m}}{M}\right)^{1/5} = \mathcal{M}_2.$$

The proof is completed. □

11.3 Neglected Gravitational Redshift

Unfortunately, in [2] (also in [3, 4, 7, 33, 148]) there is no mention about gravitational redshift of gravitational waves. Let us recall that redshift (or blueshift) of frequency of waves coming to us from the universe has three basic components:

1. A Doppler component caused by the movement of the source or the observer with respect to its neighborhood.
2. A cosmological component caused by the expansion of the universe.
3. A gravitational component caused by the change of frequency of waves in a gravitational field.

According to [2, p. 7], the luminosity distance of the considered binary system is 410^{+160}_{-180} Mpc, which is in perfect agreement with the cosmological redshift (see e.g. [284, 393])

$$z = 0.09^{+0.03}_{-0.04}, \tag{11.6}$$

that is stated by the authors. Thus for the two remaining components of the redshift it remains

$$z \approx 0. \tag{11.7}$$

Note that the corresponding Doppler redshift cannot be reliably determined, because the orientation of the orbital plane and the local movement of the binary merger are not known.

Let us also note that the gravitational redshift for the surface of a neutron star is $z \approx 0.3$ which is greater than in formula (11.6), which can be derived from (11.10) below. For the horizon of a single black hole with mass m with Schwarzschild radius

$$r = \frac{2Gm}{c^2} \tag{11.8}$$

we even have

$$z = \infty. \tag{11.9}$$

Comparing this with the relation (11.7), we find that the authors of [2] did not consider the large gravitational redshift caused by the binary black hole system.

The gravitational redshift is a direct consequence of Einstein's time dilatation. Time in a gravitational potential hole flows more slowly than outside. A photon has to spend some energy to leave a gravitational field of a mass object. Its frequency is indirectly proportional to the speed of the passing of time. Therefore, the frequency of electromagnetic waves decreases when leaving a large gravitational potential hole of the binary system. A similar phenomenon holds for gravitational waves that carry away energy and thus their frequency will decrease as well.

So let us recall the well-known formula (2.22) that can be derived from the Schwarzschild solution of Einstein's equations. It expresses the change of frequency of a photon leaving the gravitational field of a single black hole at the distance $R > r$ from its center

$$f = f_{\text{em}}\sqrt{1 - \frac{r}{R}}, \tag{11.10}$$

where r is given by (11.8), f_{em} is the emitted frequency of a photon and f is the detected frequency by a distant observer. From this and (11.3) we obtain the limiting relation (11.9) for $R \to r$.

Remark 11.2 Relations (11.8) and (11.10) can also be used to derive that the detected frequency is negligibly changed by the gravitational field at the measurement site, such as Earth, since $r = 9$ mm.

Example 11.2 Setting for instance $R = 2r$ in (11.10) (cf. [2, p. 3]), we find that $f = 2^{-1/2} f_{\text{em}}$. By (11.3) the corresponding gravitational redshift is

$$z = \sqrt{2} - 1 = 0.414 \tag{11.11}$$

which is much larger than the observed cosmological redshift in (11.6). Similarly, for $R = 3r,\ 4r$, and $5r$, we get $z = 0.225, 0.155$, and 0.118, respectively, which are also larger than (11.6).

Just before the collision of the two black holes, the spacetime between them exhibited the largest deformations. By the measured data of LIGO detectors it produced gravitational waves with an increasing recorded frequency 35–250 Hz. The distance between both the black holes was only a few Schwarzschild radii [2, p. 3]. From (11.8)–(11.11) we may deduce that the gravitational redshift of the emitted gravitational waves will be quite essential and larger than that in (11.6). Namely, a common gravitational potential hole of the two black holes is deeper than that of each of its components (see Fig. 11.2).

Example 11.3 From equality (11.2) it is obvious that the sought masses of the black holes and also the constant C of (11.4) depend on the emitted frequency $f_{\text{em}} = (z + 1)f$. Consequently, a proper determination of the total redshift z is essential. By relation (11.3) we obtain $\dot{f}_{\text{em}} = (z+1)^2 \dot{f}$, where the additional factor $(z+1)$ is due to Einstein's time dilatation. Substituting this and (11.3) into (11.2), we get the missing factor

$$(z + 1), \tag{11.12}$$

because

$$\left(f^{-11/3}\dot{f}\right)^{3/5} = (z + 1)\left(f_{\text{em}}^{-11/3}\dot{f}_{\text{em}}\right)^{3/5} \tag{11.13}$$

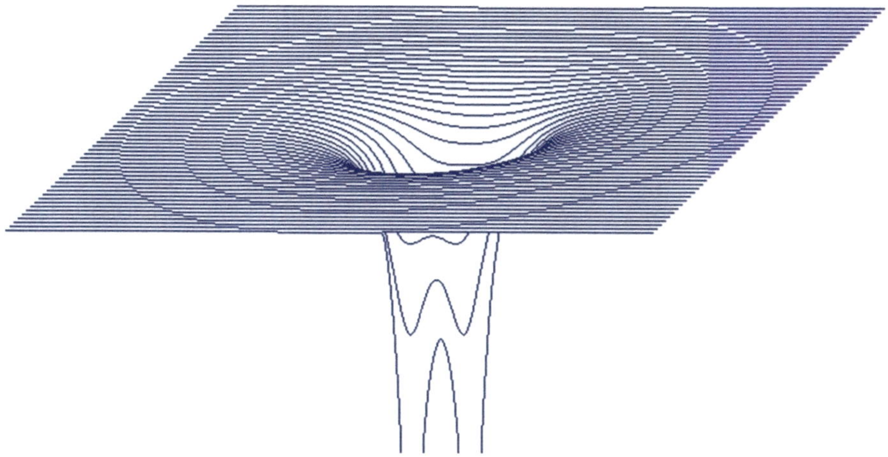

Fig. 11.2 Schematic illustration of the gravitational potential of a binary black hole system. The largest deformation of the spacetime arises close to the central saddle point. According to [2, p. 3] the highest amplitude of the detected signal was reached for the separation $R = \frac{3}{2}r$

and because the remaining factors in (11.2) are constants. In [2] only the cosmological redshift (11.6) was considered, but the total redshift is larger. For instance, from (11.6) and (11.11) we may deduce that the total redshift $z = 0.09 + 0.414$ could be slightly larger than $\frac{1}{2}$. Hence, the masses (11.1) were overestimated.

Example 11.4 Suppose for simplicity that

$$m_1 = m_2.$$

Then from (11.2) we find that the corresponding chirp mass

$$\mathcal{M} = m_1^{6/5}/(m_1 + m_1)^{1/5} = 2^{-1/5} m_1$$

linearly depends on m_1.

11.4 Other Arguments

The mechanism of the origin of the binary black hole system (11.1) is not known. According to the survey paper [71, Fig. 8], all known X-ray binaries detected in our Galaxy have components with masses less than $10M_\odot$. According to [49, 62, 256], masses of all known single stellar mass black holes are in the interval $5M_\odot$–$20M_\odot$ (e.g. Cygnus X-1 has about $20M_\odot$ and its accretion disc produces radio waves due to gravitational redshift). Hence, from a statistical point of view a system of two much larger black holes such as (11.1) seems to be quite exceptional, even though some selection effects may be present, since larger masses imply stronger signals.

Remark 11.3 By Abbott et al. [7] the mass of the larger component of the binary black hole corresponding to the event GW190521 was 85 $M_\odot$ and the total mass after coalescence was 150 $M_\odot$. The evolution path of such a large black hole binary is unknown. It is clear that many of its parameters in a common envelope of gas had to be tuned very finely.

Remark 11.4 There is a large statistically significant mass gap between all known black hole mergers and binary neutron stars, and also between single stellar black holes and binary black hole mergers, see e.g. [8, 343]. This also indicates that the gravitational redshift was ignored. Note that magnetic fields of a binary neutron star merger significantly contribute to the final coalescence, since they are very strong at short distances r (like permanent bar magnets). From Remark 9.7 we know that the magnetic potential is proportional to r^{-2} while the gravitational potential behaves like r^{-1}. The corresponding attractive forces are proportional to the gradient of these potentials.

Remark 11.5 Chen et al. in [64] have also found that the important gravitational redshift of GW150914 was not accounted for. They assume that the black hole binary system was located in close vicinity of a supermassive black hole possessing a large gravitational redshift. Their results indicate that the mass of each component of such a pair of black holes is not greater than $10M_\odot$.

Remark 11.6 The wavelength $\lambda = c/f$ corresponding to the highest detected frequency $f = 250$ Hz is equal to $\lambda = 1200$ km. This is a much larger value than the diameter of the wave zone given by the associated Schwarzschild radii of the black holes. Nevertheless, we observe that the emitted frequency $f_{\rm em} > f$ would yield a much more reliable size of the wave zone. This also shows that the gravitational redshift was missing.

Remark 11.7 According to (11.10), a distant observer will theoretically see that the fall of any mass object into a single black hole will take an infinitely long time. Actually, he should not see the plunging of this object into the black hole due to (11.9). On the other hand, a measured collision of the black holes (11.1) lasted only 40 ms. Is not that a contradiction between theory and reality? This comparison again shows that we should never equate simple theoretical models with the physical world. Moreover, since the number of emitted photons is finite, the observer will see that the fall will last only a finite time.

Remark 11.8 The smoothed signals presented from both the detectors show relatively small third and fourth amplitudes with high frequency appearing after the maximal amplitude, see [2, p. 3]. However, in principle wavelet analysis does not allow one to detect such small amplitudes with high frequencies in noisy data. The question is whether the noise removed from the signals (the so-called residual) from both the detectors is correlated or not. It appeared that it is uncorrelated except for that particular 40 ms long interval, where no noise seems to be present, see

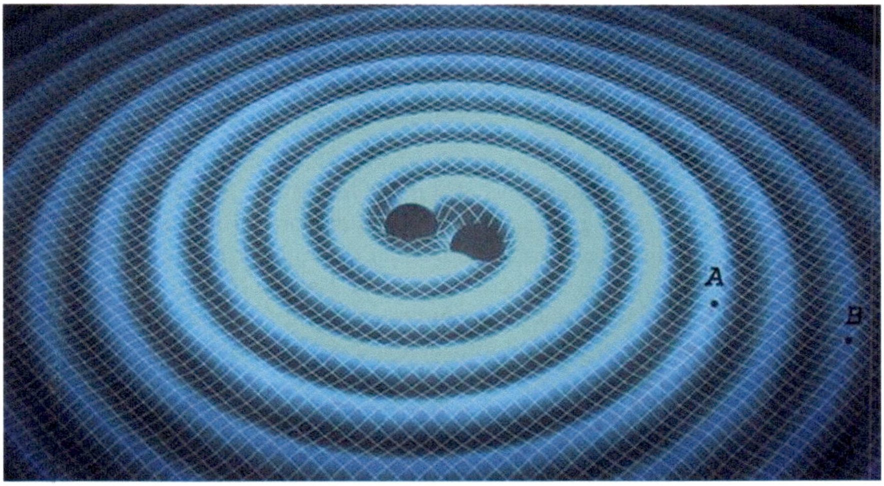

Fig. 11.3 This naive illustration implies that the orbital velocity v of binary black holes is larger than the speed of light c, see (11.14). It shows only a dipole character of gravitational waves and not their proclaimed quadrupole character

[73, 225, 258]. In this time interval, the removed noise from both the detectors is correlated. It was enough to shift the noise from the detector in Livingston about 7 ms similarly as in the detected signal. The corresponding figure can be found in [147]. In [2] such an important comparison is missing. The removed noise from the other two detected gravitational waves GW151226 and GW170104 is correlated as well. Therefore, such correlated noise could cause a bias in orbital frequencies of black holes such as (11.1).

Remark 11.9 For the time being we do not know what is the speed of gravitational waves for other media than vacuum (e.g. inside the Sun). Does this speed depend on the frequency as the speed of electromagnetic waves?

Remark 11.10 The famous Fig. 11.3, which should illustrate the propagation of gravitational waves, contradicts the General Theory of Relativity, see [201]. To see this, denote by d the distance of the two black holes and by T their orbital period. Multiply the trivial inequality

$$\pi > 2$$

by d/T. Then we immediately get a contradiction

$$v = \frac{\pi d}{T} > \frac{2d}{T} = \frac{|AB|}{T} = c, \tag{11.14}$$

where the orbital velocity v is larger than the speed of gravitational waves c (equal to the speed of light) and $|AB|$ is the distance of two consecutive maximum amplitudes of the right black hole as indicated in Fig. 11.3. However, $v \lesssim \frac{1}{3}c$ by Abbott et al. [2].

Remark 11.11 Recall that the *Archimedean spiral* is by definition a locus corresponding to the locations over time of a point moving away from a fixed point with a constant speed along a line that rotates with constant angular velocity (see [296, p. 136]). The distance between individual revolutions of this spiral in radial direction is constant. Another objection to Fig. 11.3 is that the gravitational waves should approximately form a double Archimedean spiral. Such a spiral should leave each black hole in an almost radial direction and not in the tangential direction as illustrated in the center of Fig. 11.3. Moreover, it is easy to see that Fig. 11.3 shows only a dipole and not quadrupole character of gravitational waves, in which the density of waves would be doubled. Also the exposed spacetime near the center that produced the largest amplitudes of gravitational waves, should be much more largely deformed.

11.5 Conclusions

A comprehensive and detailed study of gravitational waves and their role in astrophysics and cosmology is given in [233]. Our hypothesis that masses of binary hole mergers are overestimated is based on the following arguments:

1. The key formula (11.2) possesses several essential drawbacks which are described in Sect. 11.2.
2. In the literature on gravitational waves, no gravitational redshift is taken into account while the redshift on horizon of each black hole is infinity, compare (11.7) with (11.9).
3. One should carefully distinguish between an observed and calculated stellar black hole. As of 2022, the mass of the largest observed stellar black hole Cygnus X-1 is only about 21 $M_\odot$, whereas it is 150 $M_\odot$ for the largest calculated stellar black hole from gravitational waves, see [7].
4. No mechanism is known which would produce binary stellar black holes with masses greater than 50 solar masses, see Remark 11.3.
5. There is a large and statistically significant mass gap between all known black hole mergers and binary neutron stars, see Remark 11.4.
6. There is another large statistically significant mass gap between observed single stellar black holes and calculated binary black holes, see Remark 11.4.
7. Our hypothesis yields a more reliable size of the wave zone than in [2], see Remark 11.6.
8. Very strong gravitational fields around binary black hole mergers were described and calculated by a post-Newtonian model and not by Einstein's equations.

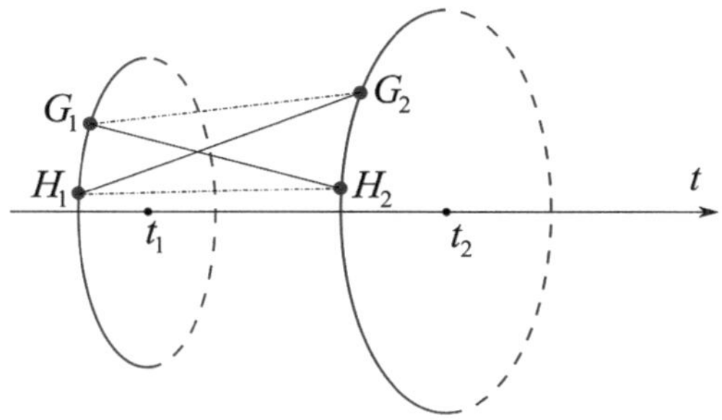

Fig. 11.4 A schematic illustration of gravitational interaction between two galaxies G and H in an expanding universe

9. A curved spacetime around physical black hole mergers is ignored.
10. A black hole on a piece of paper is a completely different object than a physical black hole.

Therefore, something has to be wrong in the current analysis of GW signals even though there are surely various selection effects. The claim that two merging black holes had masses (11.1) seems to be somewhat too strong. As shown in Sect. 11.2, the main reason is that the detected frequency f appearing in (11.2) has to be replaced by the emitted frequency $f_{\rm em}$. In other words, the chirp mass given by (11.2) has to be divided by the missing factor (11.12). We hope that this crucial mistake will be removed soon and that it will be not repeated in the coming decades.

Remark 11.12 It is not known whether the speed of gravitational waves is the same as the speed of gravitational interaction. Anyway, they are both finite. Consider two instants $t_1 < t_2$ of cosmic time. In Fig. 11.4 we observe the gravitational interaction between two galaxies G and H in an expanding universe (cf. Fig. 6.2). By G_i we denote the galaxy G at time t_i for $i = 1, 2$, and similarly for H_i. Thus, the dashed line segments G_1G_2 and H_1H_2 denote trajectories of G and H in the corresponding spacetime, respectively. Gravitational interaction does not act on objects in the past, but because of the time delay, it affects the future.

Chapter 12
Possible Distribution of Mass Inside a Black Hole

12.1 A Remarkable Coincidence

It is generally believed that the center of any physical black hole contains a point singularity with infinite mass density, see [275, 276, 278]. This seems to be only a mathematical idealization, since no physical quantity can attain infinite values due to quantum phenomena. Moreover, no point can rotate and black holes have well defined spins.

Here we introduce our own hypothesis from [166]. We suggest that the hidden center of any black hole could be occupied by a neutron-like star composed of neutrons and/or quark-gluon plasma. So there is no singularity in the center.

The maximum mass of a neutron star is at most three solar masses (cf. [6]). We show that in this case the radius of such a neutron star is approximately equal to its Schwarzschild radius (3.10). Adding a small amount of matter to this star, a black hole arises. Thus its interior could contain a compact object below the event horizon instead of the proposed point singularity.

The mass of the neutron is (see [269])

$$m = 1.675 \cdot 10^{-27}\,\mathrm{kg} \tag{12.1}$$

and its interaction diameter is given by $d = \sqrt{4\sigma/\pi} \approx 1.65 \cdot 10^{-15}$ m, where σ is the effective cross section. Assuming spherical symmetry, the density of the neutron is approximately

$$\rho_{\mathrm{n}} \approx \frac{m}{\frac{1}{6}\pi d^3} \approx 0.712 \cdot 10^{18}\,\mathrm{kg/m^3}. \tag{12.2}$$

In our Galaxy, more than 2000 neutron stars are already known. By Burgio et al. [52] most of their masses fall in a relatively narrow interval from $1.45M_\odot$ to $1.65M_\odot$, where $M_\odot = 2 \cdot 10^{30}$ kg is the mass of the Sun. Their mass is usually

M. Křížek, L. Somer, *Mathematical Aspects of Paradoxes in Cosmology*,
https://doi.org/10.1007/978-3-031-31768-2_12

Fig. 12.1 The maximum density of equal balls in the three-dimensional space is about 74 %. The maximum density of (possibly deformed) neutrons can be even higher

greater than the *Chandrasekhar limit* $1.4M_{\odot}$ for white dwarfs. However, rarely there exist neutron stars with masses close to $M_{\odot}$ (see [216]), since during the collapse of a white dwarf onto a neutron star, some amount of matter is thrown away.

The mass of each neutron star is smaller than the *Tolman-Oppenheimer-Volkoff (TOV) limit* which is approximately (see e.g. [6, 216, 223])

$$M := 3M_{\odot}. \tag{12.3}$$

As of 2022, the mass of the biggest known neutron stars (PSR J0952-0607) exceeds $2M_{\odot}$. Therefore, the above upper limit (12.3) may be further reduced in the future. Anyway, as of 2022, IGR J17091-3624 is the smallest known black hole with mass 3–7 $M_{\odot}$.

The neutron star mass density is likely not constant and increases towards the center. Thus, the mean density of a neutron star should be even higher than (12.2), i.e. denser than an atomic nucleus, cf. Fig. 12.1. It is assumed that the central part of any neutron star is occupied by a quark-gluon plasma which is a Fermi liquid satisfying the Pauli exclusion principle, see Remark 5.7. Therefore, the central quark density could be several times higher than (12.2). However, here we should take into account also Remark 3.8. Moreover, we should not identify our simple models with reality.

According to [36], hyperons appear in dense matter in the interior of a neutron star which yield approximately a double mean density $\rho = 2\rho_{\rm n}$ compared to (12.2). Then for

$$M = \frac{4}{3}\pi R^3 \rho$$

given by (12.3) we obtain that the largest possible radius of such a neutron (hyperon) star is about

$$R = \sqrt[3]{\frac{3M}{4\pi\rho}} \approx 10 \text{ km}. \tag{12.4}$$

Fig. 12.2 The lower part of the event horizon of a black hole is represented by a hemisphere. It is not made of any material substance. The interior of a black hole larger than $3M_\odot$ could contain below its event horizon a star-like object composed of dense neutron/quark matter instead of a point singularity

Recalling the definition of the Schwarzschild radius (3.10)

$$S = \frac{2GM}{c^2}, \tag{12.5}$$

we find that S corresponding to $M = 3M_\odot$ has a very similar size as (12.4), namely

$$S = 9 \text{ km}. \tag{12.6}$$

In Sect. 3.3 we derived finer estimates that take into account that the space inside the neutron star is curved and thus it has a larger relativistic volume than Euclidean volume of a massless ball with the same circumference (cf. Fig. 3.3). Adding a small amount of mass to the TOV limit in (12.3), the body becomes a black hole, but its central part below the event horizon could still remain in the state of neutron matter and/or quark-gluon plasma (see Fig. 12.2). It does not have to collapse to a point singularity (see Fig. 12.3 and [166]) due to the Pauli exclusion principle. However, for the time being we do not know whether the Pauli principle could be violated for extremely strong gravitational fields. Let us again emphasize that this does not follow from Einstein's equations, since they do not describe quantum phenomena.

12.2 Possible Distribution of Mass Inside a Black Hole

It is estimated that there are over 10^8 black holes in our Galaxy and 10^{19} in the entire observable universe. Let us denote their coordinate Schwarzschild radius (12.5) as follows

$$R_\bullet = \frac{2G}{c^2} M_\bullet \tag{12.7}$$

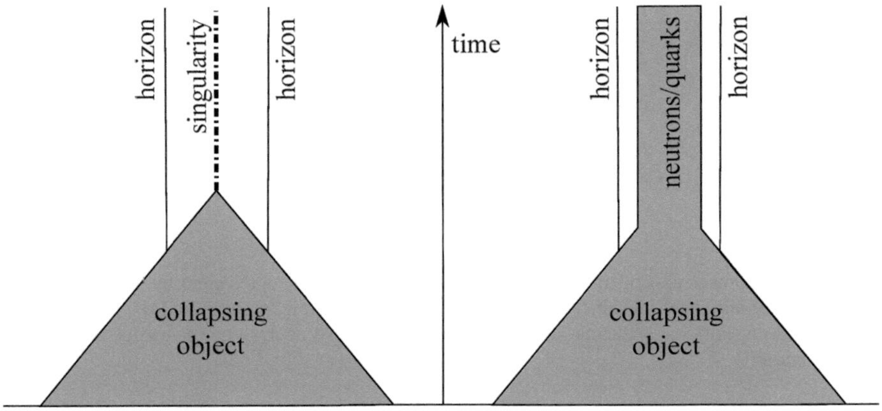

Fig. 12.3 Schematic spacetime diagrams of gravitational collapse of a star larger than $3M_\odot$ to a black hole. The time axis is vertical. The interior contains a point singularity (left) or a neutron-like star below the event horizon (right)

Table 12.1 The Schwarzschild radius $R_\bullet$ and the mean mass density $\rho_\bullet$ of various objects. The last column contains radii $R_\circ$ of objects with maximum neutron star density ρ

	$R_\bullet$ (in meters)	$\rho_\bullet$ (in kg/m^3)	$R_\circ$ (in meters)
1 kg	10^{-27}	10^{80}	$6 \cdot 10^{-7}$
Earth	0.009	10^{30}	100
$M_\odot$	3000	10^{20}	6979
$3M_\odot$	9000	10^{18}	10,000
Sgr A*	10^{10}	10^{6}	$1.4 \cdot 10^{6}$
M87*	$1.5 \cdot 10^{13}$	1	10^{8}

for a given mass $M_\bullet$. Then the corresponding mean mass density in the SI units is

$$\rho_\bullet \approx \text{const.}\frac{M_\bullet}{R_\bullet^3} \approx \text{Const.}\frac{c^6}{G^3}\frac{M_\bullet}{M_\bullet^3} \approx 10^{80}\frac{1}{M_\bullet^2}\,\text{kg/m}^3 \tag{12.8}$$

with some small (cf. (3.28)) dimensionless constants const. and Const. Consequently, the mean mass density $\rho_\bullet$ of a black hole is indirectly proportional to the square of its mass $M_\bullet$, while the mean mass density $\rho = M/V$ of a usual object is directly proportional to its mass M for a fixed volume V. In Table 12.1, we present hypothetical values of black hole mean densities of various objects.

The values in Table 12.1 associated to the central black hole Sgr A* of the Milky Way are derived e.g. in [166]. The last line of Table 12.1 corresponds to the supermassive black hole M87* in the center the supergiant elliptical galaxy M87 in the Virgo cluster. By (12.7) and (12.8) its mean mass density is $\rho_\bullet \approx 1$ kg/m^3, since this black hole is about 1500× more massive than Sgr A*. This mean density is thus like the density of air.

Remark 12.1 It is surprising that rotational axes of both photon rings of M87* and Sgr A* are directed to us (see [13]), even though the jets of M87 are directed elsewhere. The angle between the relativistic jets of M87 and the line of sight can be determined by the method described in [184, p. 310].

The limit values $M_\odot$ and $3M_\odot$ appearing in Table 12.1 were investigated in previous sections. From Table 12.1 we further observe that the Earth being a black hole would have an incredibly large mean density, 12 orders higher than the density of the neutron (12.2). Moreover, 1 kg black hole would have by (12.8) density 10^{80} kg/m^3.

Anxieties that in CERN miniature black holes could be produced are unjustified, since they would have unrealistic density. Another argument is that maximum energy of each proton in the LHC is about 10^{12} eV, while from the universe we receive protons having over 10^{20} eV and nothing happens. Namely, in our neighborhood we do not observe any miniature black holes, even though the surface of the Earth and other planets is continually bombarded by particles of cosmic rays with extremely high energies.

12.3 Other Properties

Remark 12.2 From (12.1)–(12.4) and the above discussion in Sect. 12.1 we find that each neutron star contains about 10^{57} baryons.

Remark 12.3 The existence of Hawking's micro (or mini) black holes [125] due to Hawking radiation is very improbable. By (12.8) their mean mass density would be much larger than (12.2), see the third column of Table 12.1. In fact, from (12.7) and (12.8) we get

$$\rho_\bullet \approx \frac{M_\bullet}{\frac{4}{3}\pi R_\bullet^3} = \frac{3c^2 R_\bullet}{8\pi G R_\bullet^3} = 1.6 \cdot 10^{26} \frac{1}{R_\bullet^2} \text{ kg/m}^3$$

(taking into account (3.28)). Hence, the mean mass density $\rho_\bullet$ converges to infinity as $R_\bullet \to 0$. Thus, Hawking evaporation would yield absolutely unrealistic mean densities. In fact, Stephen Hawking uses a trivial algebraic relation (12.5), but the reality can be quite different. He assumes that black holes are kept together by gravity, but does not take into account that the repulsive strong force is locally at least 40 orders of magnitude larger than the gravitational force and that Einstein's equations do not consider the strong force. Moreover, the mean density of Hawking's miniature black holes would increase with decreasing gravitation caused by radiation, which is absurd. It is more probable that for black holes of masses about $3M_\odot$ a neutron star could appear after Hawking's radiation if it exists.

Remark 12.4 We can conclude that gravity should not hold together objects with mass smaller than $\sim M_\odot$ and densities larger than the density (12.2) of the neutron or quark. On the other hand, the interior of black holes with mass larger than $3M_\odot$

Fig. 12.4 The jets of a radio source Hercules A are 1 million light years long (photo NASA)

could contain below the event horizon a star-like object composed of dense neutron or quark matter instead of the proclaimed point singularity.

If the core of a black hole rotates (cf. Fig. 12.2), then the horizon need not be exactly spherically symmetric. Its shape could resemble an ellipsoid or other axially symmetric body. The theory says that each black hole is characterized only by three parameters: its mass, spin, and electric charge (see [249, p. 876] for the no-hair theorem). But the reality can be different (see also Remark 3.6).

Remark 12.5 The jets of the radio source Hercules A in the galaxy 3C 348 are $\ell = 10^6$ ly long (1 ly $\approx 10^{16}$ m). Their shape can be roughly approximated by a huge double cone with volume V and basis area πr^2 for $r = 10^5$ ly (see Fig. 12.4). Assuming a ten times larger density $\rho = 10^{-25}$ kg/m^3 of these jets than is the mean cosmic matter density 10^{-26} km/m^3, we can roughly estimate their total mass M as follows:

$$M = \rho V = \rho 2 \frac{1}{3} \ell \pi r^2 = 10^{-25} 2 \frac{\pi}{3} 10^{22} \left(10^{21}\right)^2 \text{kg} \approx 2 \cdot 10^{64-25}\, \text{kg} = 10^9 M_\odot.$$

This is a consequence of the very strong magnetic field around the central black hole. If the magnetic field is due to a central compact body below the horizon (see Remark 3.6), it cannot be excluded that some black holes (namely quasars) should be rather considered as black & white holes. These B&W holes probably cannot be entered from the axis of jets. Note that quasars are found only at cosmological distances. Their remnants in our neighborhood are not so active now, since they have already used up most of their own energy.

12.4 Final Conclusions

Can mathematics explain the current crisis in cosmology? Our criticism of the standard cosmological model is based mainly on a rigorous use of mathematical tools and methods. We wanted to draw attention to the errors and mistakes made in contemporary cosmology. When reading any sort of cosmological literature, it is often not clear what is a definition, what is an assumption, what is a claim, what is experimentally verified fact, what is just a nice numerical simulation or artificially colored image, and what is reality (cf. also [325]). A number of cosmological definitions are vague. In addition, various quantities in the standard cosmological model and measured values are denoted by the same letter and without any warning they are sometimes arbitrarily interchanged. From heuristically derived equations important conclusions are drawn. Measured data is sometimes interpreted incorrectly, excessive extrapolations are made, various consequences are not in the form of mathematical implications, etc. Intuition can sometimes be misleading and lead to incorrect conclusions. It is a pity that there is not closer cooperation between mathematicians and cosmologists. Fortunately, if cosmologists are wrong about our universe, nothing happens.

We showed that the computational complexity of Einstein's equations is so large that one cannot calculate trajectories of two (or more) gravitationally interacting mass bodies. Their applicability is therefore quite limited. They satisfy the law of conservation of energy and momentum exactly. However, we present dozens of examples in the physical world showing a slight deviation from this law on short time intervals. Of course, we do not claim that the law of conservation of energy and momentum should not be applied. We cannot avoid it in many calculations, since it makes most situations much easier.

Revolutionary developments in physics such as Newton's theory of gravity, Special and General Relativity, and quantum mechanics came at a time when some researchers found the courage to get out of a scientific rut, and looked at natural phenomena and measured data in an entirely new way. The prevailing idea of contemporary cosmology is that 68 % of the universe is composed of some exotic dark energy, 27 % of an unknown dark matter, and only less than 5 % of normal baryonic matter. However, all models (with no exceptions!) which are used to describe the evolution of the universe were tested only on much smaller spatial and time scales. When they are applied to the entire universe, we necessarily make a considerable extrapolation without any guarantee that the result obtained is correct. The combination of measured data and model prediction is then interpreted as the presence of dark matter and dark energy, i.e., these quantities exist only by definition (see [178, p. 11]). The goal of this publication was to point out some treacherous pitfalls that cosmologists encounter if they identify the results of simple mathematical models with physical world.

While there are hundreds of books on cosmology which more or less repeat the same things in a sluggish general consensus, our book carefully analyzes many major basic cosmological statements, identifying their weaknesses and sometimes disproving supposedly well-established results. We are well aware that our book will receive criticisms from pundits in the main astrophysical stream, since it disputes many of the usual considerations. However, it is also healthy for Science to listen to some different voices from the mathematical community, since such an exercise can only reinforce the validity of the scientific method, and engender more innovative and empirically based understandings of our universe.

Bibliography

1. Aarseth S.J.: Gravitational N-body simulations. Cambridge University Press, Cambridge (2009)
2. Abbott, B.P., et al.: Observation of gravitational waves from a binary black hole merger. Phys. Rev. Lett. **116**, 061102 (2016)
3. Abbott, B.P., et al.: GW151226: observation of gravitational waves from a 22-solar-mass binary black hole coalescence. Phys. Rev. Lett. **116**, 241103 (2016)
4. Abbott, B.P., et al.: GW170104: observation of a 50-solar-mass binary black hole coalescence at redshift 0.2. Phys. Rev. Lett. **118**, 221101 (2017)
5. Abbott, B.P., et al.: Multi-messenger observation of a binary neutron star merger. Astrophys. J. Lett. **848**, L12 (2017)
6. Abbott, B.P., et al.: GW170817: observation of gravitational waves from a binary neutron star inspiral. Phys. Rev. Lett. **119**, 161101 (2017)
7. Abbott, B.P., et al.: GW190521: a binary black hole merger with a total mass of 150 $M_{\odot}$. Phys. Rev. Lett. **125**, 101102 (2020)
8. Abbott, B.P., et al.: GW190814: gravitational waves from the coalescence of a 23 solar mass black hole with a 2.6 solar mass compact object. Astrophys. J. Lett. **896**, L44 (2020)
9. Abuter, R., et al.: (Gravity Collaboration): detection of the Schwarzschild precession in the orbit of the star S2 near the Galactic centre massive black hole. Astron. Astrophys. **636**, L5 (2020)
10. Adami, C., et al.: The build-up of the Coma cluster by infalling substructures. Astron. Astrophys. **443**, 17–27 (2005)
11. Adler, J.R., Casey, B., Jacob, O.C.: Vacuum catastrophe: an elementary exposition of the cosmological constant problem. Amer. J. Phys. **63**, 620–626 (1995)
12. Ahmad, M., Alam, M.S.: Non-associativity of Lorentz transformation and associative reciprocal symmetric transformation. Int. J. Rec. Sym. Theor. Phys. **1**, 9–19 (2014)
13. Akiyama, K., et al.: First M87 event horizon telescope results. I. The shadow of the supermassive black hole. Astrophys. J. Lett. **875**, 1–17 (2019)
14. Allen, S.W., Evrard, A.E., Mantz, A.B.: Cosmological parameters from observations of galaxy clusters. Annual Rev. Astron. Astrophys. **49**, 409–470 (2011)
15. Anderle, P.: Celestial Mechanics (in Czech). Academia, Prague (1987)
16. Anderson, J.D., et al.: Shape and orientation of Mercury from radar ranging data. Icarus **124**, 690–697 (1996)
17. Anderson, J.D., et al.: Study of the anomalous acceleration of pioneer 10 and 11. Phys. Rev. D **65**, 082004 (2002)

M. Křížek, L. Somer, *Mathematical Aspects of Paradoxes in Cosmology*,
https://doi.org/10.1007/978-3-031-31768-2

18. Avdeev, N.A., Dyadina, P.I., Labazova, S.P.: Test of hybrid metric-Palatini $F(R)$-gravity in binary pulsars. J. Exp. Theor. Phys. **131**, 537–547 (2020)
19. Baade, W., Zwicky, F.: Cosmic rays from super-novae. Proc. Nat. Acad. Sci. **20**, 254–263 (1934)
20. Baade, W., Zwicky, F.: Remarks on super-novae and cosmic rays. Phys. Rev. **46**, 76–77 (1934)
21. Baierlein, R.: Two myths about special relativity. Am. J. Phys. **74**, 193–195 (2006)
22. Balázs, L.G., et al.: A possible interrelation between the estimated luminosity distances and the internal extinctions of type Ia supernovae. Astronom. Nachr. **327**, 917–924 (2006)
23. Barish, B.C., Weiss, R.: LIGO and the detection of gravitational waves. Phys. Today **52**(10), 44 (1999)
24. Baryshev, Y.V.: Paradoxes of the cosmological physics in the beginning of the 21-st century. In: Proceedings of the XXX-th Internat. Workshop on High Energy Physics – Particle and Astroparticle Physics, Gravitation and Cosmology – Predictions, Observations and New Projects, p. 297. World Scientific Publishing (2015). ArXiv 1501.01919
25. Baryshev, Y.V., Teerikorpi, P.: Fundamental Questions of Practical Cosmology: Exploring the Realm of Galaxies. Springer, Berlin (2012)
26. Barrow, J.D., Tipler, F.J.: The Anthropic Cosmological Principle. Oxford University Press, Oxford (1986)
27. Bekenstein, J.: Relativistic gravitation theory for the modified Newtonian dynamics paradigm. Phys. Rev. D **70**, 083509 (2004)
28. Bertotti, B., Farinella, P., Vokrouhlický, D.: Physics of the Solar System. Kluwer, Dordrecht (2003)
29. Binney, J., Merrifield, M.: Galactic Astronomy. Princeton University Press, Princeton (1998)
30. Birkhoff, G.D.: Relativity and Modern Physics. Harvard University Press, Cambridge (1923)
31. Biviano, A., et al.: A catalogue of velocities in the central region of the Coma cluster. Astron. Astrophys. Suppl. Ser. **111**, 265–274 (1995)
32. Blanchet, L.: Gravitational radiation from post-Newtonian sources and inspiralling compact binaries. Living Rev. Relativ. **17**(2), 187 (2014)
33. Blanchet, L., et al.: Gravitational-radiation damping of compact binary systems to second post-Newtonian order. Phys. Rev. Lett. **74**, 3515–3518 (1995)
34. Blanuša, D.: Über die Einbettung hyperbolischer Räume in euklidische Räume. Monatsh. Math. **59**, 217–229 (1955)
35. Böhringer, H., Werner, N.: X-ray spectroscopy of galaxy clusters: studying astrophysical processes in the largest celestial laboratories. Astron. Astrophys. Rev. **18**, 127–196 (2010)
36. Bombaci, I.: The hyperon puzzle in neutron stars. Nuclear Phys. News **31**, 17–21 (2021)
37. Börner, G.: The early universe. Facts and Fiction. Springer, Berlin (1993)
38. Botermann, B., et al.: Test of time dilation using stored Li+ ions as clocks at relativistic speed. Phys. Rev. Lett. **113**, 120405 (2014); Errata **114**, 239902 (2015)
39. Bouwens R.J., et al.: A candidate redshift $z \approx 10$ galaxy and rapid changes in that population at an age of 500 Myr. Nature **469**, 504–507 (2011)
40. Bowler, M.G.: Apsidal advance in SS 433? Astron. Astrophys. **510**, A28 (2010)
41. Boyer, R.H., Lindquist, R.W.: Maximal analytic extension of the Kerr metric. J. Math. Phys. **8**, 265 (1967)
42. Brander, D.: Isometric Embeddings Between Space Forms, pp. 1–48. Master Thesis, University of Pennsylvania, 2003
43. Brander, D.: Curved flats, pluriharmonic maps and constant curvature immersions into pseudo-Riemannian space forms. Ann. Global Anal. Geom. **32**, 253–275 (2007)
44. Brandts, J., Korotov, S., Křížek, M.: Simplicial Partitions with Applications to the Finite Element Method. Springer International Publishing, Cham (2020)
45. Brandts, J., Křížek, M., Zhang, Z.: Paradoxes in numerical calculations. Neural Netw. World **26**, 317–330 (2016)
46. Bretagnon, P.: Théorie du mouvement de l'ensemble des planètes. Solution VSOP82. Astron. Astrophys. **114**, 278–288 (1982)

47. Bretagnon, P., Francou, G.: Planetary theories in rectangular and spherical variables. VSOP87 solutions. Astron. Astrophys. **202**, 309–315 (1988)
48. Bristow, T.F. et al., Low Hesperian PCO2 constraints from in situ mineralogical analysis of Gale Crater, Mars. Proc. Nat. Acad. Sci. **114**, 2166–2170 (2017)
49. Broadhurst, T., Diego, J.M., Smoot, G.: Reinterpreting low frequency LIGO/Virgo events as magnified stellar-mass black holes at cosmological distances, pp. 1–16. ArXiv 1802.05273v1
50. Bruno, G.: De l'Infinito, Universo e Mondi, Venezia (1584)
51. Buitrago, F., et al.: Shaping massive galaxies: their morphology and kinematics at $z = 1 - 3$. In: Zapatero, M.R., et al. (eds.) Highlights of Spanish Astrophysics VI, Proceedings of the IX Scientific Meeting of the Spanish Astronomical Society, Madrid, pp. 154–160 (2010)
52. Burgio, G.F., et al.: Maximum mass of neutron stars with a quark core. Phys. Lett. B **526**, 19–26 (2002)
53. Burša, M., Peč, K.: Gravity Field and Dynamics of the Earth. Springer, Berlin (1993)
54. Cannon, J.W., Floyd, W.J., Kenyon, R., Parry, W.R.: Hyperbolic geometry. In: Flavors of Geometry, Mathematical Sciences Research Institute Publications, vol. 31, pp, 59–115. Cambridge University Press, Cambridge (1997)
55. Carlip, S.: Aberration and the speed of gravity. Phys. Lett. A **267**, 81–87 (2000)
56. Carpenter, E.F.: Some characteristics of associated galaxies I. A density restriction in the metagalaxy. Astrophys. J. **88**, 344 (1938)
57. Carr, B.J., Rees, M.J.: The anthropic principle and the structure of the physical world. Nature **278**, 605–612 (1979)
58. Carrera, M., Giulini, D.: Influence of global cosmological expansion on local dynamics and kinematics. Rev. Mod. Phys. **82**, 169–208 (2010)
59. Carroll, S.M., Press, W.H., Turner, E.L.: The cosmological constant. Ann. Rev. Astron. Astrophys. **30**, 499–542 (1992)
60. Carter, B.: The Significance of Numerical Coincidences in Nature, Part I. Line by line transcript of the stenciled preprint issued in 1967
61. Carter, B.: Large number coincidences and the Anthropic Principle in cosmology. In: Longair, M.S. (ed.) IAU Symposium, Confrontation of Cosmological Theories with Observational Data, vol. 63, pp. 291–298. Riedel, Dordrecht (1974)
62. Casares, J.: Observational evidence for stellar-mass black holes. In: Karas, V., Matt, G. (eds.) Proceedings of the IAU Symposium. Black Holes: From Stars to Galaxies – Across the Range of Masses, vol. 233, pp. 1–8 (2006)
63. Centrella, J., Baker, J.G., Kelly, B.J., van Meter, J.R.: Black-hole binaries, gravitational waves, and numerical relativity. Rev. Mod. Phys. **82**, 3069 (2010)
64. Chen, X., Li, S., Cao, Z.: Mass-redshift degeneracy for gravitational-wave sources in the vicinity of a supermassive black hole, 6 pp. (2017) ArXiv: 1703.10543v2
65. Chenciner, A., Montgomery, R.: A remarkable periodic solution of the three body problem in the case of equal masses. Ann. Math. **152**, 881–901 (2000)
66. Clemence, G.M.: The relativity effect in planetary motions. Rev Mod. Phys. **19**, 361–364 (1947)
67. Clowe, D., et al.: A direct empirical proof of the existence of dark matter. Astrophys. J. Lett. **648**, L109–L113 (2006)
68. Coddington, E.A., Levinson, N.: Theory of Ordinary Differential Equations. McGraw-Hill Book Company, New York (1955)
69. Colless, M., Dunn, A.M.: Structure and dynamics of the Coma cluster. Astrophys. J. **458**, 435–454 (1996)
70. Cooperstock, F.I., Faraoni, V., Vollick, D.N.: The influence of the cosmological expansion on local systems. Astrophys. J. **503**, 61–66 (1998)
71. Corral-Santana, J.M., Casares, J., Muñoz-Darias, T., Bauer, F.E., Martínez-Pais, I.G., Russell, D.M.: BlackCAT: a catalogue of stellar-mass black holes in X-ray transients. Astronom. Astrophys. **587**, A61 (2016)
72. Coxeter, H.S.M.: Regular honeycombs in hyperbolic space. In: Proceedings of International Congress of Mathematicians, vol. 3, pp. 155–169. Amsterdam (1954)

73. Creswell, J., et al.: On the time lags of the LIGO signals. J. Cosmol. Astropart. Phys. **2017**, 013 (2017)
74. Cutler, C., Flanagan, E.E.: Gravitational waves from merging compact binaries: how accurately can one extract the binary's parameters from the inspiral waveform? Phys. Rev. D **49**, 2658–2697 (1994)
75. Damjanov, I., et al.: Red nuggets at high redshift: structural evolution of quiescent galaxies over 10 Gyr of cosmic history. Astrophys. J. Lett. **739**, L44 (2011)
76. Daubechies, I.: Ten Lectures on Wavelets. CBMS Lecture Notes, vol. 61. SIAM, Philadelphia (1992)
77. Davies, G.F.: Thermal Evolution of the mantle. In Stevenson, D.J. (ed) Treatise on Geophysics, Evolution of the Earth, vol. 9, pp. 197–216. Elsevier, Amsterdam (2007)
78. Davis, T.M.: Is the universe leaking energy? Scientific American, pp. 39–47 (2010)
79. Davis, T.M., Lineweaver, C.H.: Expanding confusion: common misconceptions of cosmological horizons and the superluminal expansion of the universe. Publ. Astron. Soc. Aust. **21**, 97–109 (2004)
80. de Sitter, W.: On the relativity of inertia. Remarks concerning Einstein latest hypothesis. Proc. Kon. Ned. Acad. Wet. **19**, 1217–1225 (1917)
81. Dicke, R.H., Goldberg, H.M.: Solar oblateness and general relativity. Phys. Rev. Lett. **18**, 313–316 (1967)
82. Dickey, J.O., et al.: Lunar laser ranging: a continuing legacy of the Apollo program. Science **265**, 482–490 (1994)
83. Doppler, C.: Ueber das farbige Licht der Doppelsterne und einiger anderer Gestirne des Himmels. Abh. böhm. Ges. Wiss. **2**, 466–482 (1842)
84. Dumin, Y.V.: A new application of the Lunar laser retroreflectors: Searching for the "local" Hubble expansion. Adv. Space Res. **31**, 2461–2466 (2003)
85. Dumin, Y.V.: Testing the dark-energy-dominated cosmology by the Solar-System experiments. In: Kleinert, H., Jantzen, R.T., Ruffini, R. (eds.) Proceedings of the 11th Marcel Grossmann Meeting on General Relativity, pp. 1752–1754. World Scientific, Singapore (2008)
86. Dumin, Y.V.: The faint young Sun paradox in the context of modern cosmology. Astron. Tsirkulyar **1623**, 1–5 (2015). ArXiv: 1505.03572v1
87. Duncombe, R.L.: Relativity effects for the three inner planets. Astron. J. **61**, 174–175 (1956)
88. Eddington, A.: Space, Time and Gravitation. Cambridge University Press, Cambridge (1966)
89. Einstein, A.: Zur Elektrodynamik bewegter Körper. Ann. Phys. **322**(10), 891–921 (1905)
90. Einstein, A.: Erklärung der Perihelbewegung des Merkur aus der allgemeinen Relativitätstheorie, pp. 831–839 Sitzungsberichte der Königlich Preussischen Akademie der Wissenschaften, Berlin XLVII (1915). English translation: Explanation of the perihelion motion of Mercury from general relativity theory, by R.A. Rydin with comments by A.A. Vankov, 1–34
91. Einstein, A.: Die Feldgleichungen der Gravitation (The field equations of gravitation), pp. 844–847. Sitzungsberichte der Königlich Preussischen Akademie der Wissenschaften, Berlin, XLVIII (1915). Available at: https://www.biodiversitylibrary.org/item/92536#page/1/mode/1up
92. Einstein, A.: Die Grundlage der allgemeinen Relativitätsteorie (The foundation of the general theory of relativity). Ann. Phys. **49**, 769–822 (1916)
93. Einstein, A.: Näherungsweise Integration der Feldgleichungen der Gravitation. Sitzungsberichte K. Preuss. Akad. Wiss. **1**, 688–696 (1916)
94. Einstein, A.: Kosmologische Betrachtungen zur allgemeinen Relativitätstheorie, pp 142–152. Königlich-Preuss. Akad. Wiss. Berlin (1917). English translation in The principle of relativity. New York, Dover, 1952
95. Einstein, A.: Über Gravitationswellen. Sitzungsberichte K. Preuss. Akad. Wiss. **1**, 154–167 (1918)
96. Einstein, A., de Sitter, W.: On the relation between the expansion and the mean density of the universe. Proc. Natl. Acad. Sci. **18**, 213–214 (1932)

97. Eisenstein, D.J., Bennett, C.L.: Cosmic sound waves rule. Phys. Today **61**, 44–50 (2008)
98. Ellis, H.G.: Gravity inside a nonrotating, homogeneous, spherical body (2012). ArXiv: 1203.4750v2
99. Ewing, A.: "Black holes" in space. Sci. News Lett. **85**, 39 (1964)
100. Fahr, H.-J., Sokaliwska, M.: The influence of gravitational binding energy on cosmic expansion dynamics: new perspectives for cosmology. Astrophys. Space Sci. **339**, 379–387 (2012)
101. Feng, J.Q., Gallo, C.F.: Mass distribution in rotating thin-disk galaxies according to Newtonian dynamics. Galaxies **2**, 199–222 (2014)
102. Ferré-Mateu, A., Trujillo, I.: Superdense massive galaxies in the nearby universe. In: Bruzual, G., Charlot, S. (eds.) Proceedings of the XXVII. General Assembly of IAU, vol. S262, pp. 331–332. Kluwer, Amsterdam (2010)
103. Feynman, R.P., Leighton, R.B., Sand, M.: The Feynman lectures of physics I. Addison-Wesley Publication Company (1966)
104. Fischer, E.: The definition of density in general relativity. Internat. J. Astron. Astrophys. **7**, 303–312 (2017)
105. Florides, P.S.: A new interior Schwarzschild solution. Proc. R. Soc. Lond. A **337**, 529–535 (1974)
106. Folkner, W.M., et al.: The planetary and lunar ephemerides DE430 and DE431. IPN Progress Report, pp. 42–196 (2014)
107. Foster, J., Nightingale, J.D.: A Short Course in General Relativity, 3rd edn. Springer, New York (2006)
108. Friedman, A.: Über die Krümmung des Raumes. Z. Phys. **10**, 377–386 (1922). English translation: On the curvature of space. General Relativity and Gravitation **31**, 1991–2000 (1999)
109. Friedmann, A.: Über die Möglichkeit einer Welt mit konstanter negativer Krümmung des Raumes. Z. Phys. **21**, 326–332 (1924). English translation: On the possibility of a world with constant negative curvature of space. General Relativity and Gravitation **31**, 2001–2008 (1999)
110. Fromholz, P., Poisson, E., Will, C.M.: The Schwarzschild metric: It's the coordinates, stupid! Am. J. Phys. **82**, 295–300 (2014)
111. Gallo, C.F., Feng, J.Q.: Galactic rotation described by a thin-disk gravitational model without dark matter. J. Cosmol. **6**, 1373–1380 (2010)
112. Gamow, G.: Mr. Tompkins in paperback. Cambridge University Press, London (1965)
113. Gerber, P.: Die räumliche und zeitliche Ausbreitung der Gravitation. Z. Math. Phys. **43**, 93–104 (1898). English translation: The spatial and temporal propagation of gravity. J. Math. Phys. **43**, 93–104 (1898)
114. Gilmour, I., Sephton, M.A. (eds.): An Introduction to Astrobiology. Cambridge University Press, Cambridge (2003)
115. Glass, E.N.: Gravothermal catastrophe, an example. Phys. Rev. D **82**, 044039 (2010)
116. Gödel, K.: An example of a new type of cosmological solutions of Einstein's field equations of gravitation. Rev. Mod. Phys. **21**, 447–450 (1949)
117. Gold, T.: Rotating neutron stars as the origin of the pulsating radio sources. Nature **218**(5143), 731–732 (1968)
118. Goldberg, D.: What every computer scientist should know about floating-point arithmetic. ACM Comput. Surv. **23**, 5–48 (1991)
119. Gonzáles, V., et al.: Evolution of galaxy stellar mass functions, mass densities, and mass to light ratios from $z \sim 7$ to $z \sim 4$. Astrophys. J. Lett. **735**, L34 (2011)
120. Gradštejn, I.S., Ryžik, I.M.: Tables of Integrals, Sums, Series, and Products (in Russian). Nauka, Moscow (1991)
121. Guinan, F.E., Maloney, F.P.: The apsidal motion of the eccentric eclipsing binary DI Herculis – an apparent discrepancy with general relativity. Astron. J. **90**, 1519–1528 (1985)
122. Guth, A.H.: Inflationary universe: a possible solution to the horizon and flatness problems. Phys. Rev. D **23**, 347–356 (1981)

123. Hafele, J.C., Keating, R.E.: Around-the-world atomic clocks: predicted relativistic time gains. Science **177**, 166–168 (1972)
124. Hartmann, W.K.: Mars. Workman Publishers, New York (2003)
125. Hawking, S.W.: Gravitationally collapsed objects of very low mass. Mon. Not. R. Astron. Soc. **152**, 75–78 (1971)
126. Herdeiro, C.A.R., Lemos, J.P.S.: The black hole fifty years after: genesis of the name (2018). ArXiv: 1811.06587
127. Hewish, A., Bell, S.J., Pilkington, J.D.H., Scott, P.F., Collins, R.A.: Observation of a rapidly pulsating radio source. Nature **217**, 709–713 (1968)
128. Hilbert, D.: Über Flächen von constanter gausscher Krümmung. Trans. Am. Math. Soc. **2**, 87–99 (1901)
129. Hilbert, D.: Die Grundlagen der Physik (Erste Mitteilung). Nachr. Ges. Wissen. Göttingen, Math.-Phys. Klasse **20**, 395–407 (1915)
130. Hildebrant, H. et al.: KiDS-450: cosmological parameter constraints from tomographic weak gravitational lensing. Mon. Not. R. Aston. Soc. **465**, 1454–1498 (2017)
131. Hilditch, R.W.: An Introduction to Close Binary Stars. Cambridge University Press, Cambridge (2001)
132. Hlaváček, I., Křížek, M., Malý, J.: On Galerkin approximations of a quasilinear nonpotential elliptic problem of a nonmonotone type. J. Math. Anal. Appl. **184**, 168–189 (1994)
133. Horák, Z., Krupka, F.: Physics, vol. 2 (in Czech). SNTL, Prague (1976)
134. Hubble, E.: A relation between distance and radial velocity among extra-galactic nebulae. Proc. Nat. Acad. Sci. USA **15**, 168–173 (1929)
135. Hughes, J.P.: The mass of the Coma cluster: combined X-ray and optical results. Astrophys. J. **337**, 21–33 (1989)
136. Hulse, R.A., Taylor, J.H.: Discovery of a pulsar in a binary system. Astrophys. J. **195**, L51–L53 (1975)
137. Inoue, T.: An excess motion of the ascending node of Mercury in the observations used by Le Verrier. Celest. Mech. Dyn. Astron. **56**, 69 (1993)
138. Ives, H.E., Stilwell, G.R.: An experimental study of the rate of a moving atomic clock. J. Optic. Soc. Am. **28**, 215–226 (1938)
139. Janssen, M., Renn, J.: How Einstein found his field equations. Birkhäuser, Basel (2022)
140. Jałocha, J., Bratek, L., Kutschera, M.: Is dark matter present in NGC 4736? An iterative spectral method for finding mass distribution in spiral galaxies. Astrophys. J. **679**, 373–378 (2008)
141. Jeans, J.H.: The stability of a spherical nebula. Phil. Trans. R. Soc. A **199**, 312–320 (1902)
142. Jebsen, J.T.: Uber die allgemeinen kugelsymmetrischen Lösungen der Einsteinschen Gravitationsgleichungen im Vakuum. Arkiv för matematik, astronomi och fysik **15**(18), 1–9 (1921). On the general symmetric solutions of Einstein's gravitational equations in vacuo. General Relativity and Cosmology **37**(12), 2253–2259 (2005)
143. Jurgens, R.F., et al.: Mercury radar ranging data from 1987 to 1997. Astron. J. **124**, 486–488 (1998)
144. Kerr, R.P.: Gravitational field of a spinning mass as an example of algebraically special metrics. Phys. Rev. Lett. **11**, 237–238 (1963)
145. Kienert, H., Feulner, G., Petoukhov, V.: Faint young Sun problem more severe due to ice-albedo feedback and higher rotation rate of the early Earth. Geophys. Res. Lett. **39**, L23710 (2012)
146. Kim, A.G., et al., Implications for the Hubble constant from the first seven supernovae. Astrophys. J. Lett. **476**, L63–L66 (1997)
147. Kim, M.H.: Strange noise in gravitational-wave data sparks debate. Quanta Magazine, 6 pp. (2017)
148. Klimenko, S., et al.: Method for detection and reconstruction of gravitational wave transients with networks of advanced detectors. Phys. Rev. D **93**, 042004 (2016)
149. Kokubo, E., Ida, S., Makino, J.: Evolution of a circumterrestrial disk and formation of a single Moon. Icarus **148**, 419–436 (2000)

150. Kolesnik, Y.B., Masreliez, C.J.: Secular trends in the mean longitudes of planets derived from optical observations. Astron. J. **128**, 878–888 (2004)
151. Königsdorf, H.: Lemma 1. In: Meine ungehörigen Träume, Geschichten, Berlin, Aufbau-Verlag, Edition Neue Texte (1978)
152. Kopeikin, S., Efroimsky, M., Kaplan, G.: Relativistic Celestial Mechanics of the Solar System. Wiley, Berlin (2011)
153. Kotter, F.: Über die physikalischen Grundlagen der Einsteinschen Gravitationstheorie. Ann. Phys. **56**, 401–462 (1918)
154. Kraniotis, G.V., Whitehouse, S.B.: Compact calculation of the perihelion precession of Mercury in general relativity, the cosmological constant and Jacobi's inversion problem. Class. Quant. Grav. **20**, 4817–4835 (2003)
155. Křížek, F., Křížek, M., Šolc, J.: How massive is the black hole at the centre of our Galaxy? Obzory Mat. Fyz. Inf. **36**, 43–51 (2007)
156. Křížek, M.: Numerical experience with the three-body problem. J. Comput. Appl. Math. **63**, 403–409 (1995)
157. Křížek, M.: Numerical experience with the finite speed of gravitational interaction. Math. Comput. Simulation **50**, 237–245 (1999)
158. Křížek, M: Why do we observe apparent superluminal velocities in the universe? (in Czech). Pokroky Mat. Fyz. Astronom. **44**, 218–226 (1999)
159. Křížek, M.: On the Kepler equation (in Czech). Matematika-fyzika-informatika **19**, 449–452 (2008/2009)
160. Křížek, M.: Does a gravitational aberration contribute to the accelerated expansion of the Universe? Commun. Comput. Phys. **5**, 1030–1044 (2009)
161. Křížek, M.: Dark energy and the anthropic principle. New Astron. **17**, 1–7 (2012)
162. Křížek, M.: Zdeněk Kopal — numerical analyst. In: Rucinski, S.M., Torres, G., Zejda, M. (eds.) Proceedings of the Conference on Living Together: Planets, Host Stars and Binaries, Litomyšl, Czech Republic (2014). Astron. Soc. Pacific Conf. Ser. **496** (2015), 19–31.
163. Křížek, M.: Why is the circumference of the Sun smaller than $2\pi r$?, (in Czech). Obzory Mat. Fyz. Inf. **46**(3), 7–18 (2017)
164. Křížek, M.: Influence of celestial parameters on Mercury's perihelion shift. Bulg. Astron. J. **27**, 41–56 (2017)
165. Křížek, M.: Ten arguments against the proclaimed amount of dark matter. Gravit. Cosmol. **24**, 350–359 (2018)
166. Křížek, M.: Possible distribution of mass inside a black hole. Is there any upper limit on mass density? Astrophys. Space Sci. **364**, Article 188, 1–5 (2019)
167. Křížek, M.: Do Einstein's equations describe reality well? Neural Netw. World **29**, 255–283 (2019)
168. Křížek, M.: On paradoxes in the special theory of relativity (in Czech). Obzory Mat. Fyz. Inf. **49**, 31–40 (2020)
169. Křížek, M.: Unexpected illusions in the special theory of relativity. Adv. Stud. Theor. Phys. **14**, 335–345 (2020)
170. Křížek, M.: A critical review of paradoxes in the special theory of relativity. In: Křížek, M., Dumin, Y.V. (eds.) Proceedings of the Conference on Cosmology on Small Scales 2020, Excessive extrapolations and selected controversies in cosmology, pp. 61–78. Institute of Mathematics, Prague (2020), 61–78.
171. Křížek, M.: On the twin paradox (in Czech). Obzory Mat. Fyz. Inf. **50**(2), 1–10 (2021)
172. Křížek, M.: Relativistic perihelion shift of Mercury revisited. Astron. Nachr. **343**, e20220016, 1–7 (2022)
173. Křížek, M.: 100 years of the Friedmann equation. In: Křížek, M., Dumin, Y.V. (eds.) Proceedings of the Conference on Cosmology on Small Scales 2022, Dark Energy and the Local Hubble Expansion Problem, pp. 131–144. Institute of Mathematics, Prague (2022)
174. Křížek, M.: On extreme computational complexity of the Einstein equations. In: Neittaanmäki, P., Rantalainen, M.-L. (eds.) Proceedings of the Conference on Impact of Scientific Computing on Science and Society. Comput. Methods Appl. Sci. **58**, 21 pp. (2023)

175. Křížek, M., Brandts, J.: Manifestations of dark energy in the dynamics of the Solar system. S264 – Solar and stellar variability: Impact on Earth and planets. In: Kosovichev, A.G., Andrei, A.H., Rozelot, J.-P. (eds.) Proceedings of the IAU XXVIIth General Assembly in Rio de Janerio, August 2009, pp. 119–121. Cambridge University Press (2010)
176. Křížek, M., Brandts, J., Somer, L.: Is gravitational aberration responsible for the origin of dark energy? In: Del Valle, C.A., Longoria, D.F. (eds.) Dark Energy: Theory, Implications and Roles in Cosmology, pp. 29–57. Nova Science Publishers, New York (2012)
177. Křížek, M., Dumin, Y.V. (eds.): Cosmology on small scales 2016: Local Hubble expansion and selected controversies in cosmology. Institute of Mathematics, Czech Academy of Sciences, Prague (2016)
178. Křížek, M., Dumin, Y.V. (eds.): Cosmology on small scales 2018: Dark matter problem and selected controversies in cosmology. Institute of Mathematics, Czech Academy of Sciences, Prague (2018)
179. Křížek, M., Dumin, Y.V. (eds.): Cosmology on small scales 2020: Excessive extrapolations and selected controversies in cosmology. Institute of Mathematics, Czech Academy of Sciences, Prague (2020)
180. Křížek, M., Dumin, Y.V. (eds.): Cosmology on small scales 2022: Dark energy and the local Hubble expansion problem. Institute of Mathematics, Czech Academy of Sciences, Prague (2022)
181. Křížek, M., Gueorguiev, V.G., Maeder, A.: An alternative explanation of the orbital expansion of Titan and other bodies in the Solar system. Gravit. Cosmol. **28**, 122–132 (2022)
182. Křížek, M., Křížek, F.: Quantitative properties of the Schwarzschild metric. Publ. Astron. Soc. Bulg. **2018**, 1–10 (2018)
183. Křížek, M., Křížek, F., Somer, L.: Which effects of galactic clusters can reduce the amount of dark matter. Bulg. Astron. J. **21**, 43–65 (2014)
184. Křížek, M., Křížek, F., Somer, L.: Antigravity — its origin and manifestations. Lambert Academic Publishing, Saarbrücken (2015)
185. Křížek, M., Křížek, F., Somer, L.: Dark matter and rotation curves of spiral galaxies. Bulg. Astron. J. **25**, 64–77 (2016)
186. Křížek, M., Křížek, P.: Why has nature invented three stop codons of DNA and only one start codon? J. Theor. Biol. **304**, 183–187 (2012)
187. Křížek, M., Křížek, P., Šolc, J.: Astronomical mistakes accompanying the Prague horologe (in Czech). Čs. Čas. Fyz. **60**, 337–340 (2010)
188. Křížek, M., Mészáros, A.: On the Friedmann equation for the three-dimensional hypersphere. In: Křížek, M., Dumin, Y.V. (eds.) Proceedings of the Conference on Cosmology on Small Scales 2016, Local Hubble Expansion and Selected Controversies in Cosmology, pp. 159–178. Institute of Mathematics, Prague (2016)
189. Křížek, M., Mészáros, A.: Classification of distances in cosmology. In: Křížek, M., Dumin, Y.V. (eds.) Proceedings of Conference on. Cosmology on Small Scales 2018, Dark Matter Problem and Selected Controversies in Cosmology, pp. 92–118. Institute of Mathematics, Prague (2018)
190. Křížek, M., Mészáros, A.: A remarkable connection between the Jeans' theory of gravitational instabilities and the Newtonian limit of the General Theory of Relativity. Bulg. Astron. J. **39**, 8 (2023)
191. Křížek, M., Neittaanmäki, P.: Mathematical and Numerical Modelling in Electrical Engineering: Theory and Applications. Kluwer Academic Publishers, Dordrecht (1996)
192. Křížek, M., Pradlová, J.: On the nonexistence of a Lobachevsky geometry model of an isotropic and homogeneous universe. Math. Comput. Simulation **61**, 525–535 (2003)
193. Křížek, M., Roos, H.-G., Chen, W.: Two-sided bounds of the discretization error for finite elements. ESAIM Math. Model. Numer. Anal. **45**, 915–924 (2011)
194. Křížek, M., Šolcová, A., Somer, L.: Construction of Šindel sequences. Comment. Math. Univ. Carolin. **48**, 373–388 (2007)
195. Křížek, M., Šolcová, A., Somer, L.: Ten theorems on the astronomical clock of Prague. In: Příhonská, J., Segeth, K., Andrejsová, D. (eds.) Proceedings of the International Conference

on Presentation of Mathematics ’07, pp. 53–62. TECH Technological University, Liberec (2007)
196. Křížek, M., Somer, L.: A critique of the standard cosmological model. Neural Netw. World **24**, 435–461 (2014)
197. Křížek, M., Somer, L.: Manifestations of dark energy in the Solar system. Grav. Cosmol. **21**, 58–71 (2015)
198. Křížek, M., Somer, L.: Excessive extrapolations in cosmology. Gravit. Cosmol. **22**, 270–280 (2016)
199. Křížek, M., Somer, L.: Neglected gravitational redshift in detections of gravitational waves. In: Křížek, M., Dumin, Y.V. (eds.) Proceedings of the Conference on Cosmology on Small Scales 2018, Dark Matter Problem and Selected Controversies in Cosmology, pp. 173–179. Institute of Mathematics, Prague (2018)
200. Křížek, M., Somer, L.: Excessive extrapolations of Einstein’s equations. In: Křížek, M., Dumin, Y.V. (eds.) Proceedings of the Conference on Cosmology on Small Scales 2020, Excessive Extrapolations and Selected Controversies in Cosmology, pp. 9–34. Institute of Mathematics, Prague (2020)
201. Křížek, M., Somer, L.: Why masses of binary black hole mergers are overestimated? Galaxies **10**, Article 52, 1–8 (2022)
202. Křížek, M., Somer, L.: Anthropic principle and the Hubble-Lemaître constant. Galaxies **10**, Article 71, 1–11 (2022)
203. Křížek, M., Somer, L., Šolcová, A.: From Great Discoveries in Number Theory to Applications, Springer, Cham (2021)
204. Kroupa, P.: Star-cluster formation and evolution. In: Proceedings of the IAU S237, pp. 230–237. Cambridge University Press, Cambridge (2007)
205. Kroupa, P.: The dark matter crisis: Falsification of the current standard model of cosmology. Publ. Astron. Soc. Aust. **29**, 395–433 (2012)
206. Kroupa, P.: Galaxies as simple dynamical systems: observational data disfavor dark matter and stochastic star formation. Can. J. Phys. **93**, 169–202 (2015)
207. Kroupa, P., Pawlowski, M., Milgrom, M.: The failures of the standard model of cosmology require a new paradigm. Int. J. Mod. Phys. D **21**, 1230003 (2012)
208. Kroupa, P., Šubr, L., Jeřábková, T., Wang, L.: Very high redshift quasars and the rapid emergence of supermassive black holes. Mon. Not. Roy. Astron. Soc. **498**, 5652–5683 (2020)
209. Lainey, V., et al.: First numerical ephemerides of the Martian moons. Astron. Astrophys. **465**, 1075–1084 (2007)
210. Lainey, V., et al.: Resonance locking in giant planets indicated by the rapid orbital expansion of Titan. Nat. Astron. **4**, 1053–1058 (2020)
211. Lanczos, C.: Über eine stationäre kosmologie im sinne der Einsteinischen Gravitationstheories. Z. Phys. **21**, 73–110 (1924)
212. Landau, L.: On the theory of stars. Phys. Z. Sowjetunion **1**, 285–288 (1932)
213. Lang, K.K.: Cambridge Encyclopedia of the Sun. Cambridge University Press, Cambridge (2001)
214. Lang, K.R.: Astrophysical Formulae, vol. II. Springer, Berlin (2006)
215. Laplace, P.S.: Allgemeine geographische Ephemeriden herausgegeben von F. von Zach. **4**(1), 1–6 (1799)
216. Lattimer, J.M., Prakash, M.: The physics of neutron stars. Science **304**, 536 (2004)
217. Lelli, F., McGaugh, S.S., Schombert, J.M., Pawlowski, M.S.: One law to rule them all: The radial acceleration relation of galaxies. Astrophys. J. **836**, Article id. 152, 23 pp. (2017)
218. Lemaître, G.E.: Un Univers homogène de masse constante et de rayon croissant rendant compte de la vitesse radiale des nébuleuses extragalactiques. Ann. Soc. Sci. de Bruxelles 49–59 (1927)
219. Lemaître, G.E.: L’expansion de l’espace. Rev. Quest. Sci. **17**, 391–440 (1931)
220. Le Verrier, U.J.: Theorie du mouvement de Mercure. Ann. Obser. Imp. Paris V **76**, 1–196 (1859)

221. Li, M.-H., Liu, H.-N.: Testing the homogeneity of the Universe using gamma-ray bursts. Astron. Astrophys. **582**, A111, 1–7 (2015)
222. Li, T.-P.: Foundations of Physics and Cosmology. Science Press, Beijing (2021)
223. Linares, M., Shahbaz, T., Casares, J.: Peering into the dark side: magnesium lines establish a massive neutron star in PSR J2215+5135. Astrophys. J. **859**(54), 1–14 (2018)
224. Lineweaver, C.H., Schwartzman, D.: Cosmic thermobiology. In: Seckbach, J. (ed.) Origins, pp. 233–248. Kluwer, Dordrecht (2003)
225. Liu, H., Jackson, A.D.: Possible associated signal with GW150914 in the LIGO data. J. Cosmol. Astroparticle Phys. **1610**, 014 (2016)
226. Lopéz-Corredoira, M.: Fundamental ideas in cosmology. Scientific, Philosophical and Sociological Critical Perspectives. IOP Publishing, Bristol (2022)
227. Lorentz, H.A.: La théorie électromagnetique de Maxwell et son application aux corps mouvants. Arch. Néerl. **25**, 1–190 (1892)
228. Lyne, A.G.: A review of the double pulsar – PSR J0737-3039. Chin. J. Astron. Astrophys. **6**, 162–168 (2006)
229. Maeder, A.: An alternative to the ΛCDM model: The case of scale invariance. Astrophys. J. **834**, 194–209 (2017)
230. Maeder, A., Bouvier, P.: Scale invariance, metrical connection and the motions of astronomical bodies. Astron. Astrophys. **73**, 82–89 (1979)
231. Maeder, A.: La plus grande égnime de l'astronomie. De Newton et Einstein à l'énergie et matière noires. EDP Sciences, EDP Sciences (2023)
232. Maeder, A.M., Gueorguiev, V.G.: On the relation of the lunar recession and the length-of-the-day. Astrophys. Space Sci. **366**, Article 101 (2021)
233. Maggiore, M.: Gravitational waves. Astrophysics and Cosmology, vol. 2. Oxford Scholarship Online (2018)
234. Marchal, C.: The Three-Body Problem. Elsevier, Amsterdam (1991)
235. Martínez-Lombilla, C., Trujillo, I., Knapen, J.H.: Discovery of disc truncations above the galaxies mid-plane in Milky Way-like galaxies. Mon. Not. R. Astron. Soc. **483**, 664–691 (2019)
236. Mashhoon, B., Mobed, N., Singh, D.: Tidal dynamics in cosmological spacetimes. Classical Quant. Grav. **24**, 5031–5046 (2007)
237. McCausland, I.: Anomalies in the history of relativity. J. Sci. Explor. **13**, 271–290 (1999)
238. McGaugh, S.S.: Milky Way mass models and MOND. Astrophys. J. **683**, 137–148 (2008)
239. McLaughlin, D.E., van der Marel, R.P.: Resolved massive star clusters in the Milky Way and its satellites: Brightness profiles and a catalog of fundamental parameters. Astrophys. J. Suppl. Ser. **161**, 304–360 (2005)
240. McVittie, C.G.: The mass-particle in expanding universe. Mon. Not. R. Astron. Soc. **93**, 325–339 (1933)
241. Mészáros, P., Mészáros, A.: The brightness distribution of bursting sources in relativistic cosmologies. Astrophys. J. **449**, 9–16 (1995)
242. Meyer, Y.: Wavelets. Algorithms & Applications. SIAM, Philadelphia (1993)
243. Michell, J.: On the means of discovering the distance, magnitude, magnitude, &c. of the fixed stars, in consequence of the diminution of the velocity of their light, in case such a diminution should be found to take place in any of them, and such other data should be procured from observations, as would be farther necessary for that purpose. Phil. Trans. R. Soc. Lond. **74**, 35–57 (1784)
244. Michelson, A.A., Morley, E.W.: On the relative motion of the Earth and the luminiferous ether. Am. J. Sci. **34**, 333–345 (1887)
245. Mikulášek, Z., Krtička, J.: Foundations of Physics of Stars (in Czech), Masaryk University, Brno (2005)
246. Milgrom, M.: A modification of Newtonian dynamics as a possible alternative to the hidden mass hypothesis; Implications for galaxies; Implications of galaxy systems. Astrophys. J. **270**, 365–370, 371–383, 384–389 (1983)
247. Milne, E.A.: Relativity, Gravitation and World Structure. Clarendon Press, Oxford (1935)

248. Mirabel, I.F., Rodríguez, L.F.: A superluminal source in the Galaxy. Nature **371**, 46–48 (1994)
249. Misner, C.W., Thorne, K.S., Wheeler, J.A.: Gravitation, 20th edn. W.H. Freeman, New York (1997)
250. Mixson, J.S.: The Twin Paradox of Special Relativity Is Solved. Infinity Publishing, West Conshohocken (2014)
251. Moni Bidin, C., Carraro, G., Méndez, R.A., Smith, R.: Kinematical and chemical vertical structure of the Galactic thick disk, II. A lack of dark matter in the solar neighborhood. Astrophys. J. **751**, Article 30, 14 pp. (2012)
252. Montgomery, R.: A new solution to the three body problem. Not. Am. Math. Soc. **5**, 471–481 (2001)
253. Moriconi, M.: Special theory of relativity through the Doppler effect. Eur. J. Phys. **27**, 1409–1424 (2006)
254. Mroué, A.H., et al.: Catalog of 174 binary black hole simulations for gravitational wave astronomy. Phys. Rev. Lett. **111**, 241104 (2013)
255. Muller, J.-M. et al.: Handbook of Floating-Point Arithmetic. Birkhäuser, Basel (2009)
256. Narayan, R., McClintock, J.E.: Observational evidence for black holes, pp. 1–20. ArXiv: 1312.6698v2
257. Narlikar, J.V., Rana, N.C.: Newtonian N-body calculations of the advance of Mercury's perihelion. Mon. Not. R. Astron. Soc. **213** , 657–663 (1985)
258. Naselsky, P., Jackson, A.D., Liu, H.: Understanding the LIGO GW150914 event. J. Cosmol. Astroparticle Phys. **1608**, 029 (2016)
259. Neveu, M., Rhoden, A.R.: Evolution of Saturn's mid-sized moons. Nat. Astron. **3**, 543–552 (2019)
260. Newcomb, S.: The elements of the four inner planets and the fundamental constants of astronomy. Supp. Am. Ephemeris and Nautical Almanac for 1897, pp. 1–202. Gov. Printing Office, Washington, D.C. (1895)
261. Nobili, A.M., Will, C.M.: The real value of Mercury's perihelion advance. Nature **320**, 39–41 (1986)
262. Noerdlinger, P.D., Petrosian, V.: The effect of cosmological expansion on self-gravitating ensembles of particles. Astrophys. J. **168**, 1–9 (1971)
263. Noether, E.: Invariante Variationsprobleme. Nachr. D. König. Gesellsch. D. Wiss. Zu Göttingen, Math.-phys. Klasse, Heft **2**, 235–257 (1918)
264. Novotný, O.: Motions, Gravity Field and Figure of the Earth. Lecture Notes. University Federal da Bahia, Bahia (1998)
265. Obreschkow, D., Tissera, P., Jog., C., Carignan, C. (eds.) Galactic angular momentum. In: Proceedings of the XXXth IAU General Assembly, Focus Meeting 6, Vienna, August (2018)
266. Oort, J.: The force exerted by the stellar system in the direction perpendicular to the Galactic plane and some related problems. Bull. Astron. Inst. Neth. **6**, 249–287 (1932)
267. Pacini, F.: Rotating neutron stars, pulsars and supernova remnants. Nature **219**(5150), 145–146 (1968)
268. Park, R.S., et al.: Precession of Mercury's perihelion from ranging to the MESSENGER spacecraft. Astron. J. **153**, 121 (2017)
269. Patrignani, C., et al.: Particle Physics Booklet. University of California (2016)
270. Pauli, W.: Theory of Relativity. Dover Publication, New York (1981)
271. Peacock, J.A.: Cosmological Physics. Cambridge University Press, Cambridge (1999)
272. Peebles, P.J.E.: Principles of Physical Cosmology. Princeton University Press, New Jersey (1993)
273. Peltier, W.R,: History of earth rotation. In: Schubert, G. (ed.) Treatise on Geophysics, vol. 9. Elsevier, Amsterdam (2007). Evolution of the Earth (vol. ed. D. Stevenson)
274. Penrose, R.: The apparent shape of a relativistically moving sphere. In: Proc. Camb. Phil. Soc. **55**, 137–139 (1959)
275. Penrose, R.: Gravitational collapse and space-time singularities. Phys. Rev. Lett. **14**(3), 57–59 (1965)

276. Penrose, R.: Gravitational collapse: the role of General Relativity. Rivista del Nuovo Cimento, Numero Speziale I **252**, 252–276 (1969)
277. Penrose, R.: The Road to Reality. Vintage Books, London (2005)
278. Penrose, R.: Collected Works (6 Volume Set). Oxford University Press, Oxford (2010)
279. Perlmutter, S.: Supernovae, dark energy, and the accelerating universe. Physics Today **56**, 53–60 (2003)
280. Perlmutter, S., et al.: Measurements of the cosmological parameters Ω and Λ from the first seven supernovae at $z \geq 0.35$. Astrophys. J. **483**, 565–581 (1997)
281. Pfeiffer, H.P.: Numerical simulations of compact object binaries. Class. Quant. Grav. **29**, 124004 (2012)
282. Pick, M., Pícha, J., Vyskočil, V.: Theory of the Earth's Gravity Field. Academia, Prague (1973)
283. Pick, L., Rokyta, M.: On princess and measuring worm (in Czech: O pricenzně a píďalce). Pokroky Mat. Fyz. Astronom. **57**, 217–231 (2012)
284. Pilipenko, S.V.: Paper-and-pencil cosmological calculator, pp. 1–4 (2013). ArXiv: 1303.5961v1
285. Piña, P.E.M., et al.: No need for dark matter: resolved kinematics of the ultra-diffuse galaxy AGC 114905. Mon. Not. R. Astron. Soc. **512**, 3230–3242 (2022)
286. Pireaux, S., Rozelot, J.-P.: Solar quadrupole moment and purely relativistic gravitation contributions to Mercury's perihelion advance. Astrophys. Space Sci. **284**, 1159–1194 (2003)
287. Pitjeva, E.V.: High-precision ephemerides of planets – EPM and determination of some astronomical constants. Solar Sys. Res. **39**, 176–186 (2005)
288. Pitjeva, E.V., Pitjev, N.P.: Development of planetary ephemerides EPM and their applications. Celest. Mech. Dyn. Astron. **119**, 237–256 (2014)
289. Planck Collaboration: Planck 2013 results, I. Overview of products and scientific results. Astron. Astrophys. **571**, A1, 48 pp. (2014)
290. Planck Collaboration: Planck 2013 results, XVI. Cosmological parameters. Astron. Astrophys. **571**, A16, 66 pp. (2014)
291. Pohlen, M., et al.: Radial distribution of gas and dust in spiral galaxies. The case of M99 (NGC 4254) and M100 (NGC 4321). Astron. Astrophys. **518**, L72 (2010)
292. Poincaré, H.: Sur le problème des trois corps et les équations de la dynamique. Acta Math. **13**, 1–270 (1890)
293. Poincaré, H.: La théorie de Lorentz et le principle de réaction. Arch. Néerland. Sci. Exactes et Naturelles **5**, 252–278 (1900)
294. Poincaré, H.: Sur la dynamique de l'électron. C. R. Acad. Sci. Paris **140**, 1504–1508 (1905)
295. Rana, N.C.: An investigation of the motions of the node and perihelion of Mercury. Astron. Astrophys. **181**, 195–202 (1987)
296. Rektorys K., et al.: Survey of Applicable Mathematics, vol. I. Kluwer Academic Publishers, Dordrecht (1994)
297. Rektorys, K., et al.: Survey of Applicable Mathematics, vol II. Kluwer Academic Publishers, Dordrecht (1994)
298. Ridao, L.S., et al.: Perihelion advances for orbits of Mercury, Earth, and Pluto from extended theory of general relativity (ETGR). Can. J. Phys. **92**, 1709–1713 (2014)
299. Riess, A.G., et al.: Observational evidence from supernovae for an accelerating universe and a cosmological constant. Astron. J. **116**, 1009–1038 (1998)
300. Riess, A.G., et al.: New Hubble space telescope discoveries of Type Ia supernovae at $z >= 1$: Narrowing constraints on the early behavior of dark energy. Astrophys. J. **659**, 98–121 (2007)
301. Riess, A.G., et al.: Milky Way Cepheid standards for measuring cosmic distances and application to Gaia DR2: Implications for the Hubble constant. Astrophys. J. **861**, 126 (2018)
302. Robertson, H.P.: On the foundation of relativistic cosmology. Proc. Natl. Acad. Sci. **15**, 822–829 (1929)
303. Roseveare, N.T.: Mercury's Perihelion from Le Verrier to Einstein. Clarendon Press, Oxford (1982)

304. Roshan, M. et al.: Fast galaxy bars continue to challenge standard cosmology. Mon. Not. R. Astron. Soc. **508**, 926–939 (2021)
305. Rowan-Robinson, M.: Do type Ia supernovae prove $\Lambda > 0$? Mon. Not. R. Astron. Soc. **332**, 352–360 (2002)
306. Rubin, V.C.: The rotation of spiral galaxies. Sci. New Series **220**, 1339–1344 (1983)
307. Rubin, V.C.: Dark matter in spiral galaxies. Sci. Am. **248**, 88–101 (1983)
308. Rubin, V.C.: In: J.-P. Swings (ed.) Dark Matter in the Universe, Highlights of Astronomy, pp. 27–38. International Astronomical Union, Paris (1986)
309. Rubin, V.C.: A brief history of dark matter. In: Livio, M. (ed.) The Dark Universe: Matter, Energy, and Gravity, pp. 1–13. Cambridge University Press, Cambridge (2003)
310. Rubin, V.C., et al.: Kinematic studies of early-type stars. I. Photometric survey, space motions, and comparison with radio observations. Astrophys. J. **67**, 491–531 (1962)
311. Rubin, V.C., Ford, W.K., Thonnard, N.: Rotational properties of 21 Sc galaxies with a large range of luminosities and radii, from NGC 4605 ($R = 4$ kpc) to UGC 2885 ($R = 122$ kpc). Astrophys. J. **238**, 471–487 (1980)
312. Rydin, R.A.: The theory of Mercury's anomalous precession. In: Proceedings of the NPA, vol. 8, pp. 1–6. College Park (2011)
313. Saari, D.G., Xia Z.: Off to infinity in finite time. Not. Am. Math. Soc. **42**, 538–546 (1995)
314. Sackmann, I.J., Boothroyd, A.I., Kraemer, K.E.: Our Sun. III. Present and future. Astrophys. J. **418**, 457–468 (1993)
315. Said, S.S., Stephenson, F.R.: Solar and lunar eclipse measurements by medieval Muslim astronomers. J. Hist. Astron. **27**(28), 29–48, 259–273 (1996/1997)
316. Sanders, R.H., McGaugh, S.S.: Modified Newtonian dynamics as an alternative to dark matter. Ann. Rev. Astronom. Astrophys. **40**, 263–317 (2002)
317. Sauer, T.: The relativity of discovery: Hilbert's first note on foundations of physics. Arch. Hist. Exact Sci. **53**, 529–575 (1999)
318. Scherf, M., Lammer, H.: Did Mars possess a dense atmosphere during the first $\sim$ 400 million years? Space Sci. Rev. **217**, Article 2 (2021)
319. Schödel, R., et al.: A star in a 15.2-years orbit around the supermassive black hole at the centre of the Milky Way. Nature **419**, 694–696 (2002)
320. Schwarzschild, K.: Über das zulässige Krümmungsmaaß des Raumes. Vierteljahrsschift der Astronomischen Gesellschaft **35**, 337–347 (1900). English translation: On the permissible numerical value of the curvature of space. Abraham Zelmanov J. **1**, 64–73 (2008)
321. Schwarzschild, K.: Über das Gravitationsfeld eines Massenpunktes nach der Einsteinschen Theorie. Sitzungsber. Preuss. Akad. Wiss. Berlin, Math. Phys. 189–196 (1916). English translation: On the gravitational field of a point-mass, according to Einstein's theory. The Abraham Zelmanov Journal **1**, 10–19 (2008)
322. Schwarzschild, K.: Über das Gravitationsfeld einer Kugel aus incompressiebler Flüssigkeit nach der Einsteinschen Theorie. Sitzungsber. Preuss. Akad. Wiss. Math. Phys. 424–435 (1916). English translation: On the gravitational field of a sphere of incompressible liquid, according to Einstein's theory. The Abraham Zelmanov Journal **1**, 20–32 (2008)
323. Segeth, K.: From measured data to their mathematical description by a function and back (in Czech). Pokroky Mat. Fyz. Astronom. **60**, 133–147 (2015)
324. Shapiro, I.I.: Fourth test of general relativity. Phys. Rev. Lett. **13**(26), 789–791 (1964)
325. Sinai, Y.G.: Mathematicians and physicists = Cats and dogs? Bull. Am. Math. Soc. **43**, 563–565 (2006)
326. Smith, S.: The mass of the Virgo cluster. Astrophys. J. **83**, 23–30 (1936)
327. Šolc, M., et al.: Physics of Stars and the Universe (in Czech). SPN, Prague (1983)
328. Srinivasan, D.K., et al.: The radio frequency subsystem and radio science on the MESSENGER Mission. Space Sci. Rev. **131**, 557–571 (2007)
329. Standish, E.M.: An approximation to errors in the planetary ephemerides of the Astronomical Almanac. Astron. Astrophys. **417**, 1165–1171 (2004)
330. Standish, E.M., et al.: JPL planetary and Lunar ephemerides, DE403/LE403. JPL IOM **314.10-127**, 1–27 (2005)

331. Steiner, A.W., Lattimer, J.M., Brown, E.F.: The neutron star mass-radius relation and the equation of state of dense matter. Astrophys. J. Lett. **765**, L1–L5 (2013)
332. Stephani, H.: Relativity: An Introduction to Special and General Relativity, 3rd edn. Cambridge University Press, Cambridge (2004)
333. Stillwell, J.: The story of the 120-cell. Not. Am. Math. Soc. **48**, 17–24 (2001)
334. Stovall, K., et al.: PALFA discovery of a highly relativistic double neutron star binary (2018). ArXiv: 1802.01707v1. Accepted by Astrophys. J. Lett. 1–8.
335. Suntola, T.: The dynamic universe. Toward a unified picture of physical reality. Physics Foundations Society, The Finnish Society for Natural Philosophy. Drukatava Ltd., Latvia (2018)
336. Susobhanan, A., et al.: Exploring the effect of periastron advance in small-eccentricity binary pulsars. Mon. Not. R. Astron. Soc. **480**, 5260–5271 (2018)
337. Synge, J.L.: Relativity: The Special Theory. North-Holland Publication, Amsterdam (1956)
338. Taganov, I.N.: Irreversible Time Physics. Russian Academy of Sciences, Saint Petersburg (2016)
339. Tammann, G.A., Reindl, B.: The ups and downs of the Hubble constant. Rev. Mod. Astron. **19**, 1–29 (2006)
340. Tancredi, G., Sánchez, A., Roig, F.: A comparison between methods to compute Lyapunov exponents. Astron. J. **121**, 1171–1179 (2001)
341. Taylor, E.F., Wheeler, J.A.: Exploring Black Holes – Introduction to General Relativity. Addison Wesley, San Francisco (2000)
342. Tegmark, M.: Life 3.0: Being Human in the Age of Artificial Intelligence. Knopf, New York (2017)
343. Terrell, T.: Invisibility of Lorentz contraction. Phys. Rev. **116**, 1041–1045 (1959)
344. Thompson, T.A., et al.: A noninteracting low-mass black hole-giant star binary system. Science **366**, 637–640 (2019)
345. Tinsley, B.: Accelerating Universe revisited. Nature **273**, 208–211 (1978)
346. Tisserand, M.F.: Les Travaux de Le Verrier, vol. XV. ANN Observer, Paris (1880)
347. Titov, O.A.: Estimation of the acceleration of the Solar-system barycenter relative to a system of reference quasars. Grav. Cosmol. **55**, 95–99 (2011)
348. Tolman, R.C.: Relativity, Thermodynamics, and Cosmology. Clarendon Press/Dover, Oxford/New York (1934/1987)
349. Trujillo, I.: Origin and fate of the most massive galaxy. Highlights of Spanish Astrophysics VI. In: Zapatero, M.R. et al. (eds.) Proceedings of the IX Science Meeting, pp. 120–130. Spanish Astronomical Society, Madrid (2010)
350. Trujillo, I., et al.: Strong size evolution of the most massive galaxies since $z \sim 2$. Mon. Not. R. Astron. Soc. **382**, 109–120 (2007)
351. Tully, R.B., Fisher, J.R.: A new method of determining distances to galaxies. Astron. Astrophys. **54**, 661–673 (1977)
352. Turner, M.S.: The galactic halo. In: Gibson, B.K., Axeirod, T.S., Putrnan, M.E. (eds.) The Third Stromlo Symposium: The Galactic Halo. ASP Conference Series, vol. 165, pp. 431–452 (1999)
353. Tutukov, A.V., Fedorova, A,V.: The origin of intergalactic stars in galaxy clusters. Astron. Reports **55**, 383–391 (2011)
354. van Flandern, T.C.: A determination of the rate of change of g. Mon. Not. R. Astron. Soc. **170**, 333–342 (1975)
355. Vankov, A.: On controversies in special relativity, pp. 1–23 (2006). ArXiv: physics/0606130v1
356. Vankov, A.A.: General relativity problem of Mercury's perihelion advance revisited, pp. 1–46 (2010). ArXiv 1008.1811v1
357. van de Sande, J., et al.: The stellar velocity dispersion of a compact massive galaxy at $z = 1.80$ using X-shooter confirmation of the evolution in the mass-size and mass-dispersion relations. Astrophys. J. Lett. **736**, L9 (2011)
358. Vavryčuk, V.: Universe opacity and CMB. Mon. Not. R. Astron. Soc. **478**, 283–301 (2018)

359. Vavryčuk, V.: Universe opacity and Type Ia supernova dimming. Mon. Not. R. Astron. Soc. **489**, L63–L68 (2019)
360. Vavryčuk, V.: Considering light-matter interactions in the Friedmann equations. Proc. R. Soc. A **478**, Article 20220045, 1–25 (2022)
361. Vavryčuk, V.: Cosmological redshift and cosmic time dilation in the FLRW metric. Front. Phys. **10**, Article 826188, 1–11 (2022)
362. Vavryčuk, V.: Gravitational orbits in the expanding universe revisited. Front. Astron. Space Sci. **10**, Article 1071743, 1–18 (2023)
363. Vavryčuk, V., Kroupa, P.: The failure of testing for cosmic opacity via the distance-duality relation. Mon. Not. R. Astron. Soc. **497**, 378–388 (2020)
364. Vejchodský, T.: Three methods for two-sided bounds of eigenvalues – a comparison. Numer. Methods Partial Differ. Equ. **34**, 1188–1208 (2018)
365. Verbunt, F.: The Earth and Moon: from Halley to lunar ranking and shells, pp. 1–10. Astron Institute, Utrecht University (2002). Preprint, Corpus ID: 211732842
366. Voit, G.M.: Tracing cosmic evolution with clusters of galaxies. Rev. Mod. Phys. **77**, 207–258 (2005)
367. Walker, A.G.: On Milne's theory of world-structure. Proc. Lond. Math. Soc. **42**, 90–127 (1936)
368. Watson, J.D., Crick, F.H.C.: Genetic implications of the structure of deoxyribonucleic acid. Nature **171**, 964–969 (1953)
369. Weinberg, S.: Gravitation and Cosmology: Principles and Applications of the General Theory of Relativity. John Willey & Sons, New York (1972)
370. Weinberg, S.: The First Three Minutes, 2nd edn. Basic Books, New York (1993)
371. Weinberg, S.: Cosmology. Oxford University Press, Oxford (2008)
372. Weisberg, J.M., Taylor, J.H.: The relativistic binary pulsar B1913+16: thirty years of observations and analysis. In: F.A. Rasio, I.H. Stairs (eds.) Binary Radio Pulsars. ASP Conference Series, vol. 328, p. 25 (2005)
373. Weisskopf, V.F.: The visual appearance of rapidly moving objects. Phys. Today **13**(9), 24–27 (1960)
374. Weyl, H.: Über die statischen, kugelsymmetrischen Lösungen von Einsteins "kosmologischen" Gravitationsgleichungen. Phys. Z. **20**, 31–34 (1919)
375. Wheeler, J.A.: Our universe: the known and the unknown. Am. Sci. **56**(1), 1–20 (1968)
376. Whitney, A., et al.: Unbiased differential size evolution and the inside-out growth of galaxies in the deep CANDELS GOODS fields at $1 \leq z \leq 7$. Astrophys. J. **887**(113), 1–12 (2019)
377. Will, C.M.: The confrontation between General Relativity and experiment. Living Rev. Relat. **17**, 117 pp. (2014)
378. Williams, G.E.: Geological constraints on the Precambrian history of Earth's rotation and the Moon's orbit. Rev. Geophys. **38**, 37–60 (2000)
379. Yahalom, A.: The effect of retardation on galactic rotation curves. J. Phys. Conf. Ser. **1239**, 012006 (2019)
380. Yahalom, A.: Tully-Fisher relations and retardation theory for galaxies. Int. J. Mod. Phys. D **30**, 2142008, 8 pp. (2021)
381. Yahalom, A.: Lensing effects in galactic retarded gravity: why "Dark Matter" is the same for both gravitational lensing and rotation curves. Int. J. Mod. Phys. D **31**, 2242018, 10 pp. (2022)
382. Yahalom, A.: The weak field approximation of general relativity and the problem of precession of the perihelion for Mercury. Symmetry **15**, 39 (2023)
383. Zeldovich, Y.B., Novikov, I.D.: Relativistic Astrophysics: The Structure and Evolution of the Universe, vol. 2. University of Chicago Press, Chicago (1983)
384. Ženíšek, A.: Relativity to the Pocket (in Czech). Masaryk University, Brno (2015)
385. Zwicky, F.: Die Rotverschiebung von extragalaktischen Nebeln. Helv. Phys. Acta **6**, 110–127 (1933)
386. Zwicky, F.: On the masses of nebulae and of clusters of nebulae. Astrophys. J. **86**, 217–246 (1937)

References to Webpages (Valid as of June 2022)

387. http://www.math.cas.cz
388. www.spacetimetravel.org
389. https://en.wikipedia.org/wiki/Velocity-addition_formula
390. https://en.wikipedia.org/wiki/Interior_Schwarzschild_metric
391. https://cds.cern.ch/record/2746536
392. https://www.youtube.com/watch?v=DWF1uNb9Q1Q&feature=emb_title
393. https://ned.ipac.caltech.edu/help/cosmology_calc.html
394. https://www.ligo.org
395. https://www.gw-openscience.org
396. http://www.astro.ucla.edu/~wright/CosmoCalc.html
397. http://en.wikipedia.org/wiki/NGC_4874
398. https://users.math.cas.cz/~krizek/cosmol/
399. https://en.wikipedia.org/wiki/Stellar_classification
400. https://darkmattercrisis.wordpress.com

Index

M. Křížek, L. Somer, *Mathematical Aspects of Paradoxes in Cosmology*,
https://doi.org/10.1007/978-3-031-31768-2

Author Index

M. Křížek, L. Somer, *Mathematical Aspects of Paradoxes in Cosmology*,
https://doi.org/10.1007/978-3-031-31768-2

GPSR Compliance

The European Union's (EU) General Product Safety Regulation (GPSR) is a set of rules that requires consumer products to be safe and our obligations to ensure this.

If you have any concerns about our products, you can contact us on ProductSafety@springernature.com

In case Publisher is established outside the EU, the EU authorized representative is:

Springer Nature Customer Service Center GmbH
Europaplatz 3
69115 Heidelberg, Germany

Batch number: 10732090

Printed by Printforce, the Netherlands